CW00688854

The SIAM 100-Digit

CHALLENGE

The SIAM 100-Digit

CHALLENGE

A Study in High-Accuracy Numerical Computing

Folkmar Bornemann
Technische Universität München
Munich, Germany

Dirk Laurie
University of Stellenbosch
Stellenbosch, South Africa

Stan Wagon
Macalester College
St. Paul, Minnesota

Jörg Waldvogel
Swiss Federal Institute of Technology (ETH) Zurich
Zurich, Switzerland

With a Foreword by David H. Bailey

Society for Industrial and Applied Mathematics
Philadelphia

10 9 8 7 6 5 4 3 2 1

Library of Congress Cataloging-in-Publication Data

The SIAM 100-digit challenge : a study in high-accuracy numerical computing / Folkmar
Bornemann ... [et al.].
p. cm.
Includes bibliographical references and index.
ISBN 0-89871-561-X (pbk.)
1. Numerical analysis. I. Bornemann, Folkmar, 1967-
QA297.S4782 2004
518—dc22 2004049139

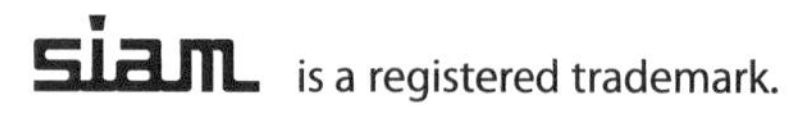

Contents

Foreword

Everyone loves a contest. I still recall when, in January/February 2002, I first read of the SIAM 100-Digit Challenge in *SIAM News*. Nick Trefethen's short article[1] introduced the 10 problems, then closed with the comment, "Hint: They're hard! If anyone gets 50 digits in total, I will be impressed." To an incorrigible computational mathematician like me, that was one giant red flag, an irresistible temptation. As it turned out I did submit an entry, in partnership with two other colleagues, but we failed to get the correct answer on at least 1 of the 10 problems and thus did not receive any award. But it was still a very engaging and enjoyable exercise.

This book shows in detail how each of these problems can be solved, as described by four authors who, unlike myself, belonged to winning teams who successfully solved all 10 problems. Even better, the book presents multiple approaches to the solution for each problem, including schemes that can be scaled to provide thousand-digit accuracy if required and can solve even larger related problems. In the process, the authors visit just about every major technique of modern numerical analysis: matrix computation, numerical quadrature, limit extrapolation, error control, interval arithmetic, contour integration, iterative linear methods, global optimization, high-precision arithmetic, evolutionary algorithms, eigenvalue methods, and many more (the list goes on and on).

The resulting work is destined to be a classic of modern computational science—a gourmet feast in 10 courses. More generally, this book provides a compelling answer to the question, "What is numerical analysis?" In this book we see that numerical analysis is much more than a collection of Victorian maxims on why we need to be careful about numerical round-off error. We instead see first hand how the field encompasses a large and growing body of clever algorithms and mathematical machinery devoted to efficient computation. As Nick Trefethen once observed [Tre98], "If rounding errors vanished, 95% of numerical analysis would remain."

As noted above, the authors of this book describe techniques that in many cases can be extended to compute numerical answers to the 10 problems to an accuracy of thousands of digits. Some may question why anyone would care about such prodigious precision, when in the "real" physical world, hardly any quantities are known to an accuracy beyond about 12 decimal digits. For instance, a value of π correct to 20 decimal digits would suffice to calculate the circumference of a circle around the sun at the orbit of the earth to within the width of an atom. So why should anyone care about finding any answers to 10,000-digit accuracy?

[1] See p. 1 for the full text.

In fact, recent work in experimental mathematics has provided an important venue where numerical results are needed to very high numerical precision, in some cases to thousands of decimal digits. In particular, precision on this scale is often required when applying integer relation algorithms[2] to discover new mathematical identities. An integer relation algorithm is an algorithm that, given n real numbers $(x_i,\ 1 \leqslant i \leqslant n)$, in the form of high-precision floating-point numerical values, produces n integers, not all zero, such that $a_1 x_1 + a_2 x_2 + \cdots + a_n x_n = 0$.

The best-known example of this sort is a new formula for π that was discovered in 1995:

$$\pi = \sum_{k=0}^{\infty} \frac{1}{16^k} \left(\frac{4}{8k+1} - \frac{2}{8k+4} - \frac{1}{8k+5} - \frac{1}{8k+6} \right).$$

This formula was found by a computer program implementing the PSLQ integer relation algorithm, using (in this case) a numerical precision of approximately 200 digits. This computation also required, as an input real vector, more than 25 mathematical constants, each computed to 200-digit accuracy. The mathematical significance of this particular formula is that it permits one to directly calculate binary or hexadecimal digits of π beginning at any arbitrary position, using an algorithm that is very simple, it requires almost no memory, and does not require multiple-precision arithmetic [BBP97, AW97, BB04, BBG04]. Since 1996, numerous additional formulas of this type have been found, including several formulas that arise in quantum field theory [Bai00].

It's worth emphasizing that a wide range of algorithms from numerical analysis come into play in experimental mathematics, including more than a few of the algorithms described in this book. Numerical quadrature (i.e., numerical evaluation of definite integrals), series evaluation, and limit evaluation, each performed to very high precision, are particularly important. These algorithms require multiple-precision arithmetic, of course, but often also involve significant symbolic manipulation and considerable mathematical cleverness as well.

In short, most, if not all, of the techniques described in this book have applications far beyond the classical disciplines of applied mathematics that have long been the mainstay of numerical analysis. They are well worth learning, and in many cases, rather fun to work with. Savor every bite of this 10-course feast.

David H. Bailey
Chief Technologist
Computational Research Department
Lawrence Berkeley National Laboratory, USA

[2] Such algorithms were ranked among *The Top 10*—assembled by Jack Dongarra and Francis Sullivan [DS00]—of algorithms "with the greatest influence on the development and practice of science and engineering in the 20th century."

Preface

This book will take you on a thrilling tour of some of the most important and powerful areas of contemporary numerical mathematics. A first unusual feature is that the tour is organized by problems, not methods: it is extremely valuable to realize that numerical problems often yield to a wide variety of methods. For example, we solve a random-walk problem (Chapter 6) by several different techniques, such as large-scale linear algebra, a three-term recursion obtained by symbolic computations, elliptic integrals and the arithmetic-geometric mean, and Fourier analysis. We do so in IEEE arithmetic to full accuracy and, at the extreme, in high-precision arithmetic to get 10,000 digits.

A second unusual feature is that we very carefully try to justify the validity of every single digit of a numerical answer, using methods ranging from carefully designed computer experiments and a posteriori error estimates to computer-assisted proofs based on interval arithmetic. In the real world, the first two methods are usually adequate and give the desired confidence in the answer. Interval methods, while nicely rigorous, would most often not provide any additional benefit. Yet it sometimes happens that one of the best approaches to a problem is one that provides proof along the way (this occurs in Chapter 4), a point that has considerable mathematical interest.

A main theme of the book is that there are usually two options for solving a numerical problem: either use a brute force method running overnight and unsupervised on a high-performance workstation with lots of memory, or spend your days thinking harder, with the help of mathematical theory and a good library, in the hope of coming up with a clever method that will solve the problem in less than a second on common hardware. Of course, in practice these two options of attacking a problem will scale differently with problem size and difficulty, and your choice will depend on such resources as your time, interest, and knowledge and the computer power at your disposal. One noteworthy case, where a detour guided by theory leads to an approach that is ultimately much more efficient than the direct path, is illustrated on the cover of this book. That diagram (taken from Chapter 1) illustrates that many problems about real numbers can be made much, much easier by stepping outside the real axis and taking a route through the complex plane.

The waypoints of our tour are the 10 problems published in the January/February 2002 issue of *SIAM News* by Nick Trefethen of Oxford University as an intriguing computing challenge to the mathematical public. The answer to each problem is a real number; entrants had to compute several digits of the answer. Scoring was simple: 1 point per digit, up to a maximum of 10 per problem. Thus a perfect score would be 100. When the dust settled several months later, entries had been received from 94 teams in 25 countries. Twenty of those teams achieved a perfect score of 100 and 5 others got 99 points. The whole fascinating story, including the names of the winners, is told in our introductory chapter, "The Story."

The contest, now known as the *SIAM 100-Digit Challenge*, was noteworthy for several reasons. The problems were quite diverse, so while an expert in a particular field might have little trouble with one or two of them, he or she would have to invest a lot of time to learn enough about the other problems to solve them. Only digits were required, neither proofs of existence and uniqueness of the solution, convergence of the method, nor correctness of the result; nevertheless, a serious team would want to put some effort into theoretical investigations. The impact of modern software on these sorts of problems is immense, and it is very useful to try all the major software tools on these problems so as to learn their strengths and their limitations.

This book is written at a level suitable for beginning graduate students and could serve as a text for a seminar or as a source for projects. Indeed, these problems were originally assigned in a first-year graduate course at Oxford University, where they were used to challenge students to think beyond the basic numerical techniques. We have tried to show the diversity of mathematical and algorithmic tools that might come into play when faced with these, and similar, numerical challenges, such as:

- large-scale linear algebra
- computational complex analysis
- special functions and the arithmetic-geometric mean
- Fourier analysis
- asymptotic expansions
- convergence acceleration
- discretizations that converge exponentially fast
- symbolic computing
- global optimization
- Monte Carlo and evolutionary algorithms
- chaos and shadowing
- stability and accuracy
- a priori and a posteriori error analysis
- high-precision, significance, and interval arithmetic

We hope to encourage the reader to take a broad view of mathematics, since one moral of this contest is that overspecialization will provide too narrow a view for one with a serious interest in computation.

The chapters on the 10 problems may be read independently. Because convergence acceleration plays an important role in many of the problems, we have included a discussion of the basic methods in Appendix A. In Appendix B we summarize our efforts in calculating the solutions to extremely high accuracies. Appendix C contains code that solves the 10 problems in a variety of computing environments. We have also included in Appendix D a sampling of additional problems that will serve as interesting challenges for readers who have mastered some of the techniques in the book.

All code in the book as well as some additional material related to this project can be found at an accompanying web page:

`www.siam.org/books/100digitchallenge`

We four authors, from four countries and three continents, did not know each other prior to the contest and came together via e-mail to propose and write this book. It took thousands of e-mail messages and exchanges of files, code, and data. This collaboration has been an unexpected benefit of our participation in the *SIAM 100-Digit Challenge.*

Notation and Terminology. When two real numbers are said to agree to d digits, one must be clear on what is meant. In this book we ignore rounding and consider it a problem of strings: the two strings one gets by truncating to the first d significant digits have to be identical. We use the symbol $\doteq$ to denote this type of agreement to all digits shown, as in $\pi \doteq 3.1415$.

When intervals of nearby numbers occur in the book, we use the notation 1.2345_{67}^{89} to denote $[1.234567, 1.234589]$.

Acknowledgments. First, we owe a special debt of thanks to John Boersma, who looked over all the chapters with an expert eye and brought to our attention many places where the exposition or the mathematics could be improved.

We are grateful to Nick Trefethen for his encouragement and advice, and to many mathematicians and members of other teams who shared their solutions and insights with us. In particular, we thank Paul Abbott, Claude Brezinski, Brett Champion, George Corliss, Jean-Guillaume Dumas, Peter Gaffney, Yifan Hu, Rob Knapp, Andreas Knauf, Danny Lichtblau, Weldon Lodwick, Oleg Marichev, Fred Simons, Rolf Strebel, John Sullivan, Serge Tabachnikov, and Michael Trott.

As long as a method of solution has been used by several contestants or can be found in the existing literature, we will refrain from giving credit to individuals. We do not suggest that any specific idea originates with us, even though there are many such ideas to be found in the book. Although each chapter bears the name of the author who actually wrote it, in every case there have been substantial contributions from the other authors.

Folkmar Bornemann, Technische Universität München, Germany

Dirk Laurie, University of Stellenbosch, South Africa

Stan Wagon, Macalester College, St. Paul, USA

Jörg Waldvogel, ETH Zürich, Switzerland

The Story

Problems worthy of attack prove their worth by hitting back.
—Piet Hein

The 100-Digit Challenge began in 2001 when Lloyd N. Trefethen of Oxford University approached SIAM with the idea of publishing 10 challenging problems in numerical computing. SIAM liked the idea and the contest was officially launched in the January/February 2002 issue of *SIAM News*. Here is the full text of Trefethen's challenge:

> Each October, a few new graduate students arrive in Oxford to begin research for a doctorate in numerical analysis. In their first term, working in pairs, they take an informal course called the "Problem Solving Squad." Each week for six weeks, I give them a problem, stated in a sentence or two, whose answer is a single real number. Their mission is to compute that number to as many digits of precision as they can.
>
> Ten of these problems appear below. I would like to offer them as a challenge to the SIAM community. Can you solve them?
>
> I will give $100 to the individual or team that delivers to me the most accurate set of numerical answers to these problems before May 20, 2002. With your solutions, send in a few sentences or programs or plots so I can tell how you got them. Scoring will be simple: You get a point for each correct digit, up to ten for each problem, so the maximum score is 100 points.
>
> Fine print? You are free to get ideas and advice from friends and literature far and wide, but any team that enters the contest should have no more than half a dozen core members. Contestants must assure me that they have received no help from students at Oxford or anyone else who has already seen these problems.
>
> Hint: They're hard! If anyone gets 50 digits in total, I will be impressed. The ten magic numbers will be published in the July/August issue of *SIAM News*, together with the names of winners and strong runners-up.
>
> — *Nick Trefethen, Oxford University*

At the deadline, entries had been submitted by 94 teams from 25 countries. The contestants came from throughout the world of pure and applied mathematics and included researchers at famous universities, high school teachers, and college students. The work was certainly easy to grade, and the results were announced shortly after the deadline: there were 20 teams with perfect scores of 100 and 5 more teams with scores of 99.

Trefethen published a detailed report [Tre02] of the event in the July/August issue of *SIAM News*, and his report nicely conveys the joy that the participants had while working on these problems. During the few months after the contest, many of the teams posted their solutions on the web; URLs for these pages can be found at the web page for this book.

Joseph Keller of Stanford University published an interesting letter in the December 2002 issue of *SIAM News*, where he raised the following point:

> I found it surprising that no proof of the correctness of the answers was given. Omitting such proofs is the accepted procedure in scientific computing. However, in a contest for calculating precise digits, one might have hoped for more.

This is indeed an important issue, one that has guided much of our work in this book. We have addressed it by providing proof for most of the problems and describing the large amount of evidence for the others, evidence that really removes any doubt (but is, admittedly, short of proof). Several responses to Keller's letter appeared in the January/February 2003 issue of *SIAM News*.

The Winners

Twenty teams scored 100 points and were deemed *First Prize Winners*:

— Paul Abbott, University of Western Australia, Nedands, and Brett Champion, Yufing Hu, Danny Lichtblau, and Michael Trott, Wolfram Research, Inc., Champaign, Illinois, USA

— Bernard Beard, Christian Brothers University in Memphis, Tennessee, USA, and Marijke van Gans, Isle of Bute, and Brian Medley, Wigan, United Kingdom ("The CompuServe SCIMATH Forum Team")

— John Boersma, Jos K. M. Jansen, Fred H. Simons, and Fred W. Steutel, Eindhoven University of Technology, the Netherlands

— Folkmar Bornemann, Technische Universität München, Germany

— Carl DeVore, Toby Driscoll, Eli Faulkner, Jon Leighton, Sven Reichard, and Lou Rossi, University of Delaware, Newark, USA

— Eric Dussaud, Chris Husband, Hoang Nguyen, Daniel Reynolds, and Christian Stolk, Rice University, Houston, Texas, USA

— Martin Gander, Felix Kwok, Sebastien Loisel, Nilima Nigam, and Paul Tupper, McGill University, Montreal, Canada

— Gaston Gonnet, ETH Zurich, Switzerland, and Robert Israel, University of British Columbia, Vancouver, Canada

— Thomas Grund, Technical University of Chemnitz, Germany

— Jingfang Huang, Michael Minion, and Michael Taylor, University of North Carolina, Chapel Hill, USA

— Glenn Ierley, Stefan L. Smith, and Robert Parker, University of California, San Diego, USA

— Danny Kaplan and Stan Wagon, Macalester College, St. Paul, Minnesota, USA

— Gerhard Kirchner, Alexander Ostermann, Mechthild Thalhammer, and Peter Wagner, University of Innsbruck, Austria

— Gerd Kunert and Ulf Kähler, Technical University of Chemnitz, Germany

— Dirk Laurie, University of Stellenbosch, South Africa

The SIAM 100-dollar, 100-digit challenge

This is to certify that

Eric Dussaud, Chris Husband, Hoang Nguyen,
Daniel Reynolds and Christiaan Stolk

were

First Prize 100–digit Winners

in this competition entered by hundreds
of contestants around the world.

Lloyd N. Trefethen
Oxford University
May 2002

Figure 1 *Each winning team received $100 and an attractive certificate.*

— Kim McInturff, Raytheon Corp., Goleta, California, USA, and Peter S. Simon, Space Systems/Loral, Palo Alto, California, USA
— Peter Robinson, Quintessa Ltd., Henley-on-Thames, United Kingdom
— Rolf Strebel and Oscar Chinellato, ETH Zurich, Switzerland
— Ruud van Damme, Bernard Geurts, and Bert Jagers, University of Twente, Netherlands
— Eddy van de Wetering, Princeton, New Jersey, USA

Five teams scored 99 points and were counted as *Second Prize Winners:*

— Niclas Carlsson, Åbo Akademi University, Finland
— Katherine Hegewisch and Dirk Robinson, Washington State University, Pullman, USA
— Michel Kern, INRIA Rocquencourt, France
— David Smith, Loyola Marymount University, Los Angeles, California, USA
— Craig Wiegert, University of Chicago, Illinois, USA

Interview with Lloyd N. Trefethen

Many aspects of the story had been known only by Trefethen himself. We are grateful that he agreed to share his views about the contest and on general issues related to numerical computing in the following interview.

Lloyd Nicholas Trefethen is Professor of Numerical Analysis at Oxford University, a fellow of Balliol College, and head of the Numerical Analysis Group at Oxford. He was born in the United States and studied at Harvard and Stanford. He then held positions at the Courant Institute, MIT, and Cornell before taking the chair at Oxford in 1997. In 1985 he was awarded the first Fox Prize in Numerical Analysis.

Trefethen's writings on numerical analysis and applied mathematics have been influential and include five books and over 100 papers. This work spans the field of theoretical and practical numerical analysis. Of more general interest are his essays, such as "The Definition of Numerical Analysis" [TB97, App.], "Maxims about Numerical Mathematics, Computers, Science, and Life" [Tre98], and "Predictions for Scientific Computing 50 years from Now" [Tre00].

You received entries from 94 teams in 25 countries, from Chile to Canada and South Africa to Finland. How many people took part?

There were teams of every size up to the maximum permitted of six—altogether, about 180 contestants. Most of these scored 40 points or more. Of course, those are just the ones I know about! There were certainly others who tried some of the problems and didn't tell me. I would hear rumors that so-and-so had been staying up late nights. Hadn't he sent me any numbers?

Were there countries that you expected entries from, but got none?

No, I wouldn't say so. I got entries from six English-speaking countries, and I did not expect to hear much from the non-English world, where *SIAM News* is less widely distributed. So it was a nice surprise to receive entries from countries all over the world, including China, Russia, Spain, Slovenia, Greece, Argentina, Mexico, and Israel.

What kind of background and training did the contestants have? Was there a significant difference between those who succeeded and those who did not?

We had contestants of every sort, from amateurs and students to world-leading mathematicians. But it has to be admitted that overall, most teams were based at universities, and the higher scoring ones usually included an expert in numerical computation with a Ph.D. A crucial ingredient of success was collaboration. More than half of our teams consisted of a single person, but only five of those singletons ended up among the winners.

Wolfram Research (Mathematica®) had a winning team and Gaston Gonnet (one of the founders of Maple®) was part of a winning team. Were there teams from other mathematical software companies?

Cleve Moler, the creator of Matlab® and Senior Scientist at MathWorks, was teaching the undergraduate course CS 138 at Stanford at the time of the Challenge, and he put problem 1 on the students' take-home final exam in March 2002. I don't know how the students did, but I can assure you that Cleve got all the digits right.

Mathematica, Maple, and Matlab were used by many teams. What other software was used, and were there any unexpected choices?

Yes, the three M's were very popular indeed. We also had plenty of C and C++ and Fortran and many other systems, including Java™, Visual Basic®, Turbo-Pascal, GMP, GSL, Octave, and Pari/GP. One contestant attacked the problems in Excel, but he was not among the high scorers.

Did you know all the answers at the time you posed the questions?

I knew eight of them to the full 10 digits. My lacunae were the complex gamma function (Problem 5) and the photon bouncing between mirrors (Problem 2).

I'm afraid that my head was so swimming with digits in those last few weeks that I did something foolish. Shortly before the contest deadline I sent a reminder to NA Digest on the web to which I added a P.S. that I thought would amuse people:

```
Hint.  The correct digits are 5640389931136747269191274224157853l
4225719123954474634207834370383758379793236749526330686862l433534.
(Not in this order.)
```

As I said, at that time I didn't even know all 100 digits! To make my little joke I took the 90 or so digits I knew and padded them out at random. It never crossed my mind that people would take this clue seriously and try to check their answers with it. They did, and several of the eventual winners found that they seemed to have too many 4s and 7s, too few 2s and 5s. I am embarrassed to have made such a mistake and worried them.

Do you have an opinion on whether "digits" in a contest such as this should mean rounded or truncated?

What a mess of an issue—so empty of deep content and yet such a headache in practice! (Nick Higham treats these matters very well in his big book *Accuracy and Stability of Numerical Algorithms* [Hig96].) No, I have no opinion on what "digits" should mean, but I very much wish I had specified a precise rule when presenting the Challenge.

Why did the list of 100-point winners grow from 18 to 20 within a few days after you posted the results?

Mainly for reasons related to just this business of rounding vs. truncation. Would you believe that there were seven teams that scored 99 points out of 100? After I circulated a list with 18 winners, two of those seven persuaded me that I had misjudged one of their answers. Two more teams failed to persuade me.

For the teams that missed a perfect score by one or two digits, can you tell us which problems were missed? Did this coincide with your estimate of the difficulty ranking? Which problems did you think were most difficult?

The troublemakers were those same two I mentioned before: the complex gamma function and the photon bouncing between mirrors. The latter is a bit surprising, since it's not hard once you realize you need extended-precision arithmetic.

Are there any other facts from the results that help in the evaluation of the relative difficulty of the problems?

As I accumulated people's responses I quickly saw which problems gave the most trouble. Along with the gamma function, another tough one was the heating of a plate (Problem 8). It was also interesting to see for which problems some people were reporting lots of extra digits, like 50 or 500. A hard problem by this measure was the infinite matrix norm (Problem 3). When the contest ended I only knew about 18 digits of the answer, as compared with 50 digits or more for most of the other problems.

What experience (either before or after the contest) makes you believe that the problems were truly hard?

The problems originated in a problem-solving course that our incoming D.Phil. students in numerical analysis at Oxford take in their first term. (I got the idea from Don Knuth, and he told me once that he got it from Bob Floyd, who got it from George Forsythe, who got it from George Pólya.) Each week I give them a problem with no hints, and, working in pairs, they have to figure out as many digits of the answer as they can. In creating the problems I did my best to tune them to be challenging for this group of highly talented but not so experienced young people.

In your SIAM News report you indicated that perhaps some of the problems should have been harder. For example, you might have changed the target time in Problem 2 from 10 seconds to 100 seconds. Are there any other such variations to the other problems that might have made them more difficult? Or was it in fact a good thing that so many teams got all 100 digits?

I think five winning teams would have been better than twenty. As you suggest, some of the problems had a parameter I tuned to try to make them hard but not too hard. Changing a few numbers in the functions of Problems 1, 4, and 9, for example, could have made them even more devilish. On the whole I underestimated the difference between a pair of new graduate students with a week's deadline and a team of experienced mathematicians with months to play with.

What was your motivation in making the Challenge public? What did you expect to get out of it? Did you in fact get out of it what you wanted?

I love this kind of hands-on computing and I wanted others to have fun too. Too often numerical analysts get lost in theory and forget how satisfying it is to compute actual numbers.

What was SIAM's reaction when you first proposed the contest?

They liked the idea from the start. Gail Corbett, the *SIAM News* editor, is wonderfully encouraging of off-beat ideas.

You excluded people who might have seen these problems as members of the problem-solving squad at Oxford. Were all the problems taken from your course, and therefore already tested on students? Did you modify any for the Challenge or were they identical to the ones you give your students?

Yes, all the problems were taken from the course, and I changed nothing but a few words here and there. I didn't dare introduce untested problems and risk making an error.

What type of problem (whether from the Challenge or not) gives your students the most difficulty?

Students these days, alas, have often not had much exposure to complex analysis, so problems in the complex plane always prove hard. Personally, I don't see how one can go far in the mathematical sciences without complex variables, but there you are.

How many digits do they usually come up with?

On a typical meeting of our course there might be four teams presenting results at the whiteboard and they might get 4, 6, 8, and 10 digits. You'd be surprised how often the results appear in just that order, increasing monotonically! I guess the students who suspect their answers aren't so accurate volunteer fast, to get their presentations over with, and those who've done well are happy to wait to the end. One year we had an outstanding student who got nearly exact solutions two weeks running—we called him "Hundred Digit Hendrik."

How do you usually decide about the correctness of the digits? How did you decide at the contest?

At home at Oxford, I have almost always computed more digits in advance than the students manage to get. For the Challenge, however, scoring was much easier. In the weeks leading up to the deadline I accumulated a big file of the teams' numerical answers to the 10 problems. I'd record all the digits they claimed were correct, one after another. Here's an extract from Problem 8 with the names changed:

```
Argonne     0.4235
Berlin      0.4240113870 3
Blanc       0.2602370772 04
Blogg       0.4240113870
Cambridge   0.4240114
Cornell     0.4240113870 3368836379743366859326
CSIRO       0.3368831975
IBM         0.4240113870 336883
Jones       0.282674
Lausanne    0.42403
Newton      0.42401139
Philips     0.4240113870
Schmidt     0.4240074597 42
Schneider   0.4240113870 336883637974336685932564512478
Smith       0.4240113870 3369
Taylor      0.4240113870
```

In the face of 80 lines of data like this, based on different algorithms run in different languages on different computers in different countries, who could doubt that the correct first 10 digits must be 0.4240113870?

What is your opinion about Joe Keller's letter in SIAM News regarding the lack of proof of correctness, and the responses it got by some of the contestants?

Joe Keller has been one of my heroes ever since I took a course from him as a graduate student. Now, as I'm sure Joe knows, it would have killed the Challenge to demand that contestants supply proofs. But this doesn't mean I couldn't have said something in the write-up afterwards about the prospects for proofs and guaranteed accuracy in numerical computation. I wish I'd done that, for certainty is an ideal we must never lose sight of in any corner of mathematics. Those responses to Joe's letter "defended" me more vigorously than I would have defended myself.

Do you think the offer of money played any role in the interest generated? Is it correct that you originally only had a $100 budget in total?

Of course it did! Any whiff of money (or sex) makes us sit up and take notice. My "budget" consisted of the Trefethen family checkbook. But do you know something funny? After I'd sent certificates to the winners, I found that some of them seemed to care quite a bit about the money for its own sake, not just as a token, even if at best they could only hope to pocket a fifth or a sixth of $100. I guess there's a special feeling in winning cash that just can't be matched by certificates and publicity.

You decided to award three $100 prizes, but then an anonymous donor stepped forward and made it possible to give each team $100. Is the donor a mathematician? Do you know him or her personally? What especially impressed the donor about the contest?

What a thrill it was one day to receive his offer out of the blue! Now that the contest is well past, the donor does not object to having his name revealed. He is William Browning, founder and President of Applied Mathematics, Inc. in Gales Ferry, Connecticut. I haven't yet met Dr. Browning, but I know from our email exchanges that he was happy to support our field in this way. In one message he wrote:

> I agree with you regarding the satisfaction—and the importance—of actually computing some numbers. I can't tell you how often I see time and money wasted because someone didn't bother to "run the numbers."

Were you surprised by the response, and if so, why?

I was surprised by people's tenacity, their determination to get all 100 digits. I had imagined that a typical contestant would spend a dozen hours on the Challenge and solve three or four problems. Well, maybe some of them started out with that plan. But then they got hooked.

How much email traffic did the contest generate?

Megabytes! Those 94 entries corresponded to about 500 email messages. People would send results in the form of 30-page documents, then send updates and improvements, ask me questions, and so on. For some weeks before the deadline I seemed to be spending all my time on this. I didn't plan for that, but it was great fun.

Were there any WWW groups formed that openly discussed the problems? Did you scan the net for those and ask them to keep quiet?

I didn't scan the net systematically, but I heard about a group in Germany that was circulating ideas on the web. I asked them to go private, which they did.

How important was the Internet for the contest? Did people find ideas, software, etc. there?

It was crucial in spreading word of the event, and in the case of the SCIMATH team of three, the participants found each other through the Internet and so far as I know have still never met face-to-face. I think the Internet helped many contestants in technical ways too. For young people these days it is simply a part of life. At Oxford we have one of the world's best numerical analysis libraries, but you don't often see the D.Phil. students in it.

Where, other than SIAM News and NA-Net, was the contest publicized?

There were half a dozen places. The two I was most aware of were Wolfram Research's MathWorld web site and a notice published in *Science* by Barry Cipra with the heading "Decimal Decathlon."

What was the biggest surprise for you on the human side of the story?

How addicted people got to these problems! I received numerous messages of thanks for the pleasure I had given people, and quite a few asked if I would be making this a regular event. (No, I won't.)

What was the biggest disappointment?

I wish there had been more entries from my adopted country of Britain.

What was the biggest surprise as to the solutions that were put forward?

One day I got a message telling me that Jean-Guillaume Dumas of the LinBox team had solved Problem 7 *exactly* with the help of 182 processors running for four days. He showed that the answer is a quotient of two 97,389-digit integers. Wow!

Did you learn any new mathematics in reviewing the solutions?

On every problem there were surprises and new ideas, for as this book shows, these problems have many links to other topics. But if I'm honest I must tell you that with a thousand pages of mathematics to evaluate in a matter of days, I didn't have time to learn many new things properly.

Some people might say that a computation to more than six significant digits is a waste of time as far as real-world applications go. What do you think?

I am very much interested in this question, and I have distilled some of my views into the notion of a *Ten-Digit Algorithm*: "Ten-digits, five seconds, and just one page." Some writings on ten-digit algorithms are in the pipeline, so if you'll forgive me, I won't elaborate here.

You said that round-off error plays a small role compared to algorithm design, and that feeling is reflected in your choice of problems, since only one required high precision. Can you expand on this point?

In my 1992 *SIAM News* essay "The Definition of Numerical Analysis" [TB97, App.] I argue that controlling rounding errors is just a small part of numerical analysis, maybe 5% or 10%. The main business of this field is the development of algorithms that converge fast, and the ideas behind these algorithms would be just as necessary even if computers could work in exact arithmetic. So yes, it seems fitting that only 1 of the 10 problems, the photon bouncing off mirrors, is one where you have to think carefully about rounding errors.

In your report in SIAM News you asked, perhaps tongue-in-cheek, whether these problems could be solved to 10,000 digits. Do you think that such work has value beyond mere digit-hunting?

Humans have always progressed by tackling challenges, whether real or artificial.

Contests like yours where a mathematician announces, "Look, here is a hard problem that I know how to solve. I wonder whether you are able to do it too," were quite common at the time of Fermat, the Bernoullis, and Euler. But we do not see many such challenges today, certainly not in numerical computing. Should there be more of them? Should a numerical analysis journal consider starting a problem section where problems in the style of the Challenge are posed?

That's a very interesting question, for you're right, there was a different style in the old days, when science was an activity of a small elite and had not been professionalized. Yes, I think it would be good to have more challenges nowadays, though journals might not be the right medium since they are so slow.

Are there any important morals to draw from the challenge?

Huckleberry Finn begins with some remarks on this subject.

What has been the impact of the Challenge?

As Mao Tse-Tung or Zhou En-Lai is supposed to have said when asked about the impact of the French Revolution, "It's too early to tell!"

What kind of activity developed after the contest? Do you know of any web pages, papers written, talks given?

Yes, there were at least a dozen talks and tech reports, for the contestants were wrapped up in these problems and wanted to share their good ideas. Of course, this book of yours is the most extraordinary result of the Challenge.

What was your reaction when you heard that a book was being planned?

I was amazed and delighted.

You have told us that the Challenge has shown us a world that seems as distant from von Neumann as from Gauss. What makes you say that?

I imagined that the Challenge would unearth 10-digit solutions that would be a kind of culmination of 50 years of progress in algorithms and software. What in fact happened seems more of a transcendence than a culmination. I think the Challenge and its amazing aftermath—the 10,000-digit mix of fast algorithms, symbolic computation, interval arithmetic, and global collaboration displayed in this book—show that we have entered a world that would be unrecognizable even to the giants of the recent past such as Turing, von Neumann, or Wilkinson. It is a world that seems hardly less distant from von Neumann, who knew about computers and jet aircraft, than from Gauss, who lived in the time of Napoleon.

Would you do it again?

Absolutely.

Comments by the Contestants

We solicited some comments from the prize-winning teams and include a summary here. The diversity of approaches is noteworthy. Several contestants programmed everything from scratch "for sport," but most teams used software packages at some point. It is clear from the responses that the primary motivation for the sometimes great effort required was simply the satisfaction one gets by solving a difficult problem. But the recognition was nice too, and so all participants, like us, are grateful to Trefethen for setting the challenge and publicizing the results.

Another general point that comes through is that problems that offered little in the way of analytical challenge were considered the least favorite. Yet all the problems can teach us something about numerical computation, and we hope that point comes through in this book.

Why do it?

— I thought it would be a good way to promote numerical analysis within the department. *(Driscoll)*

— I have done a lot of competitions before and like doing them. *(Loisel)*

— Once I read the problems over I was hooked and used my spare time until I cracked them all. *(van de Wetering)*

— The general nature of the challenge, and, for some problems, the desire to test out some new functionality in *Mathematica* that was under development. *(Lichtblau)*

— I was taking a numerical differential equations class taught by Dr. Rossi (who solved the random-walk problem). We were studying quadratures when the challenge was issued and Dr. Rossi offered an A to any student who could solve the quadrature problem to 10 digits. I couldn't. However, I did get interested in the challenge and began attending our weekly meetings. *(Faulkner)*

— My main motivation was that I have always been interested in problem solving. In this connection, I still regret that the Problems and Solutions section of *SIAM Review* was discontinued at the end of 1997, just at the moment when I went into an early retirement! The Problem section played an important role as a forum where scientists could bring their mathematical problems to the attention of a readership of widely varying expertise. I believe the editors of the journal underestimated the section's role as such a forum when they decided to drop it. All four members of our team are retired mathematicians from Eindhoven University of Technology. At the university we share a big office where we can meet regularly. After a month or so we had solved six problems and were considering submitting our solutions (recalling Trefethen's comment that he would be impressed "if anyone gets 50 digits in total"). Fortunately we did not do that but continued till we had solved all 10 problems. *(Boersma)*

Was the contest worth it?

— Yes, because of the perfect score and because the undergraduate on the team is now contemplating graduate study in numerical analysis. *(Driscoll)*

— For the satisfaction of solving the problems. *(van de Wetering)*

— It was fun and I learned quite a few things on the way. *(Kern)*

— We were able to demonstrate advantages in *Mathematica*'s arithmetic and also to show the utility of some of the sparse linear algebra under development at the time. *(Lichtblau)*

— The challenge completely redefined the way I viewed computational mathematics. At that time I had taken a course in numerical linear algebra and was taking a course in numerical DEs. I learned that just because you knew how to compute a QR decomposition didn't mean that you knew anything about numerical linear algebra. *(Faulkner)*

— Yes, definitely. It was great fun. The problems of the Challenge formed a nice mixture, to be solved by both analytical and numerical methods. As for me, I preferred the analytically oriented problems like Problems 1, 5, 6, 8, 9, and 10. Also, the use of a computer algebra package (*Mathematica* in our case) greatly contributed to the successful outcome. *(Boersma)*

What insight did you find to be most satisfying?

— I enjoyed making headway with nonnumerical (i.e., analytical) methods. It was great fun to find useful tricks and do convergence acceleration and error estimation. Also, there is something very satisfying about using independent methods and getting digits that match. *(van de Wetering)*

— The whole thing. Going from, "How can one do that?" when I first looked at the problems, to, "Yes, it can be done" at the end. *(Kern)*

— An integration by parts trick to speed the convergence for Problem 1. It turned out not to be the most elegant solution, but it blew the problem open for us. *(Driscoll)*

— I liked the infinite matrix norm problem. I wrote a routine that was able to compute Ax for any vector x without instantiating the matrix. This allowed me to use a 50,000 $\times$ 50,000 matrix in a shifted power method and still terminate in seconds. I also used a kind of hierarchical technique. I used the eigenvector for the 25,000 $\times$ 25,000 case as the initial vector in the 50,000 $\times$ 50,000 case, so very few iterations were needed. *(Loisel)*

— That I could come up with a nice solution for Problem 3, as opposed to doing it by a brute force extrapolation. In short, some thinking about this problem paid off. *(Strebel)*

— I really liked watching my professors struggle on these problems, just as we undergraduates struggle on the problems they give us. *(Faulkner)*

— I learned most from the solutions of Problems 6, 10, and 5. At first I was scared of Problems 6 and 10, because my knowledge of probability theory is rather superficial. I felt at ease again after I had found the right analytical expressions to manipulate. Steutel and I were able to generalize Problem 6 to that of a general biased two-dimensional random walk, and we even considered writing a paper on this generalized problem. However, in January 2003, Folkmar Bornemann brought to our attention that the generalized problem had been treated in papers by Henze [Hen61] and Barnett [Bar63]; so that was the end of our paper to appear! All this has now been incorporated, in a most satisfactory manner, in Chapter 6 of the present book. *(Boersma)*

Which problem was the hardest?

— Problem 5. Basically I brute-forced it and was sweating over the last digit or two. *(Driscoll)*

— Problem 5 was the one question where I had to do some research, having never tackled this type of problem. It was the last question that I tackled, so feeling I had 9 out of 10 provided enough motivation to sort this one out. *(Robinson)*

— Problems 3 and 5, the latter only because we failed to see the "correct" approach. *(Lichtblau)*

— Problem 5 was by far the hardest; it took Jos Jansen and me over a month to crack it. Our calculations led to maxima of $|f(z) - p(z)|$ which were decreasing, thus showing that we had not yet found the best polynomial. Finally we found in the literature a necessary and sufficient criterion for the polynomial $p(z)$ to be the polynomial of best approximation for $f(z)$ on $|z| = 1$. This criterion could be implemented into an algorithm that produced the best polynomial $p^*(z)$ and the associated maximum $|f(z) - p^*(z)|$ extremely rapidly. *(Boersma)*

Were any problems extremely easy for you?

— Problem 10. I'm pretty good with conformal mapping and knew right away that it was a Maple three-liner. *(Driscoll)*

Did you have a favorite problem? A nonfavorite?

— I'd say Problem 1. It's really not obvious that the limit exists. *(Kern)*

— In a strange way Problem 10 is my favorite, which is at its core not a numerical problem at all. Of course it's a physics problem. It also reminded me of the origins of Fourier theory. A fun fact is that the resulting answer is a very quickly converging series (you need only one term to get to the answer). *(van de Wetering)*

— Problems 4 and 9 provided little satisfaction. *(Robinson)*

— I liked Problem 6. It leads pretty quickly to some interesting numerics and some difficult theoretical questions. There are also effective means for solving it all along a full spectrum from pretty simple to sophisticated—direct simulation is about the only guaranteed failure. *(Driscoll)*

— My least favorite was Problem 2, as there is no clever way to solve it. *(Strebel)*

Were you confident in your digits?

— I had more than one method for most problems. But you're never 100% confident. There is always the chance of dropping a digit or misinterpreting a problem. *(van de Wetering)*

— I had either two methods to solve the problems or an independent check of the result, so I was reasonably confident. *(Strebel)*

— Problem 10 was frustrating. I felt that there must be a neat way of doing this, but could only come up with slowly converging double infinite series. In the end, I stuck with a brute force approach, but this was the question that I would have been least surprised to have got wrong. *(Robinson)*

Have you any comments on digits you missed?

— I missed the last digit for Problem 2. I must admit it took me a while to realize it was so hard. I started using the geometric modelling part of a mesh generator written by a colleague, but as it used single precision, I knew I had to go to Maple. I programmed the whole thing in Maple and explored the way the results depended on `Digits`. So I found out I needed a lot of digits, and I used `Digits := 40`. Except, for some reason, I set `Digits := 20` in the calculation I copied in my report, and there you are, last digit is wrong. *(Kern)*

Many of the teams were partially or entirely nonmathematicians. If you are not primarily a mathematician, do you feel that this sort of exercise is valuable to the applied math or scientific computation community?

— I think it is a valuable exercise for all applied sciences. It drives home the point that convergence doesn't always come easy and that when a black box spits an answer at you, you really need to look under the hood and get a clue if it was good/bad/completely wrong (which is very true in my line of work: finance). Furthermore, it is valuable to realize that solvable problems often allow for a wide variety of methods. *(van de Wetering)*

Did you find any unusual sources?

— Only for Problem 5 did I do some serious literature research. In this case, knowing that Lloyd N. Trefethen had worked in this area proved to be a disadvantage: the methods in his papers really can't produce anything close to ten digits. I wasted some hours barking up this tree. *(Driscoll)*

Have you any additional comments?

— It was a pleasure having the opportunity in my spare time to work on these challenging problems. The charm of these problems is that they cover a lot of ground, are mostly nontrivial but are expertly crafted to require little overhead: no specialized knowledge, software, or hardware is needed to crack them as posed. *(van de Wetering)*

— We had a gung-ho MATLABer (me) and a gung-ho Mapleist (DeVore). Our approaches to problems were, almost without exception, entirely unrelated. Though ultimately all the problems save Problem 2 were eminently doable in standard MATLAB, I picked up some extra respect for Maple. With regard to numbers 1 and 9 in particular, I would say that Maple can be devilishly difficult to coax into numerical integrals, but MATLAB's packaged offerings in this area aren't even on the map. *(Driscoll)*

— Overall I really enjoyed the experience and feel that I got to contribute a lot to the team. The University of Delaware Mathematics Department is a great place for undergrads, but I know that not all schools allow undergrads to be so involved. I am not a straight A student, but I could handle some of these problems, so if I have any advice to share, it is that professors should let their students get involved in learning the challenges of mathematics. You never know where the great ideas will come from. *(Faulkner)*

Relative Difficulty of the Problems

One way of evaluating the relative difficulty is to look at the statistics of how the teams performed on the problems. Table 1 shows that Problem 4 was clearly the easiest and Problem 5 the hardest. Looking at the number of teams that missed a perfect score of 10 by just 1 digit shows that Problems 1, 3, and 7 were difficult, while the mean values of the points obtained show that Problems 6 and 8 were tough. And Problem 10 was tried by the smallest number of teams, probably because of the technical difficulty of dealing with Brownian motion.

However, the general spread of results through the table is further evidence that the contest was well designed.

Table 1 *Number of teams getting k digits on problem j and the mean value of the points obtained for that problem.*

Correct digits:	0	1	2	3	4	5	6	7	8	9	10	Mean value
Problem 1 (87 teams)	5	2	1	4	4	–	3	2	–	11	55	8.2
Problem 2 (80 teams)	2	–	3	–	4	3	6	2	1	1	58	8.6
Problem 3 (78 teams)	1	1	6	1	–	–	1	–	2	15	51	8.8
Problem 4 (84 teams)	–	3	–	–	–	–	–	–	1	–	80	9.7
Problem 5 (69 teams)	5	7	5	6	5	1	1	1	1	15	22	6.3
Problem 6 (76 teams)	11	2	3	–	1	3	1	–	3	1	51	7.6
Problem 7 (78 teams)	–	1	4	1	5	–	–	–	1	14	52	8.8
Problem 8 (69 teams)	6	–	3	1	7	1	1	2	–	–	48	7.9
Problem 9 (80 teams)	3	2	3	1	2	1	1	4	6	4	53	8.4
Problem 10 (62 teams)	–	2	1	1	3	1	2	1	–	–	51	8.9

Chapter 1

A Twisted Tail

Dirk Laurie

The shortest and best way between two truths of the real domain often passes through the imaginary one.
—Jacques Hadamard [Had45, p. 123]

"Mine is a long and a sad tale!" said the Mouse, turning to Alice and sighing.
"It is a long tail certainly," said Alice, looking down with wonder at the Mouse's tail; "but why do you call it sad?"
—Lewis Carroll (*Alice's Adventures in Wonderland,* 1865)

Problem 1

What is $\lim_{\epsilon \to 0} \int_{\epsilon}^{1} x^{-1} \cos(x^{-1} \log x)\, dx$*?*

1.1 A First Look

This problem really seems to hit us with everything nasty. The graph of the integrand is appalling (see Figure 1.1; the bit near 0 has been suppressed to make the graph fit onto the page). It is difficult to believe that the required limit exists, but of course we trust Professor Trefethen not to ask the impossible. In the course of this chapter we shall be able to prove that the required limit does exist (see p. 28).

If we think of it as

$$S = \int_0^1 x^{-1} \cos(x^{-1} \log x)\, dx, \tag{1.1}$$

then there are two features, each of which by itself would be enough to make numerical integration difficult:

- The integrand is unbounded in a neighborhood of the left endpoint.
- The integrand oscillates infinitely often inside the interval of integration.

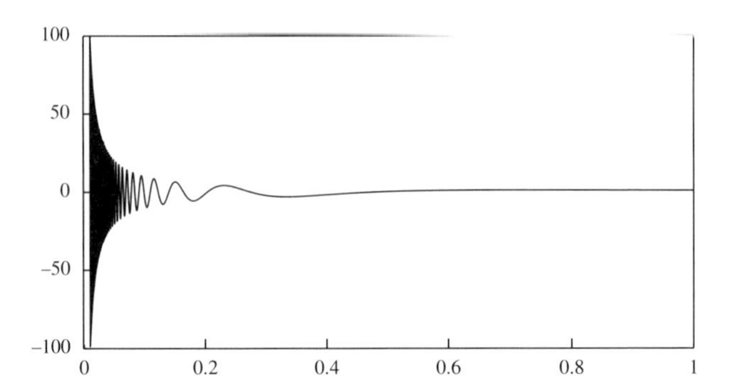

Figure 1.1. *Integrand with twisted tail.*

For the rest of this chapter, integrals such as that in (1.1) are to be thought of as improper Riemann integrals, since the corresponding Lebesgue integral does not exist.

1.2 Oscillatory Integrals in General

Let us first look in general at integrals of the form $\int_a^b g(x)\,dx$, where g has infinitely many extrema in the interval (a, b). Such integrals defeat the usual formulas (Gaussian quadrature, etc.) that assume polynomial-like behavior of g and are also quite hard for automatic integrators based on iterative subdivision.

To simplify matters, let us assume that a is the only accumulation point of the extrema of g. The basic strategy for infinitely oscillatory integrals (due to Longman [Lon56]; see also [Eva93, Chap. 4]) is usually formulated (see [PdDKÜK83, p. 80]) over an infinite range, but a finite interval can be treated similarly.

1. Choose a sequence $a_k \to a$ with $b = a_0 > a_1 > a_2 > a_3 > \cdots$. This sequence partitions $(a, b]$ as a union of subintervals $(a_k, a_{k-1}]$, $k = 1, 2, 3, \ldots$.

2. Integrate g over each of these subintervals. The idea is that the subdivision is fine enough to ensure that g does not show any unpleasant behavior when restricted to any particular subinterval; then any reasonable numerical method will work. Let

$$s_k = \int_{a_k}^{a_{k-1}} g(x)\,dx.$$

3. Sum the infinite series $\sum_{k=1}^{\infty} s_k$. The answer is the desired integral.

Note that a singularity at a (as exists in this case) does not make the problem any harder: the subdivisions cluster at a, which is always a good way of coping with a singularity. (A singularity at b, on the other hand, would have added to the difficulty of the problem.)

For this strategy to work, the terms s_k must either converge to zero very fast (which seldom happens) or have sufficiently regular behavior so that an algorithm for the extrapolation of a sequence is applicable. Some possible choices for the subdivision points a_k are:

- The zeros of g. This was Longman's original choice.
- The extrema of g.
- If $g(x) = f(x)w(x)$, where w is monotonic, it is usually easier, and just as effective, to use the extrema of f instead of those of g.

In the case of a function on $[a, \infty)$ such that the distance between successive zeros approaches a constant value h, Lyness [Lyn85] has pointed out that it is not necessary to find those zeros: if the subdivision points are equally spaced at the distance h, the method still works. Lyness does not actually mention using extrema rather than zeros, but comes close to doing so, by suggesting as subdivision points the points halfway between successive zeros of the integrand.

We need to say a bit more on the delicacy of using the zeros of g as subdivision points. The difficulty arises because (unlike the set of extrema) the set of zeros is not invariant when the graph of g is shifted up or down. There is just one correct position that leads to regularly spaced zeros; all others give a limping pattern in which too-short steps alternate with too-long steps. So we should really not be using the zeros of g, but the solutions of the equation $g(a_k) = c$ for the correct c. In many cases the choice of c is obvious: if the value of g at its extrema is always ± 1, then clearly $c = 0$ is sensible; if $\lim_{x \to a} g(x) = g_0$, then choosing $c = g_0$ is the only way to obtain an infinite sequence s_k.

In general, using the extrema is a better option. It is more robust, and the terms in the series are smaller. It is true that one cannot guarantee, as when using the zeros, that the series will alternate, but nevertheless in practice it usually does.

1.3 This Particular Oscillatory Integral

Coming back to the present case, it is obvious that we can simply use the zeros of g. Therefore, the subdivision points satisfy

$$a_k^{-1} \log a_k = -\left(k - \tfrac{1}{2}\right)\pi, \quad k = 1, 2, 3, \ldots .$$

In the case of the extrema, replace $k - 1/2$ by k. We shall have more to say on how to solve this nonlinear equation in the next section.

For $k = 1, 2, 3, \ldots, 17$, compute the zeros of the integrand and evaluate

$$s_k = \int_{a_k}^{a_{k-1}} x^{-1} \cos(x^{-1} \log x)\, dx.$$

Then do the same, using the extrema of the cosine factor of the integrand instead.

The integrand is very well behaved when restricted to a single subinterval, and almost any numerical method works: let us take Romberg integration (see Appendix A, p. 236), which has been a staple ingredient of almost all introductory numerical analysis texts since Henrici's 1964 book [Hen64]. Romberg integration gives the first 17 terms of the two series at a total cost of 2961 and 3601 function evaluations, respectively. The results (calculated on a machine which has IEEE double precision, slightly less than 16 significant digits) are shown in the left part of Table 1.1.

Table 1.1. *The first 17 terms of the series obtained by integration between zeros or extrema. Also shown are the terms obtained from series acceleration by iteratively applying Aitken's Δ^2-method.*

s_k between zeros	s_k between extrema	Accel. (betw. zeros)	Accel. (betw. extr.)
0.5494499236517820	0.3550838448097824	0.5494499236517820	0.3550838448097824
−0.3400724368128824	−0.0431531965722963	−0.2100596710635879	−0.0384770911855535
0.1905558793876738	0.0172798308777912	−0.0078881065973032	0.0076628418755175
−0.1343938122973787	−0.0093857627537754	0.0328626109424927	−0.0009933325213392
0.1043876588340859	0.0059212509664411	−0.0408139994356358	0.0001062999924531
−0.0855855139972820	−0.0040870847239680	−0.0002275288448133	−0.0000160316464409
0.0726523873218591	0.0029962119905360	0.0000467825881499	0.0000010666342681
−0.0631902437669350	−0.0022935621330502	−0.0000029629978848	−0.0000001739751838
0.0559562976074883	0.0018138408858293	0.0000004098092536	0.0000000090009066
−0.0502402983462061	−0.0014714332449361	−0.0000000298386656	−0.0000000013911841
0.0456061367869889	0.0012183087499507	0.0000000036996310	0.0000000000923884
−0.0417708676657172	−0.0010257939520507	−0.0000000002614832	−0.0000000000089596
0.0385427179214103	0.0008758961712063	0.0000000000282100	0.0000000000011693
−0.0357870041817758	−0.0007568532956294	−0.0000000000024653	−0.0000000000000529
0.0334063038418840	0.0006607085396203	0.0000000000001224	0.0000000000000079
−0.0313283690781963	−0.0005819206203306	−0.0000000000000221	−0.0000000000000007
0.0294984671675406	0.0005165328013114	−0.0000000000000018	−0.0000000000000000

Both series converge very slowly: the first like $O(k^{-1})$, the second like $O(k^{-2})$. But both series are alternating, which is theoretically significant, since one can base a convergence proof on it—see §1.8. It is also practically significant, since almost any good extrapolation algorithm will work. For example, simply iterating Aitken's Δ^2-method (see Appendix A, p. 244) gives the terms of the accelerated series shown in the right part of Table 1.1.

In both cases, the terms certainly seem to be tending to zero rapidly, and their sum is probably quite a bit more accurate than the required 10 digits. The two sums are 0.3233674316777786 and 0.3233674316777784, respectively. Distrusting the last digit, we obtain

$$S \doteq 0.323367431677778,$$

which happens to be correct to all digits shown.

1.4 Complex Integration

Knowledge of an analytic continuation of a real function off the real axis into the complex domain is a treasure beyond price. It allows numerical methods of great power and versatility.

Using Euler's identity $e^{ix} = \cos x + i \sin x$, we find that

$$S = \int_0^1 \mathrm{Re}\left(\frac{e^{(i\log x)/x}}{x}\right) dx = \int_0^1 \mathrm{Re}\left(e^{((i/x)-1)\log x}\right) dx = \int_0^1 \mathrm{Re}\left(x^{i/x-1}\right) dx.$$

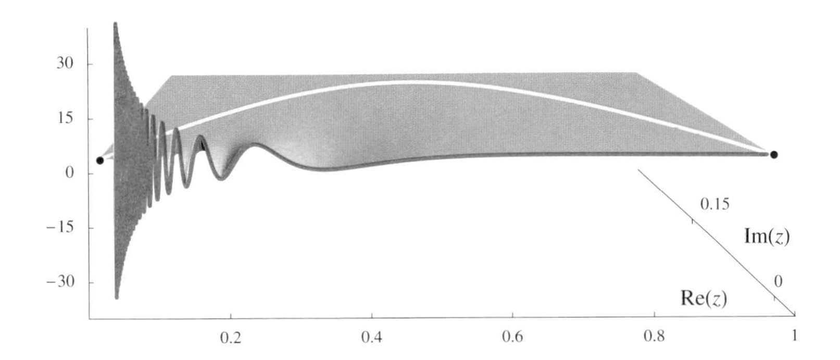

Figure 1.2. *Integrand of Problem 1 in the complex plane. The red line shows the original formulation with the integration path along the real axis, and the yellow line, the contour used for the complex integration method.*

By Cauchy's theorem, the line integral between two points in the complex plane does not depend on the path taken, as long as both paths lie in a region inside which the function is analytic. So we can replace the real variable x above by a complex variable z, taking care to select the integration path appropriately.

In this case, the integrand is very well behaved as soon as one moves into the positive quadrant away from the real axis, as the perspective view in Figure 1.2 shows. A good contour, as shown in the figure, gets away from the real axis immediately and later turns back to the point $(1, 0)$. If the contour is parametrized by $z = z(t), \quad t \in [a, b]$, where $z(a) = 0$ and $z(b) = 1$, we have

$$S = \operatorname{Re} \int_a^b z(t)^{i/z(t)-1} z'(t)\, dt. \tag{1.2}$$

This integral is an improper Riemann integral along the line $z(t) = t$, which is the original path of the problem statement, but it becomes a proper integral along a contour like the one shown if we define the integrand at 0 by continuity.

For example, take a simple parabola:

$$z(t) = t + it(1 - t), \quad z'(t) = 1 + i - 2it, \quad t \in [0, 1]. \tag{1.3}$$

Figure 1.3 shows a plot of the real part of the integrand in (1.2). The oscillation can still be seen under a microscope, but its amplitude vanishes rapidly as $t \to 0$. This function responds readily to integration by canned software.

An Octave Session

```
>> function y=func(t);
>> z=t+i*t.*(1-t); y=real(z.^(i./z-1).*(1+i*(1-2*t)));
>> endfunction
>> quad('func',0,1)

ans = 0.323367431677773
```

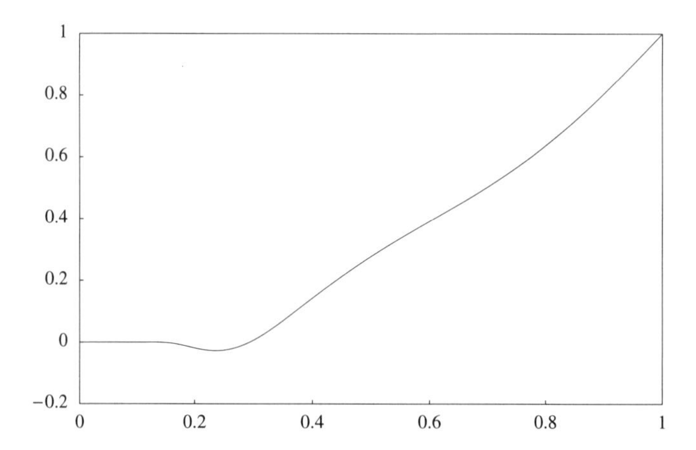

Figure 1.3. *Integrand along the contour* $z(t) = t + it(1 - t)$.

It is encouraging that the first 14 digits agree with what we had before. Let's try for more digits. Here is a PARI/GP session. We need to code the special case $t = 0$ this time because PARI/GP, unlike Octave, uses a closed quadrature formula (i.e., one that evaluates the integrand at the endpoints).

A PARI/GP Session

```
? func(t) = if(t==0,return(0),\
      z = t+I*t*(1-t); real(z^(I/z-1)*(1+I*(1-2*t)))
? intnum(t=0,1,func(t))

  0.32336743167777876139937008 68
```

And in fact (as we can confirm by computing in higher precision) we have

$$S \doteq 0.32336\,74316\,77778\,76139\,93700\,8.$$

1.5 Lambert's *W* Function

To find the partition of $(0, 1]$ needed for the subdivision method, we need to solve for a_k many equations of the form

$$a_k^{-1} \log a_k = -b_k.$$

This is not a very hard task, since the left-hand side is monotonic and one could always use bisection, but it turns out that the solution can be written explicitly in terms of a special function that, if not well known, deserves to be. Let $z = b_k$ and $w = -\log a_k$; then

$$we^w = z. \tag{1.4}$$

This equation has a unique real solution $w = W(z)$ for $z \in [0, \infty)$.

The function $-W(-z)$ was already known to Euler, who obtained it as the confluent case of a function of two variables studied by Lambert. Although the history of the function

W can be traced back to 1758, it does not appear in the *Handbook of Mathematical Functions* [AS84]. The name *Lambert's W function*, by which it is now known, was given to it in the early 1990s by the developers of Maple.[1] A delightful survey of the history, properties, and applications of Lambert's W function can be found in Corless et al. [CGH+96], which is our source for most of the following facts.[2]

Euler knew the Maclaurin series of W:

$$W(z) = z - \frac{2^1}{2!}z^2 + \frac{3^2}{3!}z^3 - \frac{4^3}{4!}z^4 + \cdots. \tag{1.5}$$

For $z_0 \neq 0$, the Taylor series of $W(z)$ in powers of $z - z_0$ is easy to find, since the derivatives of W are given by

$$\frac{d^n}{dz^n}W(z) = \frac{p_n(W(z))}{e^{W(z)\,n}(1+W(z))^{2n-1}}, \qquad n = 1, 2, 3, \ldots, \tag{1.6}$$

where the polynomials p_n satisfy $p_1(w) = 1$ and

$$p_{n+1}(w) = (1 - 3n - nw)p_n(w) + (1+w)p_n'(w), \qquad n = 1, 2, 3, \ldots.$$

Note that p_n is a polynomial of degree $n - 1$. The temptation to "simplify" (1.6) by substituting $e^{W(z)} = z/W(z)$ should be resisted, since the formula then has a quite unnecessary removable singularity at $z = 0$.

Some computing environments supply an implementation of the Lambert W function, but if not, it is an easy task to write one's own subroutine. In [CGH+96] it was suggested to solve the nonlinear equation (1.4) by an iterative method, but it is even easier to combine iteration with the Taylor series

$$W(z) = W(z_k) + \frac{1}{1!}(z - z_k)W'(z_k) + \frac{1}{2!}(z - z_k)^2 W''(z_k) + \cdots.$$

Given an approximation w_k, we compute $z_k = w_k e^{w_k}$, so that $w_k = W(z_k)$. Then keep adding terms from the Taylor series until no further improvement is possible, or until taking into account another term would not be worth the effort of computing it. The precise point at which to stop will depend on details of implementation, in particular on the relative expense of calculating e^z as compared to multiplication and division.

Note that the factor $e^{W(z_k)}$ in (1.6) has already been computed in the process of finding z_k. If the denominator $q_n(W) = e^{W(z)n}(1 + W(z))^{2n-1}$ is stored, each extra derivative can be computed by $n + 1$ multiplications and one division. Of course, the coefficients of the polynomials p_n should be precomputed up to the highest degree that will be needed.

As noted in [CGH+96], only the last iteration needs to be done to the full precision required. In our PARI/GP implementation, to evaluate W to d digits, we used eight terms of the Taylor series with an initial value accurate to $d/7.5$ digits, obtained by a recursive call. At the lowest level, the initial value $w_0 = \log(1 + z)$ was used. This is perfectly

[1]The developers of *Mathematica* chose to name the command for this function `ProductLog`.

[2]An excellent starting point when surfing for Lambert's W function is the web page `http://www.apmaths.uwo.ca/~rcorless/frames/PAPERS/LambertW/`

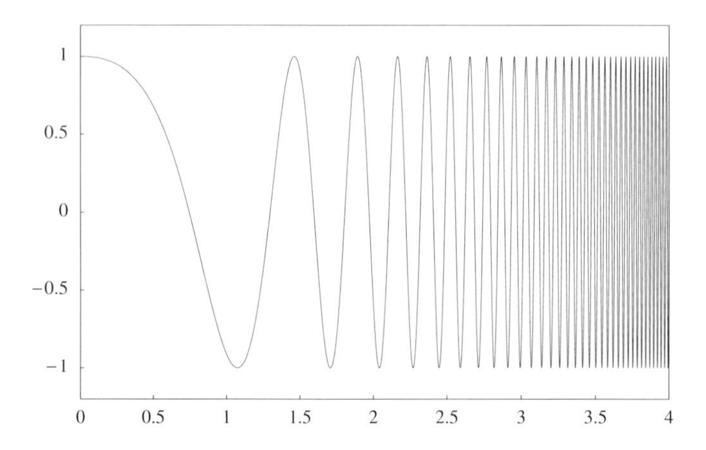

Figure 1.4. *Integrand after the transformation* $t = -\log x$.

adequate when evaluating W as a real-valued function for $z \geqslant 0$. The resulting routine, at very high precision, requires less time than two evaluations of e^z. For the subtleties involved in evaluating W as a multivalued complex function, see [CGH$^+$96].

1.6 Transformation to a Semi-Infinite Interval

As a general rule of thumb, it is easier to integrate a bounded function over an infinite interval than it is to integrate an unbounded function over a finite interval. The simplest transformation is $u = 1/x$, which leads to an integral that can be solved by contour integration. Another very obvious transformation is $t = -\log x$, so that the integral becomes

$$S = \int_0^\infty \cos(te^t)\,dt. \tag{1.7}$$

At first sight, the transformation seems to have gained nothing, since integrating between the zeros or between the extrema of the transformed integrand will produce exactly the same infinite series that we obtained before. The integrand looks only marginally less ugly than before (see Figure 1.4).

However, the transformed integrand has a simpler form and is therefore more amenable to analytic manipulation. In particular, one can use integration by parts to obtain equivalent expressions in which the amplitude of the integrand decays rapidly as t increases. For the first few terms, this can be done directly on (1.7), but if one wants the general formula, it is easier to think of the integral $\int_0^\infty \cos(xe^x)\,dx$ as $\mathrm{Re}\int_0^\infty e^{iW^{-1}(x)}\,dx$. This allows us to take advantage of a nifty trick for repeatedly integrating by parts.

Let I_0 be an integral of the form $I_0 = \int e^{cg(x)}\,dx$, where $f = g^{-1}$, i.e., $f(g(x)) = x$ for all x in the region of interest. Using the identity $g'(x)f'(g(x)) = 1$, we obtain

$$I_0 = \int f'(g(x))g'(x)e^{cg(x)}\,dx = c^{-1}f'(g(x))e^{cg(x)} - c^{-1}I_1,$$

$$\begin{aligned} I_1 &= \int \frac{d}{dx}(f'(g(x)))e^{cg(x)}\,dx = \int f''(g(x))g'(x)e^{cg(x)}\,dx \\ &= c^{-1}f''(g(x))e^{cg(x)} - c^{-1}I_2, \end{aligned}$$

$$I_2 = \int \frac{d}{dx}(f''(g(x)))e^{cg(x)}\,dx = \int f'''(g(x))g'(x)e^{cg(x)}\,dx,$$

and in general

$$\begin{aligned} \int e^{cg(x)}\,dx = -e^{cg(x)}\Big(&-c^{-1}f'(g(x)) + (-c)^{-2}f''(g(x)) + (-c)^{-3}f'''(g(x)) \\ &+ \cdots + (-c)^{-k}f^{(k)}(g(x))\Big) + (-c)^{-k}\int f^{(k+1)}(g(x))g'(x)e^{cg(x)}\,dx. \end{aligned}$$

In this case, taking $c = i$, $f(y) = W(y)$, $g(x) = xe^x$, all derivatives of f vanish at infinity. Substituting the derivatives of W at 0 that can be read off from the Maclaurin series (1.5), we are left with

$$\begin{aligned} \int_0^\infty e^{ixe^x}\,dx = i\,1^0 - i^2 2^1 + i^3 3^2 - \cdots + i^k(-k)^{k-1} \\ + i^k \int_0^\infty W^{(k+1)}(xe^x)e^x(1+x)e^{ixe^x}\,dx. \end{aligned}$$

Stopping at $k = 3$ and taking the real part, we find, with the aid of (1.6), that

$$S = 2 - \int_0^\infty \left(\frac{6}{(1+t)^3} + \frac{18}{(1+t)^4} + \frac{25}{(1+t)^5} + \frac{15}{(1+t)^6}\right) e^{-3t}\sin(te^t)\,dt.$$

The transformed integrand has an unpleasantly large slope at the origin and a nasty-looking peak, but decays quite fast (see Figure 1.5). For 16 digits, it is adequate to consider only the range $0 \leqslant t \leqslant 12$. Romberg integration is still adequate, but this integrand is not so easy, and 131,073 function evaluations are needed to confirm the result from the previous method.

1.7 Representation as a Divergent Series

If one continues the process of integration by parts indefinitely, ignoring the remainder term, an infinite series is obtained, the first few terms of which are

$$2^1 - 4^3 + 6^5 - 8^7 + \cdots. \tag{1.8}$$

This is a series of stunning simplicity, but also a violently divergent one. Divergent series like this, in which the terms have a simple analytic representation, can be formally manipulated to yield other expressions which usually are equivalent to the original quantity being sought.

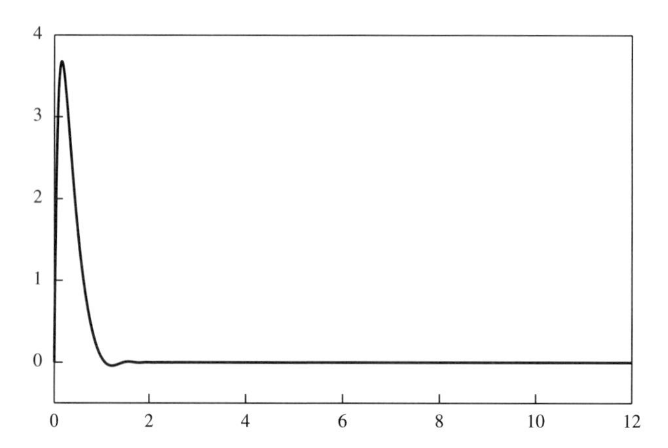

Figure 1.5. *Integrand of the remainder term after three applications of integration by parts.*

Such a technique, of course, provides only a heuristic motivation of an identity that still needs to be verified by other means.

General-purpose convergence acceleration methods (see Appendix A) make little impression on this series. However, as in the case of the contour integral method described before, the fact that we have an analytic expression for the terms allows a better method. In the case of a convergent alternating series, the identity

$$\sum_{k=1}^{\infty}(-1)^{k-1}a_k = \frac{i}{2}\int_C f(z)\csc(\pi z)\,dz$$

holds (see Theorem 3.6), provided that:

1. the integration contour C separates the complex plane into open regions Ω_R, containing the points $1, 2, 3, \ldots$, and Ω_L, containing the other integers, such that C runs counterclockwise around Ω_R;

2. $a_k = f(k)$, where f is analytic in Ω_R and decays suitably as $z \to \infty$.

The form of the divergent series (1.8) suggests that we examine the contour integral

$$\frac{i}{2}\int_C (2z)^{2z-1}\csc(\pi z)\,dz. \tag{1.9}$$

A suitable parametrized contour along which the integral converges rapidly is given by (see Figure 1.6)

$$(x, y) = (1 - \tfrac{1}{2}\cosh t, -t), \qquad -\infty < t < \infty.$$

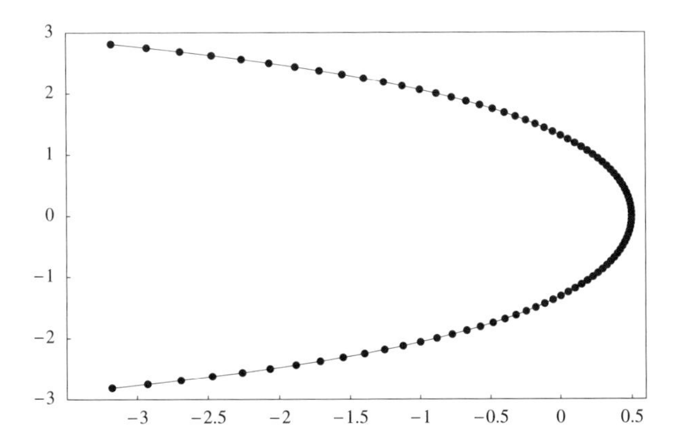

Figure 1.6. *Contour for integration of (1.9), showing points equally spaced with respect to the parametrization.*

For 16 decimal digits, it suffices to consider only the range $|t| \leqslant 2.8$. When the contour integral is evaluated by the trapezoidal rule[3] with step size h in the variable t, the results shown in Table 1.2 are obtained. It is encouraging to find that the result agrees with those obtained previously.

Table 1.2. *Approximations of contour integral (1.9) by trapezoidal sums with step size h. Also shown is the number of evaluations of the integrand.*

h^{-1}	No. of evals	Approximation to S
2	11	**0.3**34486489324265
4	23	**0.3233**98015778690
8	45	**0.323367431**981211
16	91	**0.323367431677779**
32	181	**0.323367431677779**

Although the agreement to 14 digits is highly suggestive, we do not have a proof of the validity of this method. In general, there are infinitely many analytic functions that interpolate given values at the positive integers, and only in the case of a convergent series can it be guaranteed that (1.9) is independent of the choice of interpolant. As a curiosity and a challenge, we therefore offer the *conjecture*: The improper Riemann integral (1.1) and the contour integral (1.9) are equal.

[3]See §3.6.1 of Chapter 3 for an explanation of why trapezoidal sums are exceptionally good for contour integrals of analytic functions: the convergence is *exponentially fast*.

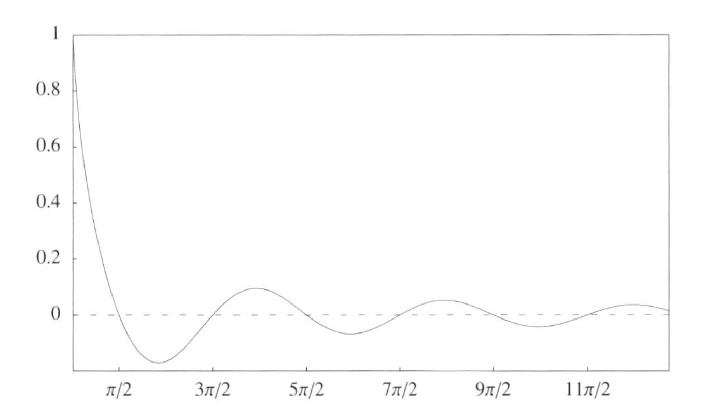

Figure 1.7. *Integrand transformed by Lambert's W function.*

1.8 Transformation to a Fourier Integral

The best-studied infinite oscillatory integrals are Fourier integrals. In §9.5 we discuss these integrals in more detail; here we show that the present problem can also be handled that way.

The essential feature of a Fourier integral is that a factor of the form $\sin(at + b)$ is present in the integrand. Here we have a factor $\cos(te^t)$, so the transformation that we need is $u = te^t$, which is to say $t = W(u)$. This gives us the formula[4]

$$S = \int_0^\infty W'(u) \cos u \, du. \tag{1.10}$$

W' is easy to compute using (1.6). As far as oscillatory integrands on the infinite interval go, this integrand is quite well behaved, as the graph in Figure 1.7 shows.

At this stage it becomes possible to prove easily that the original integral converges. Since the zeros are now equidistant, it is sufficient to show that the multiplier of $\cos u$ monotonically decreases to 0 for sufficiently large u, since in that case, integrating between the zeros gives an alternating series with monotonically decreasing terms. But $W'(u) = e^{-W(u)}/(1+W(u))$ is a product of two decreasing functions and therefore it itself decreasing.

The Ooura–Mori method [OM99] (see §9.5 for a detailed description), with the strategy of iteratively doubling the parameter M, performs as shown in Table 1.3. As can be seen from that table, both the number of function evaluations and the number of correct digits are roughly proportional to M. In view of this property, we trust all the digits except the last (which might be contaminated by round-off error) and assert that

$$S \doteq 0.32336743167777.$$

[4] *Mathematica*'s numerical integrator `NIntegrate` handles such Fourier integrals by an option that basically applies Longman's method of §1.2. Problem 1 is then solved to 13 correct digits by the following short code:

```
NIntegrate[ProductLog'[u] Cos[u], {u, 0, ∞}, Method → Oscillatory]

0.3233674316777859
```

Table 1.3. *Approximations to S by applying the Ooura–Mori method to the integral (1.10). Also shown is the number of evaluations of the integrand.*

M	No. of evals	Approximation to S
2	6	**0.3**33683545675313
4	13	**0.323**277481884531
8	26	**0.3233674**18739372
16	53	**0.323367431677**750
32	111	**0.32336743167777**9

1.9 Going for 10,000 Digits

Computation to very high precision is a specialized art best left to experts. Fortunately, those experts have produced proprietary software such as Maple and *Mathematica*, and open-source software such as GNU MP, CLN, GiNaC, PARI/GP, and doubtless many others. As far as basic arithmetic and the elementary functions are concerned, we can be confident that the experts have done their best, and while one expert may have done better than another expert, an ordinary user of those packages cannot hope to compete.

Here is an example of timings on my own workstation for a few common tasks. Starting with $x = \sqrt{3}$, Table 1.4 gives the run time in clock ticks (one clock tick is the smallest measurable unit of time, about 0.002 seconds) reported by `ginsh` (a simple interactive desk calculator interface to GiNaC) at three precision levels. Two obvious properties of that table hold true in general for all implementations of multiple-precision arithmetic:

- The cost of a d-digit arithmetic operation increases faster than d increases.
- Transcendental functions are much, much more expensive to evaluate than arithmetic functions.

The classic book of Knuth [Knu81, §4.3.3], gives an exhaustive discussion on multiprecision arithmetic.

The usual way of assessing the complexity of a d-digit calculation is to find an exponent p such that the time $t(d)$ as d is increased scales as $O(d^p)$. In the case of multiplication, the best-known fast method is that of Karatsuba, which has $d = \log_2 3 \doteq 1.6$—this is the method used by GiNaC.

Table 1.4. *Run time of various high-precision calculations.*

No. of digits	10,000	20,000	40,000
$x \times x$	1	3	12
$\sqrt{x}$	3	11	49
$1/x$	6	24	70
$\cos x$	249	1038	3749
e^x	204	694	2168

Knuth described several algorithms that can multiply two numbers in time $O(d^p)$ with $p = 1 + \epsilon$ for arbitrarily small ϵ, which is asymptotically optimal. For relatively small values of d, a value of p distinctly larger than 1 models the observed behavior of the optimal methods better. The crossover point where an optimal method is better in practice than the Karatsuba method usually is near $d = 10{,}000$.

It is also known [Bre76] that the most common transcendental functions can be evaluated in time $O(M(d)\log d)$, where $M(d)$ is the d-digit multiplication time.

The prevalence of the formula $O(d^p)$ does not imply that the actual formula for $t(d)$ is as simple as that—the big O sweeps under the carpet such multipliers as $\log d$, $\log\log d$, etc., as well as some possibly rather large constants. But the simplified formula gives us a rule of thumb: if d forms a geometric progression, then, approximately, so should $t(d)$. One can use this reasoning without actually calculating p. For example, having seen that in the case of e^x, we have $t(10{,}000) = 204$ and $t(20{,}000) = 694$, we expect that $t(40{,}000)$ should be near $694^2/204 \doteq 2361$—as indeed is the case. Also, since terms of the form $\log d$ are in effect being modelled by some power of d, this procedure tends to give an overestimate of the required time, which is not a bad thing.

Let us now assess the behavior of the more promising algorithms in terms of how fast the running time will increase as d increases.

Integration over Subintervals and Extrapolation. Assume that we can find at no cost a quadrature formula that can evaluate the integral over a subinterval to d digits using $O(d)$ points (this is optimistic, but I am going to argue that, nevertheless, this method is not competitive), and that we can get away with $O(d)$ subintervals. That gives $O(d^2)$ evaluations of the integrand.

Complex Integration. When calculating to 10,000 digits we cannot afford the luxury of an automatic quadrature routine; we need to use a routine tailored to the task at hand. The basic technique is:

1. Select the integration contour carefully.

2. Parametrize the contour in such a way that the trapezoidal rule on the parametric axis is asymptotically an optimal quadrature formula, that is, a formula in which the number of correct digits is nearly proportional to the number of function evaluations (see §3.6.1).

Details on how to do these two steps are outside the scope of this chapter, but a full discussion on a similar technique applied to other problems can be found in §§3.6 and 9.4.

After some experimentation, we recommend the following parametrized contour for the evaluation of (1.2) to d digits:

$$z = \frac{\pi e^t}{\pi e^t + 2 - 2t} + \frac{2i}{\cosh t}, \quad \log(cd) < t < 1 + \log d,$$

where $c \approx 1.53$; the trapezoidal rule with step length $2/(d+1)$ is then adequate.

Infinite Series. We cannot use this method as our main tool, but if its result happens to agree with our other well-founded methods, it is useful as a confidence booster.

Contour integration on the infinite interval typically requires step size inversely proportional to the number of digits. The double exponential decay of the integrand in (1.9) as a function of t (once from the parametrization, once from the exponentiation in the integrand) guarantees that the number of points required increases only slightly faster than the inverse of the step size, so that the number of points N actually used is modelled by $N = O(d^{1+\epsilon})$, where ϵ is small. (A more accurate model is $d \approx cN/\log N$, which when c is known, can be solved for N in terms of a branch of Lambert's W function not given by (1.5)—but that is another story.)

At each of those points we need to evaluate a hyperbolic function, a complex cosecant, and a complex power. We can expect the totality of these to be more expensive than a real-argument evaluation of $\cos(xe^x)$, but not by so much that it outweighs the advantage of $O(d^{1+\epsilon})$ versus $O(d^2)$ function evaluations.

Ooura–Mori Method. Like the previous two methods, this is another double exponential integration method, with much the same arguments being applicable as in the case of the contour integration method. It has the advantage that the argument values to the transcendental functions are real, so we expect it to be a little faster than the others.

As we have seen, the evaluation of the Lambert W function should cost no more than two evaluations of e^x.

Experimentally, one finds that $M = 1.5d$ is adequate when working to d digits.

Chapter 2

Reliability amid Chaos

Stan Wagon

If I venture to displace . . . the microscopical speck of dust . . . upon the point of my finger, . . . I have done a deed which shakes the Moon in her path, which causes the Sun to be no longer the Sun, and which alters forever the destiny of multitudinous myriads of stars. . . .
—Edgar Allen Poe (*Eureka,* 1848)

But every jet of chaos which threatens to exterminate us is convertible by intellect into wholesome force.
—Ralph Waldo Emerson ("Fate," from *The Conduct of Life,* 1860)

Problem 2

A photon moving at speed 1 *in the* x–y *plane starts at time* $t = 0$ *at* $(x, y) = (1/2, 1/10)$ *heading due east. Around every integer lattice point* (i, j) *in the plane, a circular mirror of radius* $1/3$ *has been erected. How far from* $(0, 0)$ *is the photon at* $t = 10$?

2.1 A First Look

To see even the approximate behavior of the photon, one must write a program to follow the bounces. This can be done in several ways, and we shall present an especially elegant method due to Fred Simons in §2.2. But blindly following the reflections is not at all sufficient, as one would be led astray. Using typical machine precision—16 significant digits—for the intermediate computations is not good enough to get 10 digits of the answer. We can discover this numerical instability without using high precision as follows. We can vary the initial position around a circle of radius 10^{-d} centered at $(1/2, 1/10)$ and see what sort of variation we get in the result (the distance from the final point to the origin). More precisely, we use 10 points around the circle and look at the maximum absolute difference in the answer—the distance of the photon from the origin—among the 90 pairs. The results are shown in Figure 2.1. This computation, carried out using machine precision only, shows that a change

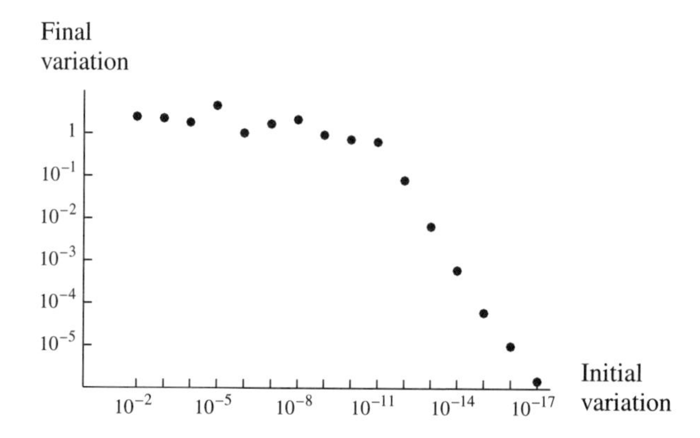

Figure 2.1. *The vertical axis represents the maximum variation in the result when the initial position varies around a circle of radius* 10^{-d} *centered at* $(1/2, 1/10)$. *This computation indicates that a computation using 16 significant digits will give only about 5 digits of the result.*

of about 10^{-17} in the initial conditions can cause the final result to vary by about 10^{-5}. This is a classic example of sensitive dependence, popularly known as the *butterfly effect*.

Since we know that round-off error due to machine precision will enter the computation quite quickly—it comes in right at the start because $1/10$ is not finitely representable in binary—we learn here that standard machine precision is simply inadequate to get more than about five digits of the answer.[5] Further, the eventual linearity of the data in Figure 2.1 leads to a prediction that about 21 digits of precision are needed to obtain 10 digits of the answer. We shall see in §2.2 that this estimate is correct.

It is interesting to note that the computations underlying Figure 2.1 are incorrect, in the sense that the computed numbers used to get the variation that is plotted are, in some cases, correct to only about 5 digits. Nevertheless, explosion in the uncertainty is a solid indicator that something is wrong. In short, machine precision can be a reliable diagnostic tool, even if it cannot provide the cure. Redoing the computation with much higher precision (50 digits, say) yields a figure that is indistinguishable from Figure 2.1. An important consequence of these observations is that software that works only in machine precision, such as MATLAB or C, will not be able to solve the problem to 10 digits (unless one uses add-ons that provide higher precision). The other nine problems in the SIAM challenge can be solved in a machine-precision environment. This is consistent with Nick Trefethen's Maxim 16 [Tre98]: "If rounding errors vanished, 95% of numerical analysis would remain."

2.2 Chasing the Photon

Consider a ray emanating from a point P in the direction of the unit vector v. If we divide the plane into unit squares centered on the lattice points, then these squares divide the ray

[5] It seems that many of the participants in the contest did not recognize this problem. Of the 82 teams that submitted solutions, 58 got 10 digits, 10 got between 6 and 9 digits, and 14 got 5 or fewer digits.

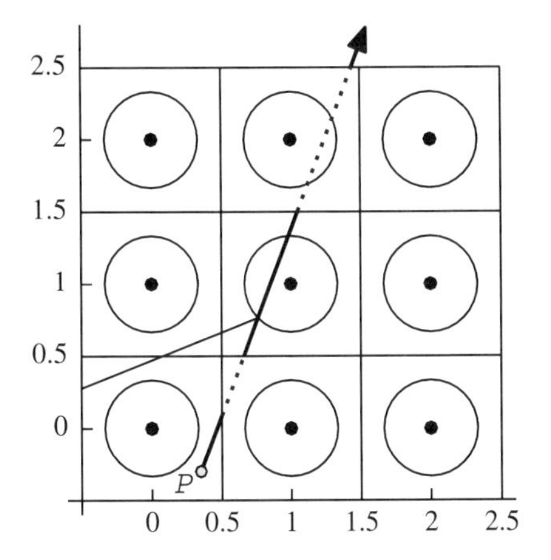

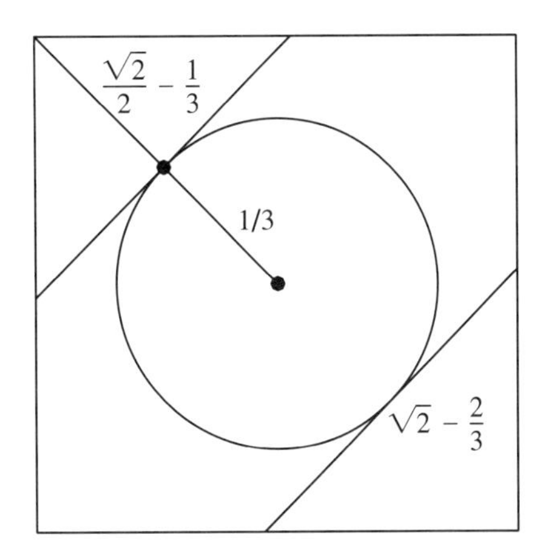

Figure 2.2. *Using lattice squares is an elegant and efficient way to organize the search for the next mirror the photon hits. The figure at right explains why: if a ray strikes a circle, its intersection with the ambient square has length at least* $\sqrt{2} - 2/3$.

into segments (Figure 2.2). For any point on the ray, it is easy to discover which square it is in by just rounding its coordinates. More important, if the ray hits the circle inside a given square, then the ray's segment inside the square has length at least $\sqrt{2} - 2/3$. We could use this number in the work that follows, but it is simpler to replace it by a rational lower bound, and $2/3$ is adequate. To repeat: if the segment inside a square has length less than $2/3$, then there is no intersection with the circle inside the square.

This observation means that if the path from P fails to hit the circle corresponding to P, then the next possible circle it can hit is the one corresponding to $P + 2v/3$. Thus we can find the intersection by considering, as long as necessary, P, $P + 2v/3$, $P + 2 \cdot 2v/3$, $P + 3 \cdot 2v/3$, and so on. As soon as we find a hit, we replace P by the point of intersection with the mirror, and v by the new direction, and carry on.

To get the new point (or determine that the circle is not hit), let m be the center of the circle corresponding to P, and choose the smallest positive root t of the quadratic equation $(P + tv - m) \cdot (P + tv - m) = 1/9$ to get the correct intersection point Q. With Q in hand, it is a simple matter to determine the new direction. Assume for a moment that the center of the circle is the origin and $Q = (a, b)$. Then the transformation taking a direction to its reflected direction is a linear transformation, A, that sends $(-a, -b)$ to (a, b) and fixes $(-b, a)$. Since we are trying to find A so that

$$A \cdot \begin{pmatrix} -a & -b \\ -b & a \end{pmatrix} = \begin{pmatrix} a & -b \\ b & a \end{pmatrix},$$

we need only a single inversion. Here is how to do it in *Mathematica*.

A *Mathematica* Session

```
(a  -b
 b   a).Inverse[(-a  -b
                 -b   a)]//Simplify//MatrixForm
```

$$\begin{pmatrix} \frac{-a^2+b^2}{a^2+b^2} & -\frac{2\,a\,b}{a^2+b^2} \\ -\frac{2\,a\,b}{a^2+b^2} & \frac{a^2-b^2}{a^2+b^2} \end{pmatrix}$$

Because $a^2 + b^2 = 1/9$, the new direction is simply

$$9\begin{pmatrix} b^2 - a^2 & -2ab \\ -2ab & a^2 - b^2 \end{pmatrix} v,$$

where we now translate and let $(a, b) = Q - m$.

Algorithm 2.1. Chasing the Reflecting Photon.

Assumptions: The photon has unit speed; the mirrors have radius $1/3$.

Input: An initial position P, a direction vector v, and a maximum time $t_{\max}$.

Output: The path of the particle from time 0 to $t_{\max}$ as the set `path` of points of reflection, together with the positions at times 0 and $t_{\max}$.

Notation: t_{rem} is the amount of time remaining, m is the center of the circle in the square containing the photon, and s is the time the photon strikes the circle, measured from the time of the preceding reflection. For a point $Q = (a, b)$, H_Q represents the matrix

$$H_Q = 9\begin{pmatrix} b^2 - a^2 & -2ab \\ -2ab & a^2 - b^2 \end{pmatrix}.$$

Step 1: Initialize: $t_{\text{rem}} = t_{\max}$, `path` $= \{P\}$.
Step 2: While $t_{\text{rem}} > 0$:
  Let $m = \text{round}(P + 2v/3)$;
  Let s = smallest positive root of $(P + tv - m).(P + tv - m) = 1/9$;
    ($s = \infty$ if there is no positive root);
  If $s < t_{\text{rem}}$: $P = P + sv$; $v = H_{p-m} \cdot v$;
    else: $s = \min(t_{\text{rem}}, 2/3)$; $P = P + sv$; end if.
Step 3: Update time and path: Let $t_{\text{rem}} = t_{\text{rem}} - s$; Append P to `path`.
Step 4: Return `path`.

Here is complete code that implements this algorithm and solves Problem 2. We start with 36 digits of the initial conditions.

A *Mathematica* Session

```
H[{a_, b_}] := 9 ( b^2 - a^2    -2 a b  )
                 ( -2 a b      a^2 - b^2 );
```

```
p = N[{1/2, 1/10}, 36]; v = {1, 0}; tRem = 10;
While[tRem > 0, m = Round[p + 2v/3];
  s = Min[Cases[t/.Solve[(p + t v - m) . (p + t v - m) == 1/9, t], _?Positive]];
  If[s < tRem, p+ = s v; v = H[p - m] .v, s = Min[tRem, 2/3]; p+ = s v];
  tRem - = s];
```

```
answer = Norm[p]
0.9952629194
```

```
Precision[answer]
10.2678536429468
```

Using 36 digits forces *Mathematica* to use its software arithmetic (high-precision arithmetic), which is based on the technique of *significance arithmetic*; this is not as pessimistic in its error estimates as interval arithmetic, but instead uses calculus-based heuristics to estimate the worst-case behavior after each computation. The result is claimed to have 10 digits of precision, and 10 digits are indeed correct.

So now we use the preceding code but with 50-digit values of the initial conditions (including the direction). The result, printed below with *Mathematica*'s tag indicating the precision, shows that the result is believed to have 32.005 correct digits.

```
0.99526291944335416089031180942672261591142088357`32.005
```

We repeat with starting precision of 100 and get the following result, with predicted error near 10^{-82}; we infer from this that the preceding result is, in fact, good to 32 digits.

```
0.9952629194433541608903118094267216210294669227341543 49
  803208858072986179622830632228394435818154`82.005
```

We learn from this that the error estimates of significance arithmetic seem to be adequate. So we can easily get 100 digits. We use 120 digits of precision to start and we get 102 digits of the answer (we will see in §2.3 how to verify these digits).

```
0.995262919443354160890311809426721621029466922734154349
  803208858072986179622830632099174981897618876031466438 9
  396105`102.005
```

Computations such as these allow us to display and compare trajectories. Figure 2.3 compares the results of two computations to time 20: one uses machine precision, the other starts with 40 digits of precision on the initial conditions, which is adequate to guarantee 10 digits of all points on the trajectory.

Mathematica's approach to high-precision arithmetic is not typical of numerical computing software. Many environments allow the user to use fixed precision for numerical computation. Such an approach will allow us to better understand the error propagation, and the results of a fixed-precision computation are shown in Table 2.1. The data in the table were obtained by simulating fixed precision in *Mathematica*; this computation would be easier in Maple, where one would use

```
> Digits := d:
```

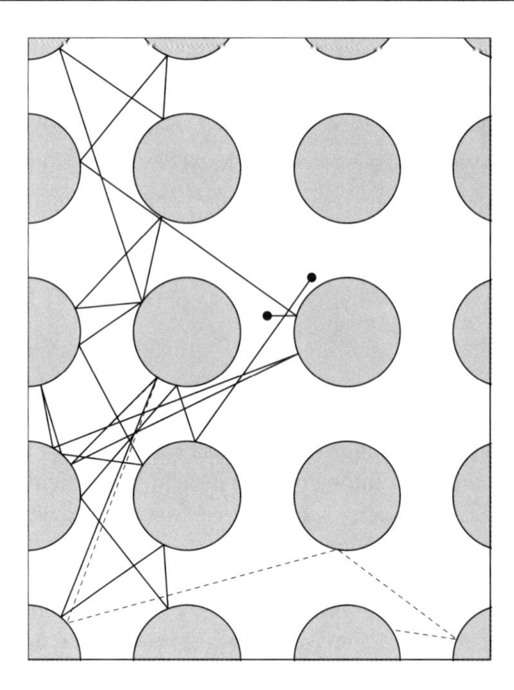

Figure 2.3. *At time 20, the machine precision trajectory (solid) is at a very different point than one following a path (dashed) computed with enough precision to guarantee correctness.*

to work with d digits of precision. In any case, it seems safe to infer from the data that the first 12 digits of the correct distance at $t = 10$ are 0.995262919443. And we note that, as predicted in §2.1, it takes a working precision of 21 digits to get 10 digits of the answer; indeed, using precision d seems to get about $d-11$ correct digits. Because the stated problem involves 17 reflections, we can say that the precision loss is about two-thirds of a digit per reflection.

2.2.1 Working Backwards

Because the motion of the photon in the direction of negative time follows exactly the same rules as in the positive direction, it is natural to investigate backwards trajectories to learn something about the number of correct digits in a numerical solution of the problem. Suppose, as in §2.1, that we are working only with machine precision. Then a natural idea is to take the computed point after 10 seconds (call it Q) and also the direction of the final segment and run the algorithm backwards to see how close the result (call it P_*) is to the starting position P.

The result of such work is $P_* = (0.5000000001419679, 0.09999035188217774)$, a point at distance $9.6 \cdot 10^{-6}$ from $(1/2, 1/10)$. In short, 10 digits were lost in the forwards-and-back computation. But what exactly can one conclude from this? Because the propagation of error inherent in the problem (independent of the algorithm) is a multiplicative process,

Table 2.1. *Results of a fixed-precision approach using precision 5 through 30.*

Precision	Result	Number of correct digits
5	3.5923	0
6	0.86569	0
7	2.386914	0
8	0.7815589	0
9	1.74705711	0
10	0.584314018	0
11	0.8272280639	0
12	1.01093541331	0
13	0.993717133054	2
14	0.9952212862076	4
15	0.99525662897655	4
16	0.995262591079377	6
17	0.9952631169165173	5
18	0.99526292565663256	7
19	0.995262917994839914	8
20	0.9952629195254311922	9
21	0.99526291946156616033	10
22	0.995262919441599585251	11
23	0.9952629194435253187805	12
24	0.99526291944336978995292	13
25	0.995262919443353261823951	14
26	0.9952629194433543857853841	15
27	0.99526291944335415781402273	16
28	0.995262919443354160804997462	19
29	0.9952629194433541607817783594	18
30	0.99526291944335416087109016456	19

it is tempting to think that 5 digits were lost in the forward computation, and then 5 more when traveling in reverse; if true, this would imply that the original computation gave the desired 10 correct digits. But this reasoning is fallacious; the true state of affairs is that 10 digits were lost in the computation of Q, and there is no further deterioration of the result when one goes backwards: yet only 10 digits are lost in the computation of P_*.

Before going more deeply into the subtleties of this, let us describe an experiment, still using only machine precision, that will show that the 5-and-5 conclusion is false. Suppose Q were in error by only about 10^{-10}. Perturb Q by 10^{-13} and run the algorithm backwards to get P_{**}. If the 5-and-5 hypothesis held, one would expect the P_{**} to be within 10^{-5} of P. But the result, using $Q + \left(10^{-13}, 10^{-13}\right)$ as the starting point, is $(0.500351, 0.115195)$, at distance 0.015 from P. So 15 digits were lost. This is consistent with the interpretation that there is already a 10-digit loss to get Q.

A backwards error analysis can shed some light on the loss of precision as the trajectory is computed. Let x denote the initial conditions (a triple: position and angle). Let $F(x)$ denote the function that takes the initial condition to its true state after 10 seconds. Let G denote the computed approximation to F using fixed precision p. Let F_b and G_b be the backwards versions of these functions, where we start at a state at time 10 and work backwards to time 0; then $F_b(F(x)) = x$ for any x. We need to make one assumption about the algorithm, that it is *numerically stable* [Hig96, §1.5]. By this we mean that the computed value is near a true time-10 value of a starting point near x; that is, $G(x) \sim_p F(x_*)$, where $x_* \sim_p x$. Here, $a \sim_m b$ means that, roughly, a and b agree to m digits.

For Problem 2 numerical stability follows from Bowen's shadowing lemma [GH83, Prop. 5.3.3] for hyperbolic dynamics, which in principle governs dispersing billiards [Tab95, Chap. 5]. A rigorous proof would have to address subtle uniformity estimates, however.

Now, when we do a computation forwards to get $y = G(x)$ and then reverse it to get $G_b(y)$, we can see how far the result is from x (using the maximum error in the three entries in x). In the machine-precision case ($p = 16$), this difference is about $10^{-4.5}$, and now we want to see what we can conclude from that. So assume that the algorithm loses d digits as time runs from 0 to 10: $G(x) \sim_{p-d} F(x)$; this should be the same when we go backwards, from 10 to 0. So $G_b(G(x)) \sim_{p-d} F_b(G(x))$. But the numerical stability assumption tells us that $G(x) \sim_p F(x_*)$ for some $x_* \sim_p x$. Therefore, $F_b(G(x)) \sim_{p-d} F_b(F(x_*)) = x_* \sim_p x$, whence $G_b(G(x)) \sim_{p-d} x$.

Despite the danger of misinterpreting backwards information, if used from the worst-case point of view it can provide useful information to give one confidence in the result. Suppose we do the entire computation forwards using fixed precision, and then backwards. We keep increasing the precision until the backwards result is within 10^{-11} of the answer. Then, we can conclude that this is also the error of the forward result at worst. This idea is generally useful in time-symmetric dynamical systems, such as ordinary differential equations.

2.3 Reliable Reflections

The answer obtained in §2.2 seems correct, but the methods are heuristic only. Interval analysis can be used to design an algorithm that yields a computer-assisted proof of correctness. The fundamentals of interval arithmetic are discussed in Chapter 4. The approach of this section is sometimes called a naive interval computation, since we will simply transform the basic numerical algorithm by replacing each operation with an interval version (as opposed to the more sophisticated interval algorithms described in Chapter 4). Recall that interval arithmetic is either built-in or in a package for *Mathematica*, Maple, Matlab, and C. But one must be careful. A first difficulty is that, for example, the *Mathematica* implementation does not include an interval version of `min`, and so one must define that oneself. More precisely, for any symbol `e`, `Min[e[{a,b}],e[{c,d}]]` in *Mathematica* returns $\min(a, b, c, d)$. But when `e` is `Interval`, one needs an interval version of `min`, for which the result is the interval $[\min(a, c), \min(b, d)]$. This is merely a technicality and easily fixed; more subtle are some of the details of the actual algorithm. And a more serious complication that makes C or MATLAB/INTLAB less suitable for this task is that very high precision is needed in the interval computations to get a final interval that determines 10 digits.

The basic idea is to use as input a small two-dimensional interval about each of the initial conditions $(1/2, 1/10)$ and $(1, 0)$ and to use interval methods to get an interval enclosure of the particle's position at time 10. If this final interval is not small enough, just restart using a smaller interval (reduce by a factor of 10) around the initial conditions. But there are subtleties:

- One must use a working precision that is large enough to handle the precision of the intervals. Using $s + 2$ digits of working precision when the initial conditions have radius 10^{-s} appears to be adequate. And this is related to another subtle point: One must be sure that the precision of the intermediate results is at least as large as the ultimate precision goal. That is, one must be sure that the precision loss when solving the quadratic equation has not built up too much. In *Mathematica*, this can be done by checking that the precision of the various computed quantities is adequate.
- When solving the quadratic equation in interval form to find the time the photon hits the current circle, one must check that the solution has no expression of the form $\sqrt{[\text{negative value, positive value}]}$. For if such an expression arose, then the mirror would not be uniquely determined by the starting interval. If this happens, we just pull the emergency brake and go back to the start of the loop, with a reduction by a factor of 10 in the size of the initial intervals.
- When checking whether the mirror-strike occurs before time runs out, one must check that the travel time along the current ray is less than the time remaining (as opposed to overlapping with the interval representing the time remaining).
- The times are intervals throughout, so one might end up with the position not at time 10, but at time $10 \pm \delta$. However, the unit speed hypothesis allows us to turn this time uncertainty into a space uncertainty, and so we can still get an interval trapping the answer.

Note that this approach can be viewed as verification of the results obtained in §2.2, yet it is also a complete algorithm by itself.

Algorithm 2.2. Using Intervals to Chase a Reflecting Photon.

Assumptions: Photon has unit speed; mirrors have radius $1/3$; interval arithmetic works for $\text{round}(\cdot)$.

Input: An initial position p, a direction vector v, a maximum time $t_{\max}$, an absolute error bound ϵ on the final position.

Output: The path of the particle from time 0 to $t_{\max}$ as the set `path` of points at the reflections, together with points at time 0 and $t_{\max}$; the last point is guaranteed to have absolute error less than ϵ.

Notation: Lower-case letters are used for numeric quantities, upper-case for intervals, script for sets of intervals; H is as in Algorithm 2.1. For an interval X, $\min(X)$ (resp.

$\max(X)$) denotes the smallest (resp. largest) number in X, $\text{diam}(X) = \max(X) - \min(X)$, and $\text{mid}(X)$ is $(\min(X) + \max(X))/2$. For a set of intervals $\mathcal{X} = \{X_i\}$, $\min(\mathcal{X})$ is $[\min_i(\min(X_i)), \min_i(\max(X))]$; $\text{diam}(\mathcal{X})$ is $\max_i \text{diam}(X_i)$. For a vector w, w_x and w_y are its x- and y-components; same for an interval vector. The intervals P, V, M, T_{rem}, $T_{\max}$ represent the position, direction, mirror-center, time remaining, and time of reflection, respectively.

Step 1: Initialize: $T_{\text{rem}} = [t_{\max}, t_{\max}]$, `path` $= \{p\}$, $s = \lfloor -\log_{10} \epsilon \rfloor$, `error` $= \infty$,
$\delta = 10^{-s}$, `wp` $= s + 2$, $u = 0$.

Step 2: While `error` $> \epsilon$:
Set the working precision to `wp` digits;
While $\min(T_{\text{rem}}) > 0$:
Let $P = ([p_x - \delta, p_x + \delta], [p_y - \delta, p_y + \delta])$;
Let $V = ([v_x - \delta, v_x + \delta], [v_y - \delta, v_y + \delta])$;
Let $M = \text{round}(P + 2V/3)$;
Use interval arithmetic to determine $\mathcal{S}$, the set of interval solutions to the quadratic equation $(P + tV - M).(P + tV - M) = 1/9$;
If $\mathcal{S}$ contains an expression of the form $\sqrt{[a, b]}$ with $a < 0 < b$, exit the inner while loop;
Let $\mathcal{T}$ be those solutions in $\mathcal{S}$ of the form $[a, b]$ with $a \geqslant 0$;
Let $\mathcal{T}$ be $\min(\mathcal{T})$ ($= [\infty, \infty]$ if $\mathcal{T}$ is empty);
Test values of T and T_{rem} and apply the appropriate case:
case $T \leqslant T_{\text{rem}}$: $P = P + TV$; $V = H_{P-M} \cdot V$; $T_{\text{rem}} = T_{\text{rem}} - T$;
case $T > T_{\text{rem}}$ and $T_{\text{rem}} \geqslant 2/3$: $T_{\text{rem}} = T_{\text{rem}} - 2/3$; $P = P + 2/3V$;
case $T > T_{\text{rem}}$ and $T_{\text{rem}} < 2/3$: $P = P + T_{\text{rem}}V$; $T_{rem} = 0$;
otherwise (incomparable intervals): exit the inner while loop.
Append $\text{mid}(P)$ to `path`;
If the precision of any of T, P, V, T_{rem} is less than $-\log_{10} \epsilon$, exit the inner while loop;
End while.
Let `error` $= \text{diam}(\{P_x + [-\max(|T_{\text{rem}}|), \max(|T_{\text{rem}}|)], P_y + [-\max(|T_{\text{rem}}|), \max(|T_{\text{rem}}|)]\})$;
Let $s = s + 1$; $\delta = 10^{-s}$; `wp` $= s + 2$;
End while.

Step 3: Return `path`.

Appendix C.5.2 contains code for `ReliableTrajectory`, an implementation of Algorithm 2.1. When running the algorithm for the first time, one has no idea what the initial interval size should be, and the program slowly decreases the size until a satisfactory initial tolerance is found. It turns out to be 10^{-40} for a 10^{-12} accuracy goal.

A *Mathematica* Session

```
Norm[Last@ReliableTrajectory[{1/2, 1/10}, {1, 0}, 10, AccuracyGoal → 12]]//
  IntervalForm
```

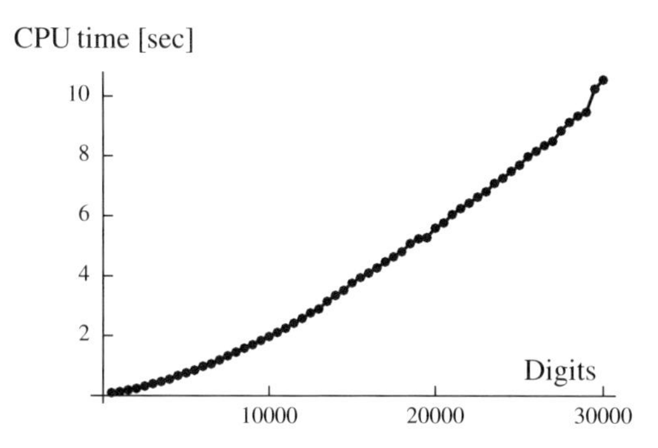

Figure 2.4. *Time needed by the interval algorithm to get d digits of the answer, for d up to 30,000.*

```
Initial condition interval radius is 10^-40.
```

$0.9952629194433^{6690}_{0390}$

From this and further work one learns that, for this *Mathematica* implementation, an interval radius of $10^{-(d+28)}$ for the initial conditions seems to be enough to get d digits of the answer; one can use this to set the starting interval sizes, thus speeding up further computations by eliminating the trial-and-error part. Such an approach allows one to get 100 digits in a fraction of a second.

```
Norm[Last@ReliableTrajectory[{1/2, 1/10}, {1, 0}, 10, AccuracyGoal → 100,
    StartIntervalPrecision → 127]]//IntervalForm
```

```
Initial condition interval radius is 10^-128.
```

$0.995262919443354160890311809426721621029466922734154349803208858072986179622830632099174981897618 87^{606190}_{599926}$

Indeed, this method works quite well even for 10,000 digits (see Appendix B). Figure 2.4 shows timing experiments up to 30,000 digits (on a 1-GHz Macintosh G4). While this might not be the fastest way of getting a large number of digits (the fixed-precision approach discussed earlier might be faster), the extra time needed by the interval method seems a small price to pay for an algorithm that eliminates the uncertainty of heuristic error estimates.

The sensitivity present in this problem is characteristic of chaotic systems and is especially well studied in certain differential equations. The problem becomes more difficult as the travel time becomes longer, but in a predictable fashion. Thus even for time 100, the interval approach has no difficulty: a starting radius of 10^{-265} turns out to be enough to get 13 digits of the answer at $t = 100$; indeed, 10^{-5460} is good enough to get the position at time $t = 2000$ (this true trajectory is shown in Figure 2.5). Of course, the possibility of knowing

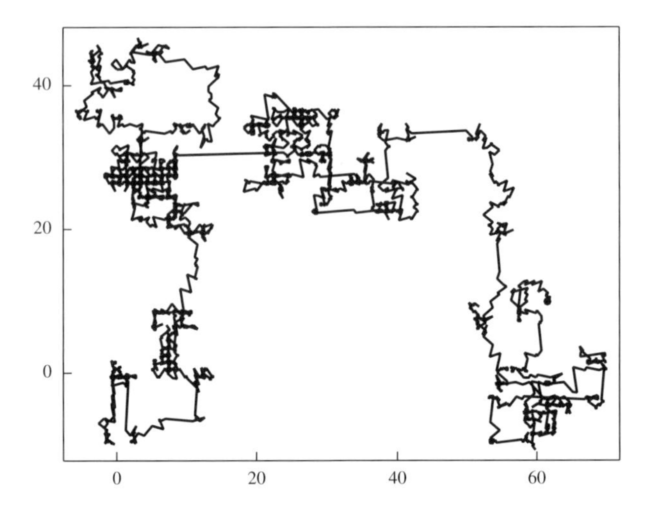

Figure 2.5. *The true trajectory of the photon up to time 2000, obtained by interval arithmetic with initial intervals having diameter of* 10^{-5460}.

a photon's initial position to such tolerances is absurd, as is the possibility of a perfectly circular mirror. But the interval algorithm presented here shows that these are problems of physics, not mathematics.

The true trajectory leads to some observations about the photon's path. One might think that the path would resemble a random walk, but in fact there are occasionally very long steps in the horizontal and vertical directions. These arise from the constraints caused by the mirrors: one can never get a very long step in a direction that is not close to vertical or horizontal.

Billiard trajectories such as those that arise in Problem 2 are quite well understood, thanks to the work of Sinai in 1970 and later researchers such as Bunimovich and Chernov. Indeed, if we view the action as taking place on a flat torus, with just a single reflecting circle, then this is exactly the first example of a dispersing system of billiards studied by Sinai [Sin70a]. It is now known that this system is ergodic (see [Tab95, Thm. 5.2.3]), and a consequence is that, for almost all initial conditions, the set of directions of the segments and the set of reflecting points on the mirror are uniformly distributed in $[0, 2\pi]$. For more on this theory, the books [Tab95] and [CM01] are a good place to start.

The graph in Figure 2.6 shows the probability distribution of the 2086 segment lengths of the true path to time 2000, which range from 0.33347 to 14.833. Using a machine-precision trajectory instead yields a similar distribution. The cusps and gaps in the distribution are interesting and arise out of the geometry of the reflections. Each segment can be classified according to the distance between the centers of the two mirrors it connects. Thus segments of type 1 will be most common, followed by type $\sqrt{2}$, $\sqrt{5}$ (knight's moves), and so on, though not all square roots of sums of two squares are represented because some mirrors are blocked by other mirrors. For the first type, any length in $[1/3, \sqrt{5}/3]$ can arise. For type $\sqrt{2}$, the possible lengths are $[\sqrt{2} - 2/3, \sqrt{14}/3]$. This explains the distribution's sudden rise at $\sqrt{2} - 2/3$. Note that the small interval from $\sqrt{5}/3 = 0.7453\ldots$ to $\sqrt{2} - 2/3 = 0.7475\ldots$

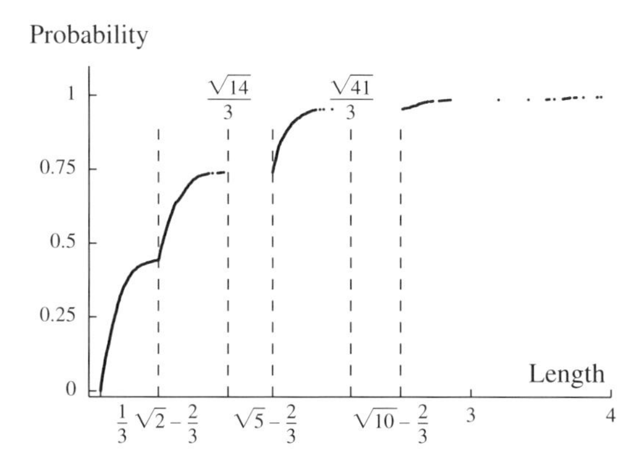

Figure 2.6. *The probability distribution of segment lengths in a verified trajectory up to time 2000. The data came from an interval computation using more than 5400 digits of precision.*

contains no lengths. The lengths of type $\sqrt{5}$ come from the interval $[\sqrt{5}-2/3, \sqrt{41}/3]$ (the upper bound is the length of the common tangent that goes from a point on the northwest side of the circle at (0,0) to the southeast side of the circle at (2,1)). And this explains the gap between $\sqrt{14}/3$ and $\sqrt{5}-2/3$. Further observations are possible if one uses the facts about uniform distribution from billiard theory alluded to above. Then it is an exercise in geometry to obtain an integral that yields the probability of a type-1 segment to be about 0.448. This would mean that 935 is the expected number of such segments among the 2086 segments in the time-2000 trajectory; in fact, there are 925 such segments, nicely consistent with the prediction. We leave more investigations along these lines—perhaps including variation in the mirror radius—to the reader.

Chapter 3

How Far Away Is Infinity?

Jörg Waldvogel

O God! I could be bounded in a nutshell,
and count myself a king of infinite space,
were it not that I have bad dreams.
—William Shakespeare (*Hamlet*,
Act 2, Scene 2, Verse 226)

Problem 3

> *The infinite matrix A with entries* $a_{11} = 1, a_{12} = 1/2, a_{21} = 1/3, a_{13} = 1/4, a_{22} = 1/5, a_{31} = 1/6$, *and so on, is a bounded operator on* ℓ^2. *What is* $\|A\|$?

In §3.1 we reduce the infinite-dimensional problem to the task of calculating a limit of finite-dimensional matrix norms. A simple estimate will be given that determines two digits. In §3.2 we calculate the matrix norms without the need of much background knowledge by using MATLAB's built-in `norm` command and approach the limit by extrapolation. This will bring us 12 digits, but without satisfactory evidence of correctness. A similar but somewhat more efficient algorithm based on the power method will help us produce 21 digits in §3.3. In §3.4 we use second-order perturbation theory to obtain a precise understanding of the error of approximation.

Striving for higher accuracy turns out to be particularly difficult for Problem 3. In fact, Trefethen's first publication of the results on his web page in May 2002 reported only 15 digits, as opposed to 40 digits on the other problems. In a later version, thanks to a method of Rolf Strebel, the missing 25 digits were added. Strebel's method, which is the subject of §3.5, is based on the Euler–Maclaurin sum formula and is the most efficient method to get the full accuracy of 16 digits in IEEE double-precision arithmetic. Though the method performs with remarkable success, the convergence rate slows down for very high accuracies beyond, say, 100 digits. We will overcome this difficulty in §3.6 by capturing infinity via complex analysis and contour integration.

3.1 A First Look

A rigorous foundation of the problem makes it necessary to recall some elementary Hilbert space theory [Rud87, Chap. 4].[6] The sequence space ℓ^2 is the set of all square-summable real-valued sequences $x = (x_1, x_2, \dots)$ and is endowed with an inner product that, for $x, y \in \ell^2$, by generalizing the Euclidean inner product on $\mathbb{R}^n$, is defined by

$$\langle x, y \rangle = \sum_{k=1}^{\infty} x_k y_k.$$

This induces the norm

$$\|x\| = \sqrt{\langle x, x \rangle},$$

which makes ℓ^2 a complete space. Complete inner-product spaces are called *Hilbert spaces*; and ℓ^2 not only is an example of such a space, but is in fact isomorphic to any separable one [Rud87, §4.19]. It is possible to generalize most of the algebraic, geometric, and analytic properties of $\mathbb{R}^n$ to Hilbert spaces, and to ℓ^2 in particular. A linear operator $A : \ell^2 \to \ell^2$ is *bounded*, and therefore continuous, if the operator norm

$$\|A\| = \sup_{x \neq 0} \frac{\|Ax\|}{\|x\|} \tag{3.1}$$

is a (finite) real number. This operator norm generalizes the *spectral norm* of a matrix in $\mathbb{R}^{n \times n}$ [Hig96, §6.2], that is, the matrix norm induced by the Euclidean vector norm.

Now, calculating the norm of the particular operator at hand looks like an infinite-dimensional optimization problem. However, as the following lemma shows, one can approximate the infinite-dimensional operator A by its n-dimensional principal submatrices A_n,

$$A = \begin{pmatrix} \boxed{\quad A_n \quad} & \vdots \\ \cdots & \ddots \end{pmatrix}, \qquad A_n \in \mathbb{R}^{n \times n}.$$

Lemma 3.1 *If* $1 \leqslant n \leqslant m$, *then*

$$\|A_n\| \leqslant \|A_m\| \leqslant \lim_{k \to \infty} \|A_k\| = \|A\| \leqslant \frac{\pi}{\sqrt{6}}.$$

Proof. Let $P_n : \ell^2 \to \operatorname{span}\{e_1, \dots, e_n\} \subset \ell^2$ be the orthogonal projection from ℓ^2 onto the n-dimensional subspace that is spanned by the first n basis sequences $(e_j)_k = \delta_{jk}$. This subspace can be identified with $\mathbb{R}^n$ and we obtain $A_n = P_n A P_n$. For $n \leqslant m$ we have

[6] Rumor has it that Hilbert himself once asked Weyl after a talk [You81, p. 312], "Weyl, you must just tell me one thing, whatever is a Hilbert space? That I could not understand." A reader who feels like this can skip to the end of Lemma 3.1.

$P_n = P_n P_m = P_m P_n$ and therefore $A_n = P_n A_m P_n$. Submultiplicativity of the operator norm, and the fact that an orthogonal projection satisfies $\|P_n\| \leqslant 1$, yields

$$\|A_n\| \leqslant \|P_n\|^2 \, \|A_m\| \leqslant \|A_m\| = \|P_m A P_m\| \leqslant \|P_m\|^2 \|A\| \leqslant \|A\|. \tag{3.2}$$

Thus, the sequence $\|A_n\|$ is monotonically increasing. We do not yet know that it is bounded, because we have not given an argument that $\|A\| < \infty$.

However, such a bound follows from the fact that the spectral norm $\|A_n\|$ is known to be bounded by the Frobenius norm $\|A_n\|_F$ (see [Hig96, Table 6.2]) and therefore

$$\|A_n\| \leqslant \|A_n\|_F = \left(\sum_{j,k=1}^{n} a_{jk}^2 \right)^{1/2} \leqslant \left(\sum_{k=1}^{\infty} k^{-2} \right)^{1/2} = \frac{\pi}{\sqrt{6}}. \tag{3.3}$$

Thus, the limit, $\lim_{n\to\infty} \|A_n\| = \sup_n \|A_n\| \leqslant \pi/\sqrt{6}$, of the increasing sequence exists. Because of the completeness of the basis sequences we know that $\lim_{n\to\infty} P_n x = x$ for all $x \in \ell^2$ and, hence,

$$\|Ax\| = \lim_{n\to\infty} \|P_n A P_n x\| \leqslant \lim_{n\to\infty} \|A_n\| \, \|x\|, \quad \text{that is,} \quad \|A\| \leqslant \lim_{n\to\infty} \|A_n\|.$$

Combined with (3.2), this finishes the proof. □

Summarizing, Problem 3 in fact asks for

$$\lim_{n\to\infty} \|A_n\|,$$

the limit of the spectral norms of the finite-dimensional principal submatrices A_n. We can safely stop talking about infinite-dimensional operators from now on: this limit will be the starting point for our computational enterprise. The precise order of convergence $\|A_n\| \to \|A\|$ will be the subject of §3.4.

Besides setting up the problem in a suitable form, Lemma 3.1 allows us to give, with proof, the first two digits of the answer:

$$1.233 \doteq \|A_3\| \leqslant \|A\| \leqslant \frac{\pi}{\sqrt{6}} \doteq 1.282, \qquad \text{that is,} \quad \|A\| \doteq 1.2.$$

In finite dimensions, the spectral norm of a matrix A_n satisfies [GL96, Thm. 2.3.1]

$$\|A_n\| = \sigma_{\max}(A_n) = \sqrt{\lambda_{\max}(G_n)}, \qquad G_n = A_n^T A_n. \tag{3.4}$$

Here $\sigma_{\max}(A_n)$ denotes the largest singular value[7] of A_n, and $\lambda_{\max}(G_n)$ denotes the largest eigenvalue of G_n.

[7]Generally, the singular values of a matrix A are the square roots of the eigenvalues of the positive semidefinite matrix $G = A^T A$.

Matrix Generation. The matrix $\tilde{A} = (1/a_{jk})$ of the reciprocal elements of A is given by the northeast to southwest arrangement of the natural numbers,

$$\tilde{A} = \begin{pmatrix} 1 & 2 & 4 & 7 & \dots \\ 3 & 5 & 8 & 12 & \dots \\ 6 & 9 & 13 & 18 & \dots \\ 10 & 14 & 19 & 25 & \dots \\ & \dots\dots\dots & & & \ddots \end{pmatrix}.$$

We take advantage of the representation of $\tilde{A}$ as the difference of a Hankel matrix and a matrix with constant columns,

$$\tilde{A} = \begin{pmatrix} 2 & 4 & 7 & 11 & \dots \\ 4 & 7 & 11 & 16 & \dots \\ 7 & 11 & 16 & 22 & \dots \\ 11 & 16 & 22 & 29 & \dots \\ & \dots\dots\dots & & & \ddots \end{pmatrix} - \begin{pmatrix} 1 & 2 & 3 & 4 & \dots \\ 1 & 2 & 3 & 4 & \dots \\ 1 & 2 & 3 & 4 & \dots \\ 1 & 2 & 3 & 4 & \dots \\ & \dots\dots & & & \ddots \end{pmatrix}.$$

Thus, the entries of A are given as

$$a_{jk} = \frac{1}{b_{j+k} - k}, \qquad j, k = 1, 2, \dots, \tag{3.5}$$

where $b_l = 1 + (l-1)l/2$. The expression (3.5) is a useful tool for efficiently generating the principal submatrices A_n as well as for coding the mapping $x \mapsto Ax$ without explicitly storing A.

A MATLAB Session

```
>> n = 1024;
>> N = 2*n-1; k = 1:N; b = 1 + cumsum(k);
>> A = 1./(hankel(b(1:n),b(n:N))-ones(n,1)*k(1:n));
```

This is more elegant[8] than the generation of A_n by means of the explicit formula

$$a_{jk} = \frac{1}{(j+k-1)(j+k)/2 - (k-1)}, \qquad j, k = 1, 2, \dots, n, \tag{3.6}$$

which is just (3.5) written out in full. Note that the decay is $a_{jk} = O((j+k)^{-2})$ for large j and k; this is faster than the decay rate of, say, the famous Hilbert matrix $H = (h_{jk})$ with $h_{jk} = 1/(j+k-1)$.

The entries a_{jk} enjoy a specific feature that will become important in §3.6: they are given by a rational function a evaluated at integer arguments, $a_{jk} = a(j,k)$, $j, k \in \mathbb{N}$. That function extends naturally as a meromorphic function to the complex u- and v-planes,

$$a(u, v) = \frac{1}{b(u+v) - v}, \qquad b(z) = 1 + (z-1)z/2, \tag{3.7}$$

with no poles at the grid points $(u, v) \in \mathbb{N}^2$.

[8]Generating the matrix by the `hankel` command used to be, in MATLAB prior to the introduction of the JIT compilation technology in Release 13, substantially faster than by the explicit formula. Now the run time of the two variants is roughly the same.

3.2 Extrapolation

In this section we present the simple method that was used by many contestants to obtain 10 to 12 digits of $\|A\|$ with some level of confidence, but without going deeper into the mysteries of the matrix A.

The method takes advantage of many packages' capability to directly compute the spectral norm $\|A_n\|$ of the principal submatrices at hand. For instance, if the matrix is named `A` in code, MATLAB calculates the norm by invoking the command `norm(A)`. Internally this is done by a singular value decomposition of A_n (see [GL96, §8.6]). The run time of using `norm(A)` scales as $O(n^3)$.

Alternatively, MATLAB offers the command `normest(A,tol)` that iteratively calculates an approximation of $\|A\|$ up to a relative error `tol`. It is based on the power method, which we will discuss in §3.3, and takes $O(n^2)$ time for a fixed accuracy `tol`.

The second column of Table 3.1 shows $\|A_n\|$ for dimensions that are a power of 2: $n = 1, 2, 4, \ldots, 4096$. The values nicely reflect the monotonicity stated in Lemma 3.1. A close look at the digits that keep agreeing from one row to the next suggests that doubling n gives a little less than one additional correct digit. From this we would infer about 10 correct digits for $n = 2048$. To be on the safe side, we go for $n = 4096$ and get

$$\|A\| \doteq 1.2742\,24152.$$

The memory needed to store the matrix A_{4096} amounts to 120 MB; calculating $\|A_{4096}\|$ by the command `norm(A)` takes 12 minutes, and by `normest(A,5e-16)`, 12 seconds.[9] Both results agree to 15 digits.

A much more efficient way of getting up to 12 digits in less than a second is based on using the values for small n and extrapolation to the limit $n \to \infty$.

A MATLAB Session *(cont. of session on previous page)*

First, let us generate and store the values of the norms for $n = 1, 2, 4, \ldots, 2^{L-1}$. We have chosen $L = 11$, corresponding to the maximum dimension $n = 1024$.

```
>> L = 11;
>> vals = [];
>> for nu = 1:L,
>>    n=2^(nu-1); An=A(1:n,1:n); vals=[vals;normest(An,5e-16)];
>> end;
```

Next, we use *Wynn's epsilon algorithm* (see Appendix A, p. 245) for approximating, by extrapolation, the limit of the data that are stored in `vals`.

```
>> L2 = 2*L-1; vv = zeros(L2,1); vv(1:2:L2) = vals;
>> for j=2:L
>>    k=j:2:L2+1-j; vv(k)=vv(k)+1./(vv(k+1)-vv(k-1));
>> end;
>> result = vv(1:2:L2);
```

The output, which is stored in the column vector `result`, contains, from top to bottom, the values of the right boundary of Wynn's triangular array. The best approximations for

[9] In this chapter all timings are for a 1.6 GHz PC.

Table 3.1. *Values of $\|A_n\|$ and extrapolation by Wynn's epsilon algorithm.*

n	`vals` = $\|A_n\|$	Wynn for $L = 10$	Wynn for $L = 11$
1	**1**.00000000000000	**1**.00000000000000	**1**.00000000000000
2	**1**.18335017655166	**1.2**9446756234379	**1.2**9446756234379
4	**1.2**5253739751680	**1.274**09858051212	**1.274**09858051212
8	**1.27**004630585408	**1.27422415**478429	**1.27422415**478429
16	**1.27**352521545013	**1.2742241**6116628	**1.2742241**6116628
32	**1.274**11814436915	**1.274224152**78695	**1.27422415282**204
64	**1.2742**0913129766	**1.2742241528**5119	**1.27422415282**075
128	**1.27422**212003778	**1.274224152**71828	**1.27422415282**239
256	**1.27422**388594855	**1.27422415**371348	**1.2742241528**1540
512	**1.2742241**1845808	**1.2742241**1845808	**1.2742241528**9127
1024	**1.2742241**4844970	—	**1.2742241**4844970
2048	**1.274224152**26917	—	—
4096	**1.274224152**75182	—	—

$\|A\|$ are generally located just below the central element. Table 3.1 gives the results of the session just run with $L = 11$, and of another run with $L = 10$. The run time was less than half a second. In the less expensive case of the extrapolation, $L = 10$, the entries for $n = 32$, 64, and 128 agree to 10 digits. In the case $L = 11$ these entries agree to 12 digits; therefore, at this point we can confidently report

$$\|A\| \doteq 1.2742\,24152\,82. \tag{3.8}$$

Thus, extrapolation yields two more digits in a run time that is faster by a factor of 40 when compared to the use of the simple approximation by $\|A_{4096}\|$. Similar results are obtained with other extrapolation techniques. In Appendix A, pp. 247 and 252, the reader will find, as yet another example, applications of *Levin's U* and the *rho* algorithm to Problem 3.

This looks like a quick exit from Problem 3. However, to get more evidence for the calculated digits, we need higher accuracy, which this approach cannot easily provide.

3.3 The Power Method

One of the bottlenecks of the method in §3.2 is the enormous amount of memory needed to store the matrix A_n for larger n: 120 MB are needed for $n = 4096$. This leads us to ask for an iterative method of calculating the norm $\|A_n\|$ that addresses A_n only by the matrix-vector products $x \mapsto Ax$ and $x \mapsto A^T x$. Such a method is readily obtained if we recall (3.4), that is,

$$\|A_n\|^2 = \lambda_{\max}(G_n), \qquad G_n = A_n^T A_n.$$

The largest eigenvalue of the symmetric positive semidefinite matrix G_n can efficiently be calculated by the *power method* [GL96, §8.2.1], a basic method that can be found in virtually any textbook on elementary numerical analysis.

Algorithm 3.1. Power Method for Calculating $\lambda_{\max}(G_n)$.

Take an initial vector $x^{(0)}$ of norm 1

for $\nu = 1$ **to** $\nu_{\max}$ **do**

$y^{(\nu)} \quad = G_n x^{(\nu-1)};$

$\lambda^{(\nu-1)} = \left(x^{(\nu-1)}\right)^T y^{(\nu)};$

if $\lambda^{(\nu-1)}$ is sufficiently accurate **then exit**;

$x^{(\nu)} \quad = y^{(\nu)}/\|y^{(\nu)}\|;$

end for

Convergence Theory. This simple method has a correspondingly simple convergence theory.

Theorem 3.2 [GL96, Thm. 8.2.1]. *If the largest eigenvalue $\lambda_1 = \lambda_{\max}(G_n)$ of a symmetric, positive semidefinite $n \times n$ matrix G_n is* simple, *that is, if the n eigenvalues of G_n can be ordered as $\lambda_1 > \lambda_2 \geqslant \lambda_3 \geqslant \cdots \geqslant \lambda_n \geqslant 0$, and if the initial vector $x^{(0)}$ is in general position, that is, not orthogonal to the eigenvector belonging to λ_1, there is a constant $c > 0$ such that*

$$|\lambda^{(\nu)} - \lambda_{\max}(G_n)| \leqslant c\rho^\nu, \qquad \nu = 1, 2, \ldots,$$

with the contraction rate $\rho = (\lambda_2/\lambda_1)^2$.

There are two slick ways of showing that the matrix at hand, $G_n = A_n^T A_n$, does in fact have a largest eigenvalue that is simple. The second method has the advantage of giving a quantitative bound for the contraction rate ρ.

Method 1. We observe that all entries of G_n are strictly positive. Therefore, the *Perron–Frobenius theory* of nonnegative matrices [HJ85, §8.2] is applicable. In particular, there is a general theorem [HJ85, Thm. 8.2.5] of Perron stating that for positive matrices the eigenvalue of largest modulus, that is, the *dominant* eigenvalue, is always simple.

Method 2. We recall the bound (3.3) on $\|A_n\|_F$, which—given the representation [GL96, form. (2.5.7)] of the Frobenius norm as the sum of the squares of the singular values—can be expressed as

$$\|A_n\|_F^2 = \lambda_1 + \lambda_2 + \cdots + \lambda_n \leqslant \frac{\pi^2}{6}.$$

Because of $\lambda_1 = \|A_n\|^2 \geqslant \|A_3\|^2 \doteq 1.52$ we obtain for $n \geqslant 3$,

$$\lambda_2 \leqslant \frac{\pi^2}{6} - \lambda_1 \leqslant 0.125, \qquad \rho = \left(\frac{\lambda_2}{\lambda_1}\right)^2 \leqslant 6.8 \cdot 10^{-3}.$$

In fact, a numerical calculation shows that for large n the second largest eigenvalue of G_n is $\lambda_2 \doteq 0.020302$, which yields—with (3.8), that is, $\lambda_1 = \|A_n\|^2 \doteq 1.6236$—the contraction rate

$$\rho \doteq 1.5635 \cdot 10^{-4}. \tag{3.9}$$

This corresponds to a gain of $-\log_{10} \rho \doteq 3.8$ digits per iteration step, which actually has been observed in our numerical experiments.

Application to Problem 3. We use (3.5) to efficiently implement the matrix-vector multiplication step in Algorithm 3.1, that is

$$y^{(\nu)} = G_n x^{(\nu-1)}, \qquad G_n = A_n^T A_n.$$

There is no need to explicitly set up the matrix G_n. Instead we factor the matrix-vector product into two separate steps, $\tilde{x}^{(\nu)} = A_n x^{(\nu-1)}$ and $y^{(\nu)} = A_n^T \tilde{x}^{(\nu)}$, which are given by

$$\begin{aligned} \tilde{x}_j^{(\nu)} &= \sum_{k=1}^{n} a_{jk} x_k^{(\nu-1)} = \sum_{k=1}^{n} \frac{x_k^{(\nu-1)}}{b_{j+k} - k}, \qquad j = 1, \ldots, n, \\ y_k^{(\nu)} &= \sum_{j=1}^{n} a_{jk} \tilde{x}_j^{(\nu)} = \sum_{j=1}^{n} \frac{\tilde{x}_j^{(\nu)}}{b_{j+k} - k}, \qquad k = 1, \ldots, n. \end{aligned} \tag{3.10}$$

To minimize the influence of roundoff errors it is advisable [Hig96, p. 90] to accumulate all sums[10] from small to large terms, that is, from $j = n$ to $j = 1$ and from $k = n$ to $k = 1$. The price paid for not storing the matrix A_n is minimal: instead of one scalar multiplication in the inner loop of each matrix-vector product, two index additions and one division need to be carried out. Since by (3.9) the contraction rate ρ is essentially independent of the dimension n, the run time of approximating $\|A_n\|$ to a fixed accuracy scales as $O(n^2)$.

The reader will find an implementation `OperatorNorm(x,tol)` of these ideas coded in PARI/GP in Appendix C.2.1 and coded in MATLAB in Appendix C.3.1. It takes as input an initial vector `x` and an absolute tolerance `tol` for $\|A_n\|$ and outputs the approximation to $\|A_n\|$ as well as the final vector $x^{(k)}$ of the power method. This final vector allows for a *hierarchical* version of the iterative procedure: since the eigenvector of G_n belonging to the dominant eigenvalue is an approximation of a fixed vector in the sequence space ℓ^2, we can profitably use this final vector of dimension n, padded by zeros, to start the power method for dimension $2n$. Compared to a fixed initial vector, such as the first basis vector, this reduces the run time by about a factor of 2.

We now extend Table 3.1 up to the dimension $n = 32768$. To increase the confidence in the correctness of the solution we go to extended precision, using PARI/GP's default precision of 28 decimal digits.

[10]The sums involve only nonnegative terms if the initial vector $x^{(0)}$ was chosen to be nonnegative.

A PARI/GP Session *(compare with MATLAB session on p. 51)*

```
? dec=28;  default(realprecision,dec); tol=10^(2.0-dec);
? L=16; vals=vector(L); res=[1.0,1.0]; vals[1]=res[1];
? for(nu=2,L, x0=concat(res[2],0*res[2]); res=OperatorNorm(x0,tol);\
        vals[nu]=res[1]);
? L2=2*L-1; vv=vector(L2,j, if(j%2==1, vals[(j+1)/2], 0));
? for(j=2,L, forstep(k=j, L2+1-j, 2,\
        vv[k]=vv[k]+1/(vv[k+1]-vv[k-1])));
? result=vector(L,j, vv[2*j-1])
```

The entries of the vectors `vals` and `result` are shown in the second and third columns of Table 3.2.[11] The run time was less than eight hours.

Table 3.2. *Values of $\|A_n\|$ and extrapolation by Wynn's epsilon algorithm.*

n	`vals` = $\|A_n\|$	Wynn for $L = 16$
..	....	....
64	**1.2742**09131297662	**1.2742241528212283**43690360386
128	**1.27422**2120037776	**1.27422415282122818 8**076956823
256	**1.27422**3885948554	**1.274224152821228188 21**3398344
512	**1.2742241**18458080	**1.274224152821228188212**253584
1024	**1.2742241**48449695	**1.274224152821228188 21**3171143
2048	**1.274224152**269170	**1.274224152821228188 2**05973369
4096	**1.274224152**751815	**1.274224152821228188 2**99867942
8192	**1.2742241528**12522	**1.27422415282122818**5212556813
16384	**1.27422415282**0138	**1.2742241528212283**35619616358
32768	**1.274224152821**091	**1.274224152821**091776178588977

The entries for $n = 256$, 512, and 1024 agree to 21 digits. Therefore, we can confidently report

$$\|A\| \doteq 1.2742\,24152\,82122\,81882\,1.$$

This is about as far as we can get by the straightforward technique of handling the principal submatrices A_n and extrapolating to the limit as $n \to \infty$ by a general-purpose method. In §3.5 we will start using conclusive information about the *sequence* of submatrices, or equivalently, about the infinite matrix A.

A Dead End: Calculating a Closed Form for $A^T A$. We conclude this section with a digression to study a rather surprising closed form for the elements

$$g_{jl} = \sum_{k=1}^{\infty} a_{kj} a_{kl}$$

[11] It is comforting to observe that Tables 3.2 and 3.1 agree to all the digits given.

of the infinite matrix $G = A^T A$. Such a closed form suggests using the principal submatrices $\hat{G}_n$ of G instead of the matrices $G_n = A_n^T A_n$ to run the power method. However, as we shall see, this does not seem to be of any use in the actual calculation of $\|A\|$. A reader with no further interest in higher transcendental functions and symbolic computation might wish to skip the rest of this section and go to §3.4.

By representing the coefficient a_{jk} as in (3.6) we obtain $g_{jl} = \sum_{k=1}^{\infty} R_{jl}(k)$, where

$$R_{jl}(k) = \frac{4}{\big(k^2 + (2j-1)k + (j-1)(j-2)\big)\big(k^2 + (2l-1)k + (l-1)(l-2)\big)},$$

a rational function with denominator of degree 4 in k. An elegant technique for evaluating sums of rational functions over equally spaced points is described in [AS84, §6.8]:[12] the expansion of R_{jl} in partial fractions directly yields the value of the sum in terms of the psi (or digamma) function, $\psi(z) = \Gamma'(z)/\Gamma(z)$, and its derivatives, called polygamma functions.

The partial-fraction expansion of R_{jl} is determined by its poles, located at

$$k_{1,2} = \frac{1}{2}\left(1 - 2j \pm \sqrt{8j-7}\right), \qquad k_{3,4} = \frac{1}{2}\left(1 - 2l \pm \sqrt{8l-7}\right).$$

Now, the term g_{jl} will be a linear combination of the values $\psi(1-k_\nu)$, $\psi'(1-k_\nu)$, $\psi''(1-k_\nu), \ldots, \psi^{(\mu_\nu - 1)}(1-k_\nu)$, where μ_ν is the multiplicity of the pole k_ν. Some computer algebra systems, such as *Mathematica* and Maple, offer implementations of this algorithm.

A Maple Session

```
> j:=1: l:=1:
> j1:=2*j-1: j2:=(j-1)*(j-2): l1:=2*l-1: l2:=(l-1)*(l-2):
> gjl:=sum(4/(k^2+j1*k+j2)/(k^2+l1*k+l2),k=1..infinity);
```

$$g_{11} = \frac{4}{3}\pi^2 - 12.$$

A few more results are, after some beautification:

$$g_{12} = \frac{2}{9}\pi^2 - \frac{43}{27}, \quad g_{22} = \frac{4}{27}\pi^2 - \frac{31}{27}, \quad g_{24} = \frac{137}{1350}, \quad g_{17} = \frac{26281}{396900},$$

$$g_{13} = 4\gamma - 2 + \left(2 + \frac{8}{\sqrt{17}}\right)\psi\left(\frac{7-\sqrt{17}}{2}\right) + \left(2 - \frac{8}{\sqrt{17}}\right)\psi\left(\frac{7+\sqrt{17}}{2}\right),$$

where γ is Euler's constant,

$$g_{35} = -\frac{2}{17}\left(\psi\left(\frac{7-\sqrt{17}}{2}\right) + \psi\left(\frac{7+\sqrt{17}}{2}\right)\right)$$
$$+ \frac{2}{17}\left(\left(1 + \frac{4}{\sqrt{33}}\right)\psi\left(\frac{11-\sqrt{33}}{2}\right) + \left(1 - \frac{4}{\sqrt{33}}\right)\psi\left(\frac{11+\sqrt{33}}{2}\right)\right).$$

[12] According to [Nie06, §24] the technique dates back to a short note of Appell from 1878.

From the point of view of closed forms the case g_{33} is of particular interest. Maple yields

$$g_{33} = \frac{4}{17}\left(\psi'\left(\frac{7-\sqrt{17}}{2}\right) + \psi'\left(\frac{7+\sqrt{17}}{2}\right)\right) + \frac{8\sqrt{17}}{289}\left(\psi\left(\frac{7-\sqrt{17}}{2}\right) - \psi\left(\frac{7+\sqrt{17}}{2}\right)\right).$$

The *Mathematica* results are detailed below.

A *Mathematica* Session

```
g[j_, l_] :=
  Sum[4/((k^2 + (2 j - 1) k + (j - 1) (j - 2)) (k^2 + (2 l - 1) k + (l - 1) (l - 2))), {k, 1, ∞}] //
  FullSimplify
```

```
g[3, 3]
```

$$-\frac{9}{4} - \frac{4}{289}\pi \operatorname{Sec}\left[\frac{\sqrt{17}\,\pi}{2}\right]^2 \left(-17\pi + \sqrt{17}\operatorname{Sin}\left[\sqrt{17}\,\pi\right]\right)$$

John Boersma communicated to us a proof that the two expressions have the same value indeed. Moreover, he proved the remarkable fact that g_{jj} is generally expressible in terms of elementary functions:[13]

$$g_{jj} = \frac{4\pi^2}{3(2q+1)^2} - \frac{4}{(2q+1)^2}\left(\sum_{m=1}^{j-q-1}\frac{1}{m^2} + \sum_{m=1}^{j+q}\frac{1}{m^2}\right) - \frac{8}{(2q+1)^3}\sum_{m=j-q}^{j+q}\frac{1}{m},$$

if $j = (q^2+q+2)/2$ for some $q = 0, 1, 2, \ldots$, and otherwise

$$g_{jj} = -4\sum_{m=0}^{j-1}\frac{1}{(m^2+m+2-2j)^2} + \frac{4\pi}{(8j-7)^2}\sec^2(\sqrt{8j-7}\,\pi/2)\Big((8j-7)\pi - \sqrt{8j-7}\,\sin(\sqrt{8j-7}\,\pi)\Big).$$

The partial-fraction algorithm shows that any entry g_{jl} of the infinite matrix G can be computed as a closed-form expression in terms of the functions ψ and ψ'. As elegant as this might be, this approach turns out to be a dead end:

- In contrast to the entries a_{jl} of the matrix A the entries g_{jl} of G take much more effort to calculate for given indices j and l. This is, however, not compensated by a faster rate of decay.

[13] Courtesy of J. Boersma; his proof can be found on the web page for this book.

- Recall from Table 3.1 that

$$\|A\| - \|A_{128}\| \doteq 2.0328 \cdot 10^{-6}.$$

Consider, on the other hand, the principal submatrix $\hat{G}_{128}$ of G of dimension $n = 128$, whose generation involves substantial symbolic and numerical effort. Its dominant eigenvalue, as computed by MATLAB, is found to be

$$\lambda_{\max}(\hat{G}_{128}) \doteq 1.6236447743.$$

Therefore,

$$\|A\| - \sqrt{\lambda_{\max}(\hat{G}_{128})} \doteq 0.9485 \cdot 10^{-6},$$

a rather modest gain of accuracy.

3.4 Second-Order Perturbation Theory

A closer look at Table 3.2 suggests that the norms of the principal submatrices A_n of dimension n approximate the norm of the operator A up to an error of order $O(n^{-3})$. We will use second-order perturbation theory to explain this behavior. As a by-product we obtain a correction term that can be used to solve Problem 3 to 12 correct digits in IEEE double precision.

The following result is a special case of a second-order perturbation expansion that Stewart originally established for the *smallest* singular value of a matrix. It holds true, in fact, for the largest singular value with exactly the same proof.

Theorem 3.3 [Ste84]. *Let*

$$\tilde{A} = \begin{pmatrix} A & B \\ C & D \end{pmatrix} = \begin{pmatrix} A & 0 \\ 0 & 0 \end{pmatrix} + E$$

be a block matrix of dimension $m \times m$. Let u and v be the left and right singular vectors belonging to the largest singular value of the $n \times n$ principal submatrix A,[14] $n \leqslant m$. For $\|E\| \to 0$, there holds the asymptotic expansion

$$\|\tilde{A}\|^2 = \|A\|^2 + \|u^T B\|^2 + \|Cv\|^2 + O(\|E\|^3).$$

For the problem at hand we apply this perturbation theorem to the block partitioning

$$A_m = \begin{pmatrix} A_n & B_{n,m} \\ C_{n,m} & D_{n,m} \end{pmatrix} = \begin{pmatrix} A_n & 0 \\ 0 & 0 \end{pmatrix} + E_{n,m}, \qquad n \leqslant m.$$

Using once again the bound of the spectral norm by the Frobenius norm, we get

$$\|E_{n,m}\|^2 \leqslant \|E_{n,m}\|_F^2 = \|A_m\|_F^2 - \|A_n\|_F^2 \leqslant \sum_{k>n^2/2} k^{-2} = O(n^{-2}).$$

[14]That is, u and v are vectors of norm 1 that satisfy $AA^T u = \sigma_{\max}^2(A)u$ and $A^T Av = \sigma_{\max}^2(A)v$. Algorithm 3.1 actually allows us to calculate them: $x^{(\nu)}$ and $Ax^{(\nu)}$, normalized to have norm 1, approximate u and v up to an error of order $O(\rho^\nu)$.

In the same way we obtain $\|B_{n,m}\| = O(n^{-1})$ and $\|C_{n,m}\| = O(n^{-1})$. Thus, if we denote by u_n and v_n the left and right singular vectors of A_n that belong to the largest singular value, Theorem 3.3 results in the error estimate

$$\begin{aligned}\|A_n\|^2 \leqslant \|A_m\|^2 &= \|A_n\|^2 + \|u_n^T B_{n,m}\|^2 + \|C_{n,m}v_n\|^2 + O(n^{-3}) \\ &\leqslant \|A_n\|^2 + \|B_{n,m}\|^2 + \|C_{n,m}\|^2 + O(n^{-3}) = \|A_n\|^2 + O(n^{-2}). \quad (3.11)\end{aligned}$$

Thus, the argument so far proves an error estimate of order $O(n^{-2})$ only—instead of the numerically observed order $O(n^{-3})$. We must have given away too much in using the submultiplicativity bounds

$$\|u_n^T B_{n,m}\| \leqslant \|B_{n,m}\| = O(n^{-1}), \qquad \|C_{n,m}v_n\| \leqslant \|C_{n,m}\| = O(n^{-1}).$$

In fact, a closer look at the singular vectors u_n and v_n reveals, experimentally, that their kth component decays of order $O(k^{-2})$ as $k \to \infty$.[15] *Assuming* this to be true, we obtain after some calculations the refined estimates

$$\|u_n^T B_{n,m}\| = O(n^{-3/2}), \qquad \|C_{n,m}v_n\| = O(n^{-3/2});$$

hence $\|A_m\|^2 = \|A_n\|^2 + O(n^{-3})$. Since all these estimates are uniform in m we can pass to the limit and obtain the desired error estimate

$$\|A\| = \|A_n\| + O(n^{-3}).$$

We *conjecture* that the actual decay rate of the components of the singular vectors improves the error term $O(n^{-3})$ in (3.11) in the same way to $O(n^{-4})$, that is,

$$\|A_m\|^2 = \|A_n\|^2 + \|u_n^T B_{n,m}\|^2 + \|C_{n,m}v_n\|^2 + O(n^{-4}).$$

If we combine with $\|A\|^2 = \|A_m\|^2 + O(m^{-3})$ and take

$$m_n = \lceil n^{4/3} \rceil,$$

we finally get an improved approximation of the conjectured order $O(n^{-4})$:

$$\|A\| = \left(\|A_n\|^2 + \|u_n^T B_{n,m_n}\|^2 + \|C_{n,m_n}v_n\|^2\right)^{1/2} + O(n^{-4}). \quad (3.12)$$

The improved order of convergence, as compared to $\|A_n\|$, is paid for with a moderate increase in the computational cost from $O(n^2)$ to $O(m_n \cdot n) = O(n^{7/3})$. Therefore, the run time needed to achieve a given error ϵ of approximation improves from $O(\epsilon^{-2/3})$ to $O(\epsilon^{-7/12})$.

Table 3.3 shows some numerical results[16] obtained with MATLAB, which are consistent with the conjectured order $O(n^{-4})$. The run time was about 10 seconds; 12 digits are correct for $n = 2048$. Because of round-off errors in the last two digits, Wynn's epsilon algorithm cannot improve upon these 12 digits.

[15]This reflects most probably the proper decay rate of the corresponding left and right singular vectors of the *infinite* matrix A.

[16]The code can be found on the web page for this book.

Table 3.3. *Values of $\|A_n\|$ and the improved approximation (3.12).*

n	$\|A_n\|$	$\left(\|A_n\|^2 + \|u_n^T B_{n,m_n}\|^2 + \|C_{n,m_n} v_n\|^2\right)^{1/2}$
128	**1.27422**212003778	**1.27422413**149024
256	**1.27422**388594855	**1.27422415**144773
512	**1.27422411**845808	**1.27422415**273407
1024	**1.27422414**844970	**1.274224152 8**1574
2048	**1.27422415**226917	**1.27422415282**070

3.5 Summation Formulas: Real Analysis

As an alternative to the acceleration of convergence of the sequence $\|A_n\|$, we now pursue the idea of passing to the limit $n \to \infty$ earlier, at each iteration step of the power method (Algorithm 3.1). Equivalently, this means to apply the power method to the infinite matrix A. In doing so we have to evaluate infinite-dimensional matrix-vector products such as Ax and $A^T x$, that is, series such as

$$\sum_{k=1}^{\infty} a_{jk} x_k \quad \text{and} \quad \sum_{k=1}^{\infty} a_{kj} x_k. \tag{3.13}$$

A simple truncation of the sum at index n would lead us back to the evaluation of $\|A_n\|$. Thus, we need a more sophisticated method to approximate these sums. Essentially, such methods can be justified on the ground of the actual rate of decay of the terms of the sum.

In the course of the power method, the vectors x at hand are approximations of the left and right singular vectors belonging to the largest singular value of A. From the discussion of §3.4 we have good reasons to assume that $x_k = O(k^{-2})$ as $k \to \infty$. The explicit formula (3.6) implies that, for j fixed, the matrix entries a_{jk} and a_{kj} decay as $O(k^{-2})$, too. Further, we observe that all terms of the sum will be nonnegative, if the initial vector for the power method was chosen to be nonnegative. Hence, a good *model* of the sums (3.13) is

$$\zeta(4) = \sum_{k=1}^{\infty} k^{-4} = \frac{\pi^4}{90}, \tag{3.14}$$

for which we will test our ideas later on. For sums of this type, we will derive some *summation formulas*

$$\sum_{k=1}^{\infty} f(k) \approx \sum_{k=1}^{n} w_k \cdot f(c_k), \tag{3.15}$$

where—in analogy to quadrature formulas—the nonnegative quantities w_k are called *weights* and the c_k are the *sampling points*. We will take the freedom to view $f(k)$ as a function of k that naturally extends to noninteger arguments $c_k \notin \mathbb{N}$. This point of view is natural indeed for the coefficients $a_{jk} = a(j,k)$ at hand, with $a(j,k)$ the rational function given in (3.7).

Now, if we approximate the matrix-vector products that arise in the power method by the summation formula (3.15), the components x_k will inherit this property. This way, the main step $y^{(\nu)} = A^T A x^{(\nu-1)}$ of the power method (see (3.10)) transforms to

$$\begin{aligned} \tilde{x}^{(\nu)}(c_j) &= \sum_{k=1}^{n} w_k \cdot a(c_j, c_k) \cdot x^{(\nu-1)}(c_k), \qquad j = 1, \dots, n, \\ y^{(\nu)}(c_k) &= \sum_{j=1}^{n} w_j \cdot a(c_j, c_k) \cdot \tilde{x}^{(\nu)}(c_j), \qquad j = 1, \dots, n. \end{aligned} \tag{3.16}$$

Upon introducing the diagonal matrix $W_n = \text{diag}(w_1, \dots, w_n)$ and the matrix $T_n = (a(c_j, c_k))_{jk}$, we observe that the transformed power iteration (3.16) calculates, in fact, the dominant eigenvalue of the matrix

$$\tilde{G}_n = T_n^T W_n T_n W_n. \tag{3.17}$$

By means of the similarity transformation $B \mapsto W_n^{1/2} B W_n^{-1/2}$ we conclude that this eigenvalue is, also, the dominant eigenvalue of the matrix

$$W_n^{1/2} \tilde{G}_n W_n^{-1/2} = (W_n^{1/2} T_n W_n^{1/2})^T \, (W_n^{1/2} T_n W_n^{1/2}) = \tilde{A}_n^T \tilde{A}_n.$$

Hence, we may actually calculate the *norm* $\|\tilde{A}_n\|$ of the transformed matrix

$$\tilde{A}_n = W_n^{1/2} T_n W_n^{1/2} = \left(\frac{\sqrt{w_j w_k}}{(c_j + c_k - 1)(c_j + c_k)/2 - (c_k - 1)} \right)_{j,k=1,\dots,n}. \tag{3.18}$$

We conjecture, and the results of §3.4 as well as our numerical experiments confirm, that

$$\lim_{n \to \infty} \|\tilde{A}_n\| = \|A\|,$$

with the same order of approximation as the underlying summation formula (3.15). A proof of this fact is an open problem that we leave as a challenge to the reader.

Strebel's Summation Formula. To get a working algorithm we need a summation formula that is of higher order than simple truncation at n, which, for the problem at hand, is of order $O(n^{-3})$. We will follow the ideas that were communicated to us by Rolf Strebel and construct a method that is, at least for the evaluation of $\zeta(4)$, *provably* of order $O(n^{-7})$.

A summation formula of the type (3.15) should shorten the range of indices from $1, \dots, \infty$ to $1, \dots, n$ without introducing too much error. In the case of an *integral*, a substitution by means of a one-to-one mapping $\phi : [1, n) \to [1, \infty)$ would do the job *exactly*,

$$\int_1^\infty f(x)\, dx = \int_1^n \phi'(\xi)\, f(\phi(\xi))\, d\xi.$$

If we view the sum as a kind of "approximation" to the integral, we might try

$$\sum_{k=1}^{\infty} f(k) \approx \sum_{k=1}^{n-1} \phi'(k)\, f(\phi(k)).$$

Now, the precise relation between equally spaced sums and integrals is described by the following theorem. For the convenience of presentation we have shifted the lower index of summation to $k = 0$.

Theorem 3.4 (Euler–Maclaurin sum formula [Hen77, §11.11]). *Let n, m be positive integers and f a function that is $2m$ times continuously differentiable on the interval $[0, n]$. Then*

$$\sum_{k=0}^{n} f(k) = \frac{1}{2}\big(f(0)+f(n)\big)+\int_0^n f(x)\,dx+\sum_{k=1}^{m} \frac{B_{2k}}{(2k)!}\big(f^{(2k-1)}(n)-f^{(2k-1)}(0)\big)+R_{2m},$$

with the remainder bounded by

$$|R_{2m}| \leqslant \frac{|B_{2m}|}{(2m)!}\int_0^n |f^{(2m)}(x)|\,dx.$$

Here, the quantities B_{2k} denote the Bernoulli numbers.

The terms in this formula that appear in addition to the sum $\sum_{k=0}^{n} f(k)$ and the desired integral $\int_0^n f(x)\,dx$ will restrict the possible choices of a suitable mapping ϕ. One such possibility is given in the next lemma.

Lemma 3.5 (R. Strebel). *Let n be a positive integer and $f_n(x) = (x+n)^{-\alpha}$, $\alpha > 1$. The function*

$$\phi_n(\xi) = \frac{n^{1+\beta}(n-\xi)^{-\beta}}{\beta} - \frac{n}{\beta} - \frac{(1+\beta)\xi^2}{2n}, \qquad \beta = \frac{6}{\alpha - 1},$$

is strictly increasing and maps $[0, n)$ onto $[0, \infty)$. Then

$$\sum_{k=0}^{\infty} f_n(k) = \sum_{k=0}^{n-1} \phi_n'(k) \cdot f_n(\phi_n(k)) + O(n^{-3-\alpha}).$$

Proof. We write $\tilde{f}_n(\xi) = \phi_n'(\xi) \cdot f_n(\phi_n(\xi))$ for short. The mapping ϕ_n was chosen such that, as $\xi \to 0$,

$$\phi_n(\xi) = \xi + O(\xi^3).$$

Therefore

$$\tilde{f}_n(0) = f_n(0), \qquad \tilde{f}_n'(0) = f_n'(0).$$

A short calculation shows that

$$\tilde{f}_n'''(0) = c_{1,\alpha}\, n^{-3-\alpha}, \qquad f_n'''(0) = c_{2,\alpha}\, n^{-3-\alpha},$$

with some constants $c_{1,\alpha}$ and $c_{2,\alpha}$ that depend on α only. Moreover, ϕ_n was also chosen such that, as $\xi \to n$,

$$\tilde{f}_n(\xi) = c_{3,\alpha}(n-\xi)^5 + o((n-\xi)^5),$$

with a constant $c_{3,\alpha}$. Hence

$$\tilde{f}_n(n) = \tilde{f}_n'(n) = \tilde{f}_n''(n) = \tilde{f}_n'''(n) = \tilde{f}_n^{(4)}(n) = 0.$$

Finally, we observe that

$$\int_0^\infty |f_n^{(4)}(x)|\,dx = O\left(\int_0^\infty (x+n)^{-4-\alpha}\,dx\right) = O(n^{-3-\alpha})$$

and

$$\int_0^n |\tilde{f}_n^{(4)}(\xi)|\,d\xi \leqslant n \cdot \max_{\xi\in[0,n]} |\tilde{f}_n^{(4)}(\xi)| = n \cdot O(n^{-4-\alpha}) = O(n^{-3-\alpha}).$$

If we apply the Euler–Maclaurin sum formula twice, we thus obtain

$$\begin{aligned}\sum_{k=0}^\infty f_n(k) &= \frac{1}{2} f_n(0) + \int_0^\infty f_n(x)\,dx - \frac{B_2}{2!} f_n'(0) + O(n^{-3-\alpha}) \\ &= \frac{1}{2}\tilde{f}_n(0) + \int_0^n \tilde{f}_n(\xi)\,d\xi - \frac{B_2}{2!}\tilde{f}_n'(0) + O(n^{-3-\alpha}) = \sum_{k=0}^{n-1} \tilde{f}_n(k) + O(n^{-3-\alpha}),\end{aligned}$$

which is the assertion. □

We arrive at *Strebel's summation formula* for $f(x) = x^{-\alpha}$, $\alpha > 1$, which is of the desired form

$$\sum_{k=1}^\infty f(k) = \sum_{k=1}^n w_k \cdot f(c_k) + O(n^{-3-\alpha}), \tag{3.19}$$

by splitting, with $m = \lceil n/2 \rceil$, $\sum_{k=1}^\infty f(k) = \sum_{k=1}^{m-1} f(k) + \sum_{k=0}^\infty f_m(k)$ and applying Lemma 3.5 to the second term. Then, the weights are

$$w_k = \begin{cases} 1, & 1 \leqslant k \leqslant m-1, \\ \phi_m'(k-m), & m \leqslant k \leqslant 2m-1, \end{cases} \tag{3.19–1}$$

and the sampling points are

$$c_k = \begin{cases} k, & 1 \leqslant k \leqslant m-1, \\ m + \phi_m(k-m), & m \leqslant k \leqslant 2m-1. \end{cases} \tag{3.19–2}$$

If $n = 2m$, we additionally define $w_n = 0$. In Appendix C.3.2 the reader will find an implementation as a MATLAB procedure that is called by

```
[w,c] = SummationFormula(n,alpha).
```

Table 3.4. *Values of $\|\tilde{A}_n\|$ for two different summation formulas.*

n	$\|\tilde{A}_n\|$ with (3.19)	$\|\tilde{A}_n\|$ with (3.20)
4	**1.2**84206027130483	**1.2**19615351739390
8	**1.274**196943864618	**1.2**63116205917547
16	**1.27422**3642340573	**1.27420**7431536352
32	**1.2742241**47506024	**1.274224152**001268
64	**1.274224152**770219	**1.274224152821228**
128	**1.2742241528**20767	**1.274224152821228**
256	**1.27422415282122**4	**1.274224152821228**
512	**1.274224152821228**	**1.274224152821228**
1024	**1.274224152821228**	**1.274224152821228**

We test it on the evaluation of $\zeta(4)$, using $n = 2$, 20, and 200:

A MATLAB Session

```
>> zeta4 = pi^4/90;
>> alpha = 4; f = inline('1./x.^alpha','x','alpha');
>> error = [];
>> for n = [2,20,200]
>>     [w,c] = SummationFormula(n,alpha);
>>     error = [error; abs(zeta4 - w*f(c,alpha)')];
>> end
>> error

error = 8.232323371113792e-002
        1.767847579436932e-008
        1.554312234475219e-015
```

The result nicely reflects that the error is, for $\alpha = 4$, of order $O(n^{-7})$.

Application to Problem 3. Now that we have a good summation formula in hand, we take the transformed matrix $\tilde{A}_n$, as defined in (3.18), and calculate its norm for various n using the power method of §3.3, applied here to $\tilde{A}_n^T\tilde{A}_n$ instead of $A_n^T A_n$. The results of a run[17] in MATLAB up to dimension $n = 1024$ are shown in the second column of Table 3.4; the run time was less than a second. The data are consistent with the conjectured $O(n^{-7})$ order of approximation. The reduction of the dimension is remarkable: whereas $\|A_{32768}\|$ gives 13 correct digits only, $\|\tilde{A}_{512}\|$ is correct to 16 digits—even if evaluated in the realm of IEEE double-precision arithmetic.

An Exponential Summation Formula. For much higher accuracy, the $O(n^{-7})$ order of approximation is yet not good enough. Following the ideas of the proof of Lemma 3.5 one might try to eliminate all terms $B_{2k}f^{(2k-1)}(0)/(2k)!$ in the Euler–Maclaurin sum formula

[17] The code can be found on the web page for this book.

for $\sum_{k=0}^{\infty} f(k)$ by using a transformation

$$\tilde{f}_n(\xi) = (1 + n^{-1}\phi'_{\exp}(\xi/n))\, f(\xi + \phi_{\exp}(\xi/n)),$$

where $\phi_{\exp}(\xi)$ is a function that vanishes with all its derivatives at $\xi = 0$ and that grows fast to infinity for $\xi \to 1$. In fact, we can then show that

$$\sum_{k=0}^{\infty} f(k) = \sum_{k=0}^{n-1} (1 + n^{-1}\phi'_{\exp}(k/n))\, f(k + \phi_{\exp}(k/n)) + R_{2m} - \tilde{R}_{2m}, \tag{3.20}$$

where R_{2m} and $\tilde{R}_{2m}$ denote the remainder terms of the Euler–Maclaurin formulas for the two sums. The point is that the summation formula (3.20) holds for *all* m. An analysis of the error term $R_{2m} - \tilde{R}_{2m}$ would try to find a particular index m_n, depending on n, that minimizes the error. However, such an analysis turns out to be very involved, even for the simple function $f(x) = (x+1)^{-\alpha}$, $\alpha > 1$. Strebel did some numerical experiments and obtained very promising results for the particular choice

$$\phi_{\exp}(u) = \exp\left(\frac{2}{(1-u)^2} - \frac{1}{2u^2}\right). \tag{3.20–1}$$

In Appendix C.3.2 the reader will find an implementation as a Matlab procedure that is called by

```
[w,c] = SummationFormula(n,'exp').
```

Indeed, for Problem 3 this formula is a further improvement, even in IEEE double-precision arithmetic. The results of a run[18] in MATLAB up to dimension $n = 1024$ are shown in the third column of Table 3.4; the run time was once more less than 1 second. To get 16 correct digits, we need go only to dimension $n = 64$, which takes under 0.1 second.

We have implemented the summation formula in PARI/GP and applied it to Problem 3 using high-precision arithmetic.[18] Table 3.5 shows the number of correct digits for various

Table 3.5. *Number of correct digits of* $\|\tilde{A}_n\|$ *with summation formula (3.20);* `dec` *is the mantissa length of the high-precision arithmetic.*

n	`dec`	No. of correct digits	Run time
100	30	25	1.9 sec
200	50	40	13 sec
400	75	55	92 sec
600	75	66	3.5 min
800	100	75	10 min
1000	100	82	27 min

[18]The code can be found on the web page for this book.

n; "correctness" was assessed by comparing with the result obtained for $n = 1200$. Though the formula performs with remarkable success—we get 25 digits in less than 2 seconds as compared to 8 hours for 21 digits in §3.3—the convergence rate slows down for larger n. In the next section we will turn to summation formulas derived from complex analysis that enjoy *reliable* exponential convergence rates.

3.6 Summation Formulas: Complex Analysis

In this section we describe a general technique for evaluating sums as contour integrals. Because there are excellent methods known for numerical quadrature, this will result in a very efficient algorithm to deal with the infinite sums (3.13) that have emerged in the application of the power method to the infinite matrix A.

The technique is a consequence of the residue theorem of complex analysis and is therefore restricted to sums of terms that depend analytically on the index. The process is better known in the opposite direction: the evaluation of a contour integral by a sum of residues. Summation by contour integration has earlier been used by Milovanović [Mil94], among others. A particularly useful result of this approach to summation, general enough to serve our purposes, is given by the following theorem.

Theorem 3.6 *Let $f(z)$ be a function that is analytic in a domain of the complex plane. Further, suppose that for some $\alpha > 1$, $f(z) = O(z^{-\alpha})$ as $z \to \infty$. Let C be a contour in the domain of analyticity of f that is symmetric with respect to the real axis, passes from infinity in the first quadrant into the fourth quadrant, and has, on its left (with respect to its orientation), all the positive integers, no other integers, and no boundary points of the domain of analyticity of f. Then*

$$\sum_{k=1}^{\infty} f(k) = \frac{1}{2\pi i} \int_C f(z) \cdot \pi \cot(\pi z)\, dz, \tag{3.21}$$

$$\sum_{k=1}^{\infty} (-1)^k f(k) = \frac{1}{2\pi i} \int_C f(z) \cdot \pi \csc(\pi z)\, dz. \tag{3.22}$$

Proof.[19] Consider the 1-periodic meromorphic function $\pi \cot(\pi z)$, whose poles are all simple and located at the integers $z = n$, $n \in \mathbb{Z}$. From the Laurent series at $z = 0$, namely,

$$\pi \cot(\pi z) = \frac{1}{z} - \frac{\pi^2}{3} z + O(z^3),$$

we conclude that the residues are all 1.

Let C_1 be the arc of the contour C between a point P and its complex conjugate $\bar{P}$; let C_2 be the circular arc of radius r joining $\bar{P}$ and P counterclockwise (see Figure 3.1). If we restrict ourselves to the radii $r_n = n + 1/2$, $n \in \mathbb{N}$, the function $\pi \cot(\pi z)$ remains

[19] We restrict ourselves here to (3.21). The formula (3.22) for an alternating series can be proved in analogy if one notes that at the pole $z = n$, $n \in \mathbb{Z}$, the residue of the function $\pi \csc(\pi z)$ is $(-1)^n$.

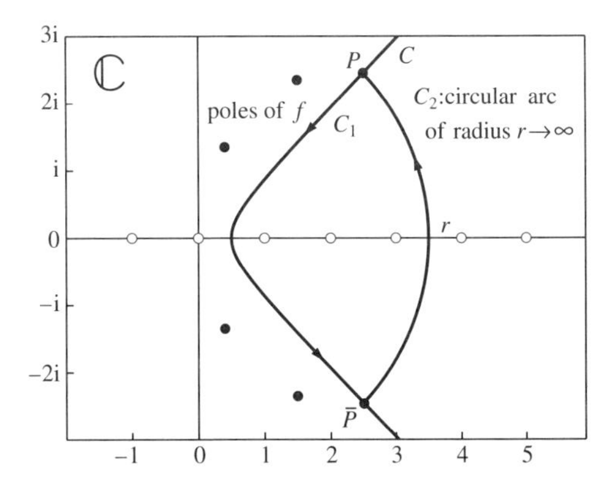

Figure 3.1. *Contours in the proof of Theorem 3.6 ($n = 3$).*

uniformly bounded when $|z| = r_n$, $z \in \mathbb{C}$, and $n \to \infty$. Now, the residue theorem [Hen74, Thm. 4.7a] gives

$$\sum_{k=1}^{n} f(k) = \frac{1}{2\pi i} \int_{C_1} f(z) \cdot \pi \cot(\pi z)\, dz + \frac{1}{2\pi i} \int_{C_2} f(z) \cdot \pi \cot(\pi z)\, dz.$$

Passing to the limit $n \to \infty$ yields the conclusion, because

$$\left| \int_{C_2} f(z) \cdot \pi \cot(\pi z)\, dz \right| = \operatorname{length}(C_2) \cdot O(r_n^{-\alpha}) = O(r_n^{1-\alpha}) \to 0.$$

Here we have used the asymptotic decay $f(z) = O(z^{-\alpha})$, $\alpha > -1$. □

Theorem 3.6 enables us to transform infinite sums into integrals. We first choose an appropriate contour C, parametrized by the real variable t through the complex-valued function $Z(t)$. Since we are working with tools of complex analysis, it is best to use contours C that are *analytic curves*; that is, the function Z must itself be an analytic function of the real variable t. For ease of numerical evaluation we prefer curves that can be expressed in terms of elementary functions. Using the parametrization, the integral in (3.21) transforms to

$$\sum_{k=1}^{\infty} f(k) = \int_{-\infty}^{\infty} F(t)\, dt, \qquad F(t) = \frac{1}{2i} f(Z(t)) \cdot \cot(\pi Z(t)) \cdot Z'(t). \tag{3.23}$$

Example. Let us transform the sum (3.14), which defines $\zeta(4)$, into an integral. We choose the contour C parametrized through the function $Z(t) = (1 - it)/2$, apply Theorem 3.6 to the function $f(z) = z^{-4}$, and obtain

$$\zeta(4) = \frac{1}{2\pi i} \int_C \frac{\pi \cot(\pi z)}{z^4}\, dz = 16 \int_{-\infty}^{\infty} \frac{t(1 - t^2) \tanh(\pi t/2)}{(1 + t^2)^4}\, dt. \tag{3.24}$$

Any numerical quadrature method used to approximate the integral (3.23) will result in a summation formula for f of the form (3.15). However, because of the intermediate function Z, the weights and sampling points will now be complex.

Note that once the contour $\mathcal{C}$ has been chosen, the parametrization of $\mathcal{C}$ can still be modified by using a new parameter τ that is related to t by an analytic transformation $t = \Phi(\tau)$ with $\Phi'(\tau) > 0$. This additional flexibility is the main advantage of the summation by contour integration over the real-analysis method of §3.5. A proper choice of the parametrization will help us reduce the number of terms in the resulting summation formula considerably.

3.6.1 Approximation of Contour Integrals by Trapezoidal Sums

We approximate the integral $S = \int_{-\infty}^{\infty} F(t)\,dt$ in (3.23) by its trapezoidal sum $T(h)$ of step size $h > 0$,

$$T(h) = h \sum_{j=-\infty}^{\infty} F(j \cdot h).$$

In many texts on numerical analysis the trapezoidal rule is treated as the "ugly duckling" among algorithms for approximating definite integrals. It turns out, however, that the trapezoidal rule, simple as it is, is among the *most powerful* algorithms for the numerical quadrature of analytic functions. Among the first who pointed out that exceptional behavior of the trapezoidal rule for infinite intervals were Milne (in an unpublished 1953 note; see [DR84, p. 212]) and Bauer, Rutishauser, and Stiefel [BRS63, pp. 213–214]. Later, Schwartz [Sch69] and Stenger [Ste73] applied the trapezoidal rule to more general analytic integrals; see also the book by Davis and Rabinowitz [DR84, §3.4]. We also mention the Japanese school starting with the work of Iri, Moriguti, and Takasawa [IMT70], which is also based on the trapezoidal rule and is now known as the IMT method. Their ideas were further developed into the *double-exponential formulas* by Takahasi and Mori [TM74] and by Mori [Mor78]. Full generality for handling analytic integrals is achieved by combining the trapezoidal rule (applied to integrals over $\mathbb{R}$) with analytic transformations of the integration parameter (see [Sch89, Chap. 8]). Applications to multidimensional integrals over Cartesian-product domains (rectangles, strips, quadrants, slabs, etc.) are discussed in [Wal88].

Truncation of Trapezoidal Sums. The *infinite* sums $T(h)$ are inherently difficult to compute if the integrand decays too slowly, such as $O(|t|^{-\alpha})$ as $t \to \pm\infty$. An obvious problem is the large number of terms that would have to be included in the sum; more serious, however, is the estimation of the remainder of a truncated sum.

Typically such a sum is truncated by disregarding all terms that are considered "too small." Let us formalize this idea by specifying a threshold or truncation tolerance $\epsilon > 0$ and define the truncated trapezoidal sum

$$T_\epsilon(h) = h \sum_{j\in\mathbb{Z}:|F(j\cdot h)|\geqslant\epsilon} F(j \cdot h). \tag{3.25}$$

Then an approximation of or a useful bound on the remainder $R_\epsilon(h) = T(h) - T_\epsilon(h)$ is needed. To get a rough idea of the principal problems involved, we consider, as a model, the truncation of the integral

$$\int_1^\infty t^{-\alpha}\,dt, \quad \alpha > 1,$$

at the threshold ϵ, that is, at the point of integration $t_\epsilon = \epsilon^{-1/\alpha}$. The remainder is

$$R_\epsilon = \int_{t_\epsilon}^{\infty} t^{-\alpha}\, dt = \frac{\epsilon^{(\alpha-1)/\alpha}}{\alpha - 1},$$

which can be much larger than the truncation tolerance ϵ. For instance, we obtain $R_\epsilon = \sqrt{\epsilon}$ for $\alpha = 2$. Therefore, in the case of slowly decaying integrands the truncation by a threshold does not lead to accurate results. Instead, we propose to enhance the decay of the integrand by introducing a new integration variable. For integrals along the real axis the transformation,

$$t = \sinh(\tau), \qquad dt = \cosh(\tau)\, d\tau, \tag{3.26}$$

proves to be appropriate. The integral becomes

$$S = \int_{-\infty}^{\infty} G(\tau)\, d\tau, \qquad G(\tau) = F(\sinh(\tau))\cosh(\tau).$$

If $F(t)$ decays as a power of t, $G(\tau)$ decays *exponentially*. Consider, again as a model, the truncation by a threshold ϵ of a typical integral with an exponentially decaying integrand,

$$\int_0^{\infty} ae^{-\alpha\tau}\, d\tau, \quad \alpha > 0, \ a > 0.$$

The point of truncation is $t_\epsilon = \log(a/\epsilon)/\alpha$, and the remainder is

$$R_\epsilon = \int_{t_\epsilon}^{\infty} ae^{-\alpha\tau}\, d\tau = \epsilon/\alpha.$$

Thus, we see that a truncation tolerance $\epsilon = \alpha \cdot \texttt{tol}$ suffices to accumulate the trapezoidal sum of an exponentially decaying integrand to the accuracy `tol`.

The reader will find a MATLAB implementation of these ideas as the routine

```
TrapezoidalSum(f, h, tol, level, even)
```

in Appendix C.3.2. The routine assumes that the integrand `f` decays monotonically in the tail where the threshold `tol` applies. The nonnegative integer `level` tells the routine how often the sinh transformation (3.26) has to be applied recursively. We will call the method used with `level` = 1 a *single-exponential formula*, with `level` = 2 a *double-exponential formula*. If the switch `even` is set to `'yes'`, only half of the sum is accumulated for symmetry reasons. For numerical accuracy, the sum is accumulated from the smaller to the larger values (assuming monotonicity). Therefore, the terms have to be stored until the smallest one has been computed.

Discretization Error. The error theory of the trapezoidal rule is ultimately connected to the Fourier transform of F:

$$\hat{F}(\omega) = \int_{-\infty}^{\infty} F(t) e^{-i\omega t}\, dt.$$

In fact, the *Poisson summation formula* [Hen77, Thm. 10.6e] expresses $T(h)$ as a corresponding trapezoidal sum, with step size $2\pi/h$, of the Fourier transform $\hat{F}$:

$$T(h) = h \sum_{j=-\infty}^{\infty} F(j \cdot h) = \sum_{k=-\infty}^{\infty} \hat{F}\left(k \cdot \frac{2\pi}{h}\right).$$

Note that the sum over k has to be interpreted as a *principal value*, that is, as the limit of the sum from $-N$ to N as $N \to \infty$. This, however, becomes relevant only if $\hat{F}$ decays slowly. Now, we observe that the term $k = 0$ of the trapezoidal sum of the Fourier transform is the integral at hand:

$$\hat{F}(0) = \int_{-\infty}^{\infty} F(t)\,dt = S.$$

Therefore, the Poisson summation formula yields the error formula

$$E(h) = T(h) - S = \hat{F}(2\pi/h) + \hat{F}(-2\pi/h) + \hat{F}(4\pi/h) + \hat{F}(-4\pi/h) + \cdots. \tag{3.27}$$

Hence, the decay rate of $E(h)$ for $h \to 0$ is determined by the asymptotic behavior of the Fourier transform $\hat{F}(\omega)$ as $\omega \to \pm\infty$. In many specific cases this asymptotics can be found by the saddle point method (method of steepest descents); see, e.g., [Erd56] or [Olv74] for the theory or [GW01, pp. 495ff] for a worked example.

The error formula (3.27) yields an especially nice and simple result if we assume that F is analytic in the strip $|\mathrm{Im}(t)| < \gamma_*$, $\gamma_* > 0$, and that $F(x + iy)$ is integrable with respect to x, uniformly in $|y| \leqslant \gamma$ for any $\gamma < \gamma_*$. Then the modulus of the Fourier transform is known to decay exponentially [RS75, Thm. IX.14]:[20]

$$|\hat{F}(\omega)| = O\left(e^{-\gamma|\omega|}\right) \qquad \text{as} \quad \omega \to \pm\infty,$$

where $0 < \gamma < \gamma_*$ can be chosen arbitrarily. If we plug this into (3.27) we obtain the error estimate[21]

$$E(h) = O\left(e^{-2\pi\gamma/h}\right) \qquad \text{as} \quad h \to 0; \tag{3.28}$$

that is, we get *exponential* convergence: halving the step size results in doubling the number of correct digits. Note that exponential convergence is much better than the higher order convergence $O(h^{2m})$ achieved in Romberg integration with $m - 1$ steps of Richardson extrapolation applied to the trapezoidal rule (see Appendix A, p. 236).

Example. Let us illustrate the use of the truncated trapezoidal sum (3.25) and the sinh transformation (3.26) for the integral (3.24), which resulted from the application of the method of contour integration to the sum defining $\zeta(4)$. Since the integrand decays slowly as $O(|t|^{-5})$ as $t \to \pm\infty$, we have to apply the sinh transformation (3.26) at least once. We start with the single-exponential formula (`level` $= 1$), using the step sizes $h = 0.3$ and $h = 0.15$.

[20]The more difficult L^2-version of this result is one of the classic Paley–Wiener theorems [PW34, §3, Thm. IV], stated in their seminal monograph on Fourier transforms in the complex domain.

[21]For a different proof see [DR84, p. 211].

A MATLAB Session

```
>> f = inline('z^(-4)','z');
>> Z = inline('1/2-i*t/2','t');
>> dZ = inline('-i/2','t');
>> F_Summation = inline('real(f(Z(t))*cot(pi*Z(t))*dZ(t)/2/i)',...
>>       't','f','Z','dZ');
>> tol = 1e-16; level = 1; even = 'yes'; s = [];
>> for h = [0.3 0.15]
>>       s = [s;
>>        TrapezoidalSum(F_Summation,h,tol,level,even,f,Z,dZ)];
>> end
>> s

s = 1.082320736262448e+000
    1.082323233711138e+000
```

For $h = 0.3$, using 73 terms of the trapezoidal sum, 6 digits of the result are correct; whereas all 16 digits are correct for $h = 0.15$ using 143 terms of the trapezoidal sum. The errors of several runs with h in the range from 1 to 1/6 are shown (dashed line) in the left part of Figure 3.2; they nicely reflect the exponential convergence. The corresponding number of terms used in the truncated trapezoidal sums are shown in the right part of Figure 3.2.

A look at the right integral in (3.24) shows that the singularities of the integrand are, *before* the sinh transformation, located at $(2k+1)i$, $k \in \mathbb{Z}$. *After* the transformation the singularities τ_s of the integrand satisfy

$$\sinh(\tau_s) = (2k+1)i, \qquad \text{which implies} \qquad \operatorname{Im}(\tau_s) = \frac{(2m+1)\pi i}{2}$$

for some $m \in \mathbb{Z}$; see the left part of Figure 3.3. Hence, the transformed integrand is analytic in the strip $|\operatorname{Im}(\tau)| < \gamma_*$ with $\gamma_* = \pi/2$ and the error estimate (3.28) specializes to $E(h) = O(e^{-\beta/h})$, for all $\beta < \pi^2$. Thus, an increase of $1/h$ by, say, 1 yields, in the asymptotic regime, a gain in accuracy of approximately $\pi^2/\log(10) \doteq 4$ digits. Indeed, the dashed

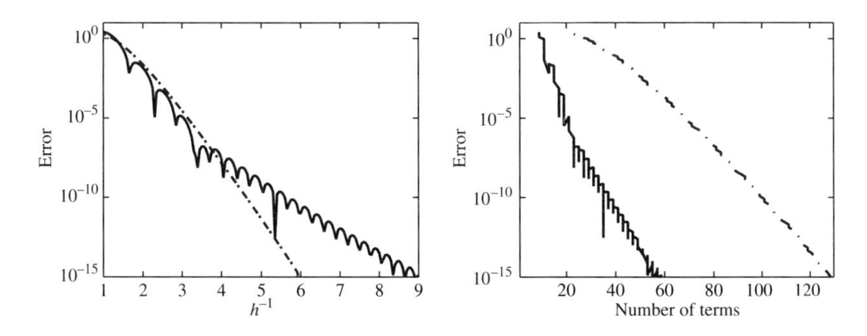

Figure 3.2. *Error of trapezoidal sum vs. $1/h$ (left), and vs. the number of terms (right). Results are shown for a single-exponential (dashed line) and a double-exponential (solid line) formula.*

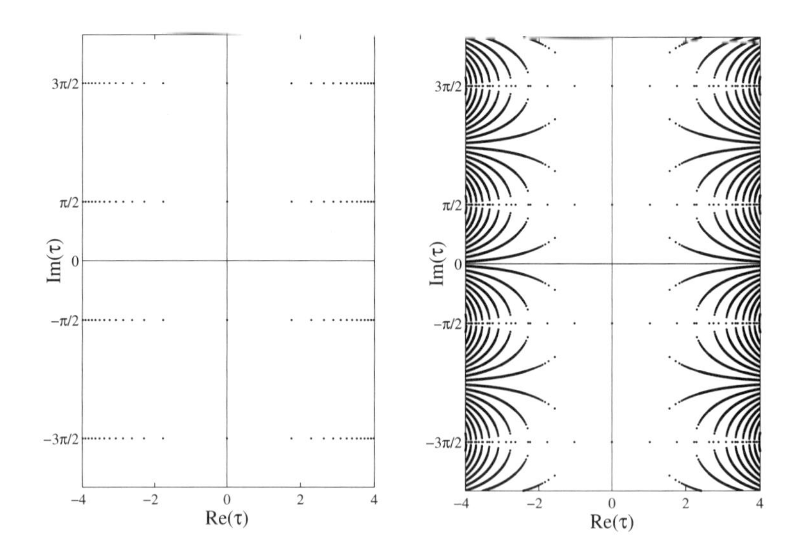

Figure 3.3. *Proliferation of singularities in the parameter plane for a single-exponential (left) and a double-exponential (right) formula. There are many more singularities outside the rectangles shown.*

line in the left part of Figure 3.2 has a slope of approximately -4 in its linear asymptotic regime.

It is tempting to enhance the decay of the integrand by a double-exponential parametrization, that is, by applying the sinh transformation (3.26) once more.[22] This would simplify the computation of the trapezoidal sums by making the number of terms in the truncated sum $T_\epsilon(h)$ essentially independent of the threshold ϵ. Such a repeated application of the sinh transformation has to be used with care, however.

The right part of Figure 3.3 shows the singularities of the integrand of the example at hand after such a second sinh transformation. We observe that the integrand is no longer analytic in a strip containing the real axis. In contrast, the domain of analyticity now has the shape of a "funnel" of exponentially decaying width. Although the formula (3.28) for the discretization error is not applicable, integrands with such funnel domains of analyticity can still be useful: for the integral at hand, the solid line in the left part of Figure 3.2 shows that the convergence is apparently still exponential, even if the decay has slowed down. Therefore, smaller step sizes have to be used to get the same accuracy as with the singly transformed integrand. This potential increase of the computational effort is compensated by the much faster decay of the integrand, which allows much earlier truncation of the trapezoidal sum. So let us compare the two methods with respect to the number of terms that were actually used: the right part of Figure 3.2 shows that, at the same level of accuracy, the double application of the sinh transformation requires fewer terms by a factor of 2.5. Thus the computational cost of the sinh transformations is about the same for both the single-exponential and the double-exponential formula. However, the double-exponential

[22] To do so, one puts `level` = 2 in the MATLAB session of p. 71.

formula saves the factor of 2.5 in the evaluations of the original integrand. This will become a major advantage in the final solution of Problem 3, for which the transformations (and the transcendentals of the summation formula) are applied only once, but a large number of integrals has to be calculated during the power iteration.

Summary. For the convenience of the reader we finally write down the summation formula of the type (3.15) that is obtained by applying the trapezoidal rule to the contour integral (3.23). If we choose a contour parametrized through $Z(t)$, a reparametrization $\Phi(\tau)$, a step size h, and a truncation point T, we get

$$\sum_{k=1}^{\infty} f(k) \approx \sum_{k=-m}^{m} w_k f(c_k), \qquad m = \lfloor T/h \rfloor,$$

with the (complex-valued) sampling points and weights

$$c_k = Z(\Phi(kh)), \qquad w_k = \frac{h}{2i} \cot(\pi c_k) \cdot Z'(\Phi(kh)) \cdot \Phi'(kh).$$

The discussion so far has shown that for certain f there are suitable choices that make the error exponentially small in the number of terms, that is, $n = 2m + 1$.

3.6.2 Application to Problem 3

As in §3.5, having a good summation formula in hand, we can use the power iteration in the form (3.16) to approximate $\|A\|^2$. That is, we calculate the dominant eigenvalue of the transformed matrix $\tilde{G}_n$, which we defined in (3.17).[23] To make the method work we need only choose a contour, a reparametrization, and a truncation point. Then we apply the procedure for various step sizes.

Choosing the Contour. The sample points enter the transformed power iteration (3.16) via the expression $a(c_j, c_k)$ in a twofold operational sense: as a means to efficient summation as well as a new index for the resulting vectors. Therefore, the poles of the underlying analytic functions depend on the contour itself. In fact, these poles lie on the loci given by the zeros $v(t)$ and $u(t)$ of the denominator $b(u + v) - v$ of $a(u, v)$, as given by (3.7), while one of the variables is running through $Z(t)$:

$$b(u(t) + Z(t)) - Z(t) = 0, \qquad b(Z(t) + v(t)) - v(t) = 0.$$

These are quadratic equations and we obtain four connected branches for the loci of the poles (see Figure 3.4).

Because of the dependence of the poles on the chosen contour, the simplest such choice, which was good for our model $\zeta(4) = \sum_{k=1}^{\infty} k^{-4}$, namely $Z(t) = \sigma - it$ with

[23] As in §3.5 we can argue that this value is also the dominant eigenvalue of the matrix $\tilde{A}_n^T \tilde{A}_n$, where the transformed matrix $\tilde{A}_n$ is defined in (3.18). However, because now $\tilde{A}_n$ has *complex* entries, it is *not* the case that this eigenvalue is equivalently given by $\|\tilde{A}_n\|^2$—a value that in the complex case is equal to the dominant eigenvalue of $\tilde{A}_n^H \tilde{A}_n$ instead.

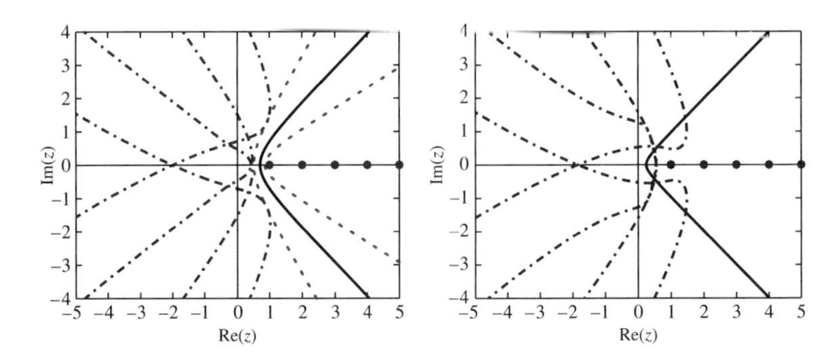

Figure 3.4. *Loci of poles (dashed lines) for the contour (3.29) (solid line) with a good choice of the parameter* $\sigma = 1/\sqrt{2}$ *(left) and a bad choice* $\sigma = 1/4$ *(right). The dotted lines on the left show the boundaries of the image* $z = Z(t)$ *(with* $\sigma = 1/\sqrt{2}$*) of the strip* $|\mathrm{Im}(t)| \leqslant 0.25$ *in the parameter plane.*

$0 < \sigma < 1$, does not satisfy the conditions of Theorem 3.6. Instead we use the right branch of a rectangular hyperbola $\mathrm{Re}(z^2) = \sigma^2$, suitably parametrized to yield exponentially decaying integrands for the trapezoidal rule:

$$Z(t) = \sigma(\cosh(t) - i\sinh(t)), \qquad Z'(t) = -i \cdot \overline{Z(t)}. \tag{3.29}$$

The particular choice $\sigma = 1/\sqrt{2}$ yields the equivalent representation $Z(t) = \cosh(t - i\pi/4)$; the loci of poles are shown on the left part of Figure 3.4 and satisfy the conditions of Theorem 3.6. As shown on the right part of Figure 3.4, the choice of the parameter $\sigma = 1/4$ leads to intersections of the loci of poles with the contour. Such a choice is not admissible.

Choosing the Parametrization. With $\sigma = 1/\sqrt{2}$, and without further reparametrization, the domain of analyticity extends to the strip $|\mathrm{Im}(t)| < \gamma_*$, where $\gamma_* \approx 0.25$, as can be read off from the dotted lines on the left of Figure 3.4. Therefore, we expect errors that are exponentially small in the reciprocal step size h^{-1}. Although practical for low accuracies, such as calculations in IEEE double precision, this parametrization turns out to be rather slow for higher accuracies. Some experimentation led us to the particular choice

$$t = \Phi(\tau) = \tau + \frac{\tau^3}{3}. \tag{3.30}$$

Choosing the Truncation Point. From §3.5 we know that the decay rate of the terms of the sums in the power iteration is the same as for $\zeta(4) = \sum_{k=1}^{\infty} k^{-4}$. Thus, with the parametrization (3.30) a good point of truncation for the truncation tolerance ϵ is at

$$T = \log^{1/3} \epsilon^{-1}. \tag{3.31}$$

Results in IEEE Double-Precision Arithmetic. The results of a MATLAB run[24] with the choices (3.29) for the contour, (3.30) for the reparametrization, and (3.31) for the truncation

[24] The code can be found on the web page for this book.

Table 3.6. *Values of* $\lambda_{\max}^{1/2}(\tilde{G}_n)$, *truncation tolerance* $\epsilon = 10^{-16}$.

h	n	$\sigma = 1/\sqrt{2}$	$\sigma = 1/4$
0.64	11	**1**.165410725532445	$9.202438226420232 \cdot 10^{-17}$
0.32	21	**1.2**52235875316941	$5.344162844425345 \cdot 10^{-17}$
0.16	41	**1.27**3805685815999	2.445781134567926
0.08	83	**1.27422413**7562002	1.722079300161066
0.04	167	**1.274224152821228**	4.210984571814455

point are shown in Table 3.6. The run with the suitable contour parameter $\sigma = 1/\sqrt{2}$, shown in the third column, nicely exhibits the exponential convergence: the number of correct digits doubles if we double the dimension n; 16 correct digits were obtained in less than a second. The run with the bad contour parameter $\sigma = 1/4$, shown in the fourth column, proves that it is important to satisfy the conditions of Theorem 3.6: all the digits are garbage. Thus a careful theoretical study turns out to be indispensable for the method at hand.

Results in High-Precision Arithmetic. For experiments with higher accuracies we have implemented this approach in PARI/GP with the above choices (3.29), (3.30), (3.31), and $\sigma = 1/\sqrt{2}$. Because the contour is symmetric with respect to the real axis, the sums (3.16) of the transformed power method also enjoy a symmetry with respect to complex conjugation. Taking advantage of this symmetry allowed us to cut the computational effort in half.

Table 3.7 shows the number of correct digits for various runs; "correctness" was assessed by comparing with a result that was obtained in a month-long computation with a predicted accuracy of 273 digits. Such a prediction can be obtained as follows.

At the end of §3.6.1 we considered reparametrizations that enforce a stronger decay of the integrand and observed that they certainly yield shorter trapezoidal sums, but at the expense of a slower convergence rate caused by the proliferation of singularities. Experiments show that for Problem 3 as well, these two effects tend to balance each other in the sense that different parametrizations differ only by a constant factor in the overall computational effort. Now, for the case of no reparametrization the computational effort to obtain d correct digits is easily estimated: the exponential convergence yields the reciprocal step size $h^{-1} = O(d)$, the exponential decay results in the truncation point $T = O(d)$, and the power iteration needs $O(d)$ iterations. Thus, we get the dimension $n = O(T/h) = O(d^2)$ and[25]

$$\text{no. of } d\text{-digit operations} = O\big(\text{no. of power iterations} \cdot n^2\big) = O\big(d^5\big).$$

On the other hand, for the parametrization (3.30) we have the truncation point given in (3.31), namely, $T = O(d^{1/3})$. If we assume the same asymptotic operation count $O(d^5)$ as above, we get the reciprocal step size $h^{-1} = O(d^{5/3})$. In fact, the data of Table 3.7 are

[25] Since these operations are basically multiplications, the run time scales, using Karatsuba multiplication, as $O(d^{6.58\ldots})$ and, with an FFT-based fast multiplication, as $O(d^{6+\kappa})$, where $\kappa > 0$ is arbitrary.

Table 3.7. *Number of correct digits of* $\lambda_{\max}^{1/2}(\tilde{G}_n)$ *with* $\epsilon = 10^{2-\mathtt{dec}}$; `dec` *is the mantissa length of the high-precision arithmetic.*

$1/h$	n	`dec`	No. of correct digits	Run time
40	321	28	25	9.8 sec
65	577	38	36	43 sec
90	863	48	47	2.3 min
110	1119	57	56	5.0 min
160	1797	86	76	32 min
240	2991	105	98	2.0 h
384	5315	144	131	13 h
640	9597	183	180	71 h

consistent with this asymptotic behavior; a simple fit yields

$$d \approx 51.4 \sinh\left(\frac{3}{5}\,\mathrm{arcsinh}\left(\frac{h^{-1}}{47.7}\right)\right). \tag{3.32}$$

This empirical law allows us to predict the accuracy for a given step size h.

As efficient as this reliably exponentially convergent method is, the complexity of the problem is still too large to allow us to calculate 10,000 digits, as we have done for each of the other nine problems: the dimension of the approximating problem would be $n \approx 3 \cdot 10^7$. Calculating the dominant eigenvalue of a full matrix of that dimension to 10,000 digits is beyond imagination: infinity is too far away.

Chapter 4

Think Globally, Act Locally

Stan Wagon

In order to find all common points, which are the solutions of our nonlinear equations, we will (in general) have to do neither more nor less than map out the full zero contours of both functions. Note further that the zero contours will (in general) consist of an unknown number of disjoint closed curves. How can we ever hope to know when we have found all such disjoint pieces?

— W. H. Press et al. (*Numerical Recipes* [PTVF92])

Problem 4

What is the global minimum of the function

$$e^{\sin(50x)} + \sin(60e^y) + \sin(70\sin x) + \sin(\sin(80y)) \\ - \sin(10(x+y)) + (x^2+y^2)/4 \text{ ?}$$

4.1 A First Look

Let $f(x, y)$ denote the given function. On a global scale, f is dominated by the quadratic term $\left(x^2 + y^2\right)/4$, since the values of the other five summands lie in the intervals $[1/e, e]$, $[-1, 1]$, $[-1, 1]$, $[-\sin 1, \sin 1]$, and $[-1, 1]$, respectively. Thus the overall graph of f resembles a familiar paraboloid (Figure 4.1). This indicates that the minimum is near $(0, 0)$. But a closer look shows the complexity introduced by the trigonometric and exponential functions. Indeed, as we shall see later, there are 2720 critical points inside the square $[-1, 1] \times [-1, 1]$. From this first look, we see that a solution to the problem breaks into three steps:

1. Find a bounded region that contains the minimum.
2. Identify the rough location of the lowest point in that region.
3. Zoom in closer to pinpoint the minimum to high precision.

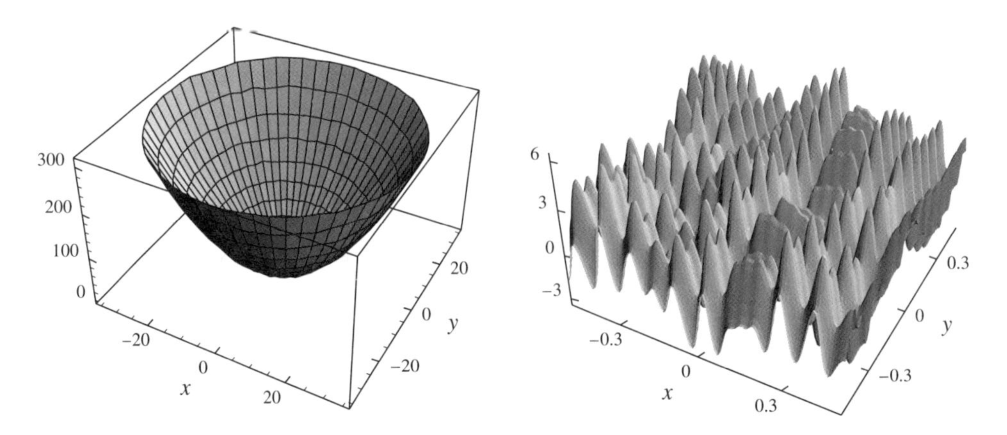

Figure 4.1. *Two views of the graph of $f(x, y)$. The scale of the left-hand view masks the complexity introduced by trigonometric and exponential terms. It is this complexity that makes finding the global minimum a challenge.*

Step 1 is easy. A quick computation using a grid of modest size yields the information that the function can be as small as -3.24. For example, one can look at the 2601 values obtained by letting x and y range from -0.25 to 0.25 in steps of 0.01.

A *Mathematica* Session

For later convenience, we define `f` so that it accepts as inputs either the two numeric arguments or a single list of length 2.

```
f[x_, y_] := ⅇ^Sin[50 x] + Sin[60 ⅇ^y] + Sin[70 Sin[x]] + Sin[Sin[80 y]] -
    Sin[10 (x + y)] + (x^2 + y^2) /4;
f[{x_, y_}] := f[x, y];
```

```
Min[Table[f[x, y], {x, -0.25, 0.25, 0.01}, {y, -0.25, 0.25, 0.01}]]
-3.246455170851875
```

This upper bound on the minimum tells us that the global minimum must lie inside the circle of radius 1 centered at the origin, since outside that circle the quadratic and exponential terms are at least $1/e + 1/4$ and the four sines are at least $-3 - \sin 1$, for a total above -3.23. And step 3 is easy once one is near the minimum: standard optimization algorithms, or root-finding algorithms on the gradient of f, can be used to locate the minimum very precisely. Once we are close, there is no problem getting several hundred digits. The main problem is in step 2: How can we pinpoint a region that contains the correct critical point, and is small enough that it contains no other? Figure 4.2, a contour plot of $f(x, y)$ with the contour lines suppressed, shows what we are up against.

In fact, a slightly finer grid search will succeed in locating the proper minimum; several teams used such a search together with estimates based on the partial derivatives of f to show that the search was fine enough to guarantee capture of the answer. But we will

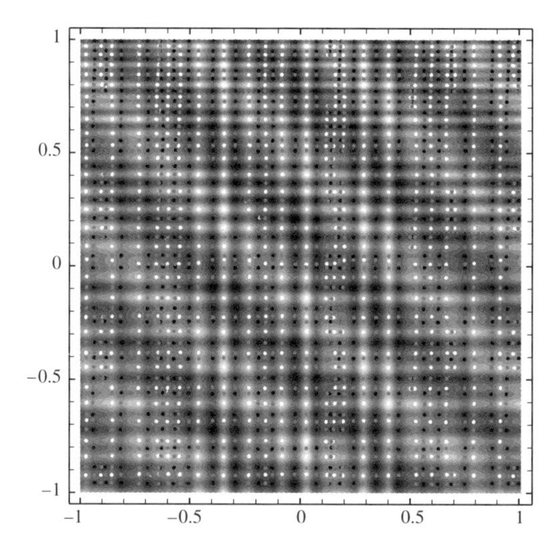

Figure 4.2. *A density plot of $f(x, y)$ with increasing values going from black to white. The local minima are shown as black dots (693), the local maxima as white dots (667), and the saddles (1360) as gray dots. There are 2720 critical points in this square, computed by the method of §4.4.*

not discuss such methods here, focusing instead on general algorithms that do not require an analysis of the objective function.

4.2 Speedy Evolution

While a grid search can be effective for this particular problem (because the critical points do have a rough lattice structure), it is easy to design a more general search procedure that uses randomness to achieve good coverage of the domain. An effective way to organize the search is to use ideas related to evolutionary algorithms. For each point in the current generation, n random points are introduced, and the n best results of each generation (and its parents) are used to form the new generation. The scale that governs the generation of new points shrinks as the algorithm proceeds.

Algorithm 4.1. Evolutionary Search to Minimize a Function.
Inputs: $f(x, y)$, the objective function;
R, the search rectangle;
n, the number of children for each parent, and the number of points in the new generation;
ϵ, a bound on the absolute error in the location of the minimum of f in R;
s, a scaling factor for shrinking the search domain.

Output: An upper bound to the minimum of f in R, and an approximation to its location.

Notation: `parents` = current generation of sample points, `fvals` = f-values for current generation.

Step 1: Initialize: Let z be the center of R; `parents` = $\{z\}$; `fvals` = $\{f(z)\}$;
$\{h_1, h_2\}$ = side-lengths of R.

Step 2: The main loop:
While $\min(h_1, h_2) > \epsilon$,
For each $p \in$ `parents`, let its children consist of n random points in a rectangle around p; get these points by using uniform random x and y chosen from $[-h_1, h_1]$ and $[-h_2, h_2]$, respectively;
Let `newfvals` be the f-values on the set of all children;
Form `fvals` $\cup$ `newfvals`, and use the n lowest values to determine the points from the children and the previous parents that will survive;
Let `parents` be this set of n points; let `fvals` be the set of corresponding f-values;
Let $h_i = s \cdot h_i$ for $i = 1, 2$.

Step 3: Return the smallest value in `fvals`, and the corresponding parent.

This algorithm is nicely simple and can be programmed in just a few lines in a traditional numerical computing environment. It generalizes with no problem to functions from $\mathbb{R}^n$ to $\mathbb{R}$. The tolerance is set to 10^{-6} in the *Mathematica* code that follows because the typical quadratic behavior at the minimum means that it should give about 12 digits of precision of the function value. Note that it is possible for some children to live outside the given rectangle, and therefore the final answer might be outside the rectangle. If that is an issue, then one can add a line to restrict the children to be inside the initial rectangle.

A *Mathematica* Session

Some experimentation will be needed to find an appropriate value of n. We use $n = 50$ here, though this algorithm will generally get the correct digits even with n as low as 30.

```
h = 1; gen = {f[#], #}&/@{{0, 0}};
```

```
While[h > 10^-6,
  new = Flatten[Table[#[[2]] + Table[h(2 Random[] - 1), {2}], {50}]&/@gen,
     1]; gen = Take[Sort[Join[gen, {f[#], #}&/@new]], 50];
  h = h/2];
```

```
gen[[1]]
{-3.30686864747396, {-0.02440308163632728, 0.2106124431628402}}
```

Figures 4.3 and 4.4 use a logarithmic scale to show the steady convergence of the points in each generation to the answer. Figure 4.4 shows how the points swarm into the correct one from all directions.

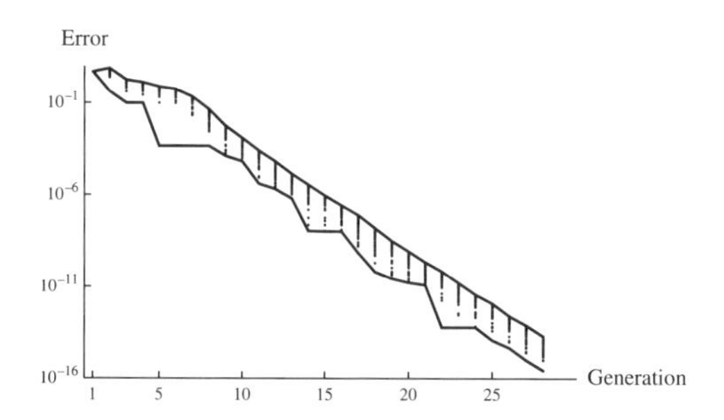

Figure 4.3. *The base-10 logarithm of the f-value error for each point in each generation of a genetic search, in a run with $n = 50$ and tolerance of 10^{-8} in the x–y coordinates. The convergence to the correct result is nicely steady.*

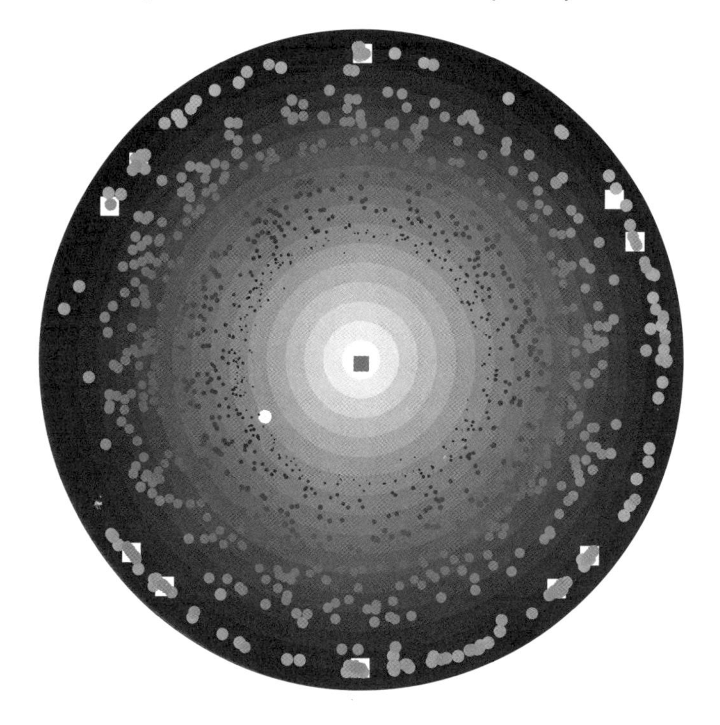

Figure 4.4. *Survival of the fittest: the results of a genetic algorithm with a high working precision and accuracy goal of 10^{-13}; the computation runs through 45 generations with 50 members of each. The scale is logarithmic, with each gray ring representing another power of 10 away from the true minimum location (center). Points of the same shade correspond to the same generation, so the innermost black points are the last generation (28th). They are within 10^{-6} of the true best; the point with the lowest f-value (which is not the point closest to the answer!) is the one atop a white disc. The outermost points—the first generation—have f-values between -1.3 and 4.2. But by the 9th generation all points are in the vicinity of the correct critical point. The white squares mark the location of the 20 lowest local minima; the clustering near those points at the early generations is visible.*

For more confidence one would compute a sequence of values as n increases. When $n = 70$, the algorithm solves the problem with high probability (992 successes in 1000 runs). This approach has the virtue of being easy to program and very fast, and so long as the resolution is fine enough to find the correct local minimum, it will converge to the answer even if one requests many digits (and uses appropriate working precision). Of course, it comes with no guarantee, and repeated trials will usually be necessary to gain confidence in the correctness of the result (the next section will present an algorithm that eliminates the uncertainty inherent in a random search). Searches such as this have limitations—a move to higher dimensions puts stress on the number of search points; a microscopic downward spike would be impossible to find—but such issues are problematic for just about any optimization method. Given the speed and minimal programming effort, this genetic approach is a fine tool for an initial investigation into an optimization problem.

An obvious way to speed things up is to switch at some point to a method that is designed to quickly converge to the nearest local minimum. It can be hard to decide, in an automated fashion, when one is close enough to the right minimum. However, for the problem at hand, switching to either a minimization routine (Brent's method is popular) or, better, a root-finder that finds the zero of the gradient speeds things up a lot if one wants a large number of digits. Once we have a seed that we believe in, whether from a random search or a more reliable interval algorithm as in §4.3, using Newton's method on the gradient is very fast and yields 10,000 digits in under a minute (see Appendix B).

If one wants to use a random search, there are some canned algorithms that can be used. For example, the `NMinimize` function in *Mathematica* can solve a variety of optimization problems, including problems with constraints and in any number of variables. But there are several methods available for it, and several options for each, so, as with all complicated software, it can be tricky finding a combination that works. Simulated annealing and the Nelder–Mead method are available, but they do not seem very good on this sort of continuous problem (they tend to end up at a nonglobal minimum). An evolutionary algorithm called *differential evolution* does work well (for more on this method, see §5.2) and will reliably (85 successes in 100 random trials) solve the problem at hand when called in the following form.

A *Mathematica* Session

```
NMinimize[{f[x, y], x^2 + y^2 ≤ 1}, {x, y},
  Method → {"DifferentialEvolution", "SearchPoints" → 250}]
```

```
{-3.3.3068686474752402,
  {x → -0.024403079743737212, y → 0.21061242727591697}}
```

This gives 14 digits of the answer and causes the objective function to be evaluated only about 6000 times (the simpler evolutionary algorithm presented earlier can get 15 digits with about 120,000 function evaluations).

Another approach is to use a more comprehensive global optimization package. One such, based on random search and statistically based reasoning, is *MathOptimizer*, a commercial package for *Mathematica* developed by Janos Pintér.[26]

[26] `http://www.dal.ca/~jdpinter/m_f_m.html`

4.3 Interval Arithmetic

Let R be the square $[-1, 1] \times [-1, 1]$, which we know must contain the answer. Random search methods can end up at a local but nonglobal minimum; the best algorithm will be one that is guaranteed to find the global minimum. Such an algorithm arises from a subdivision process. Repeatedly subdivide R into smaller rectangles, retaining only those rectangles that have a chance of containing the global minimum. The identification of these subrectangles can be done by estimating the size of the function and its derivatives on the rectangle. This point of view is really one of interval arithmetic. In interval arithmetic, the basic objects are closed intervals $[a, b]$ on the real line; extended arithmetic is generally used, so that endpoints can be $\pm\infty$. One can design algorithms so that elementary functions can be applied to such intervals (or, in our case, pairs of intervals), and the result is an interval that contains the image of the function on the domain in question. However, the resulting interval is not simply the interval from the minimum to the maximum of the function. Rather, it is an enclosing interval, defined so as to make its computation easy and fast; it will generally be larger than the smallest interval containing all the f-values.

For example, it is easy to determine an enclosing interval for $\sin([a, b])$: just check whether $[a, b]$ contains a real number of the form $\frac{\pi}{2} + 2n\pi$ $(n \in \mathbb{Z})$. If so, the upper end of the enclosing interval is 1; if not, it is simply $\max(\sin a, \sin b)$. Finding the lower end is similar. The exponential function is monotonic, so one need look only at the endpoints. For sums, one follows a worst-case methodology and adds the left ends and then the right ends. So if $g(x) = \sin x + \cos x$, then this approach will not yield a very tight result, as $[-2, 2]$ is only a loose bound on the actual range of g, $[-\sqrt{2}, \sqrt{2}\,]$. This phenomenon is known as *dependence*, since the two summands are treated as being independent when they are not. Nevertheless, as the intervals get smaller and the various pieces become monotonic, the interval computations yield tighter results. Products are similar to sums, but with several cases. There are several technicalities that make the implementation of an interval arithmetic system tedious and difficult to get perfectly correct: one critical point is that one must always round outwards. But *Mathematica* has interval arithmetic built in, and there are packages available for other languages, such as INTPAKX[27] for Maple, INTLAB[28] for MATLAB, and the public-domain SMATH library[29] for C. Thus one can use a variety of environments to design algorithms that, assuming the interval arithmetic and the ordinary arithmetic are implemented properly, will give digits that are verified to be correct. For the algorithms we present here, we will assume that a comprehensive interval package is available.

One simple application of these ideas is to the estimation exercise used in §4.1 to ascertain that the square R contains the global minimum. The interval output of f applied to the half-plane $-\infty < x \leqslant -1$ is $[-3.223, \infty]$, and the same is true if the input region—the half-plane—is rotated 90°, 180°, or 270° around the origin. This means that in these four regions—the complement of R—the function is greater than -3.23, and so the regions can be ignored. Here is how to do this in *Mathematica*, which has interval arithmetic implemented for elementary functions.

---

[27] `http://www.mapleapps.com/powertools/interval/Interval.shtml`

[28] `http://www.ti3.tu-harburg.de/~rump/intlab/`

[29] `http://interval.sourceforge.net/interval/C/smathlib/README.html`

A *Mathematica* Session

```
f[Interval[{-∞, -1.}], Interval[{-∞, ∞}]]
Interval[{-3.223591543636455, ∞}]
```

Now we can design an algorithm to solve Problem 4 as follows (this approach is based on the solution of the Wolfram Research team, the only team to use interval methods on this problem, and is one of the basic algorithms of interval arithmetic; see [Han92, Chap. 9] and [Kea96, §5.1]). We start with R and the knowledge that the minimum is less than -3.24. We then repeatedly subdivide R, retaining only those subrectangles T that have a chance of containing the global minimum. That is determined by checking the following three conditions. Throughout the interval discussion we use the notation $h[T]$ to refer to an enclosing interval for $\{h(t) : t \in T\}$.

(a) $f[T]$ is an interval whose left end is less than or equal to the current upper bound on the absolute minimum.

(b) $f_x[T]$ is an interval whose left end is negative and right end is positive.

(c) $f_y[T]$ is an interval whose left end is negative and right end is positive.

For (a), we have to keep track of the current upper bound. It is natural to try condition (a) by itself; such a simple approach will get one quickly into the region of the lowest minimum, but the number of intervals then blows up because the flat nature of the function near the minimum makes it difficult to get sufficiently tight enclosing intervals for the f-values to discard them. The other two conditions arise from the fact that the minimum occurs at a critical point and leads to an algorithm that is more aggressive in discarding intervals. Conditions (b) and (c) are easy to implement (the partial derivatives are similar to f in complexity) and the subdivision process then converges quickly to the answer for the problem at hand. While a finer subdivision might sometimes be appropriate, simply dividing each rectangle into four congruent subrectangles is adequate.

In the implementation we must be careful, as it is important to always improve the current upper bound as soon as that is possible. In the algorithm that follows, this improvement is done for the entire round of same-sized rectangles when a_1 is updated. The algorithm is given here for the plane, but nothing new is needed to apply it to n-space; later in this section we will present a more sophisticated algorithm for use in all dimensions.

Algorithm 4.2. Using Intervals to Minimize a Function in a Rectangle.
Assumptions: An interval arithmetic implementation that works for f, f_x, and f_y is available.

Inputs: R, the starting rectangle;
$f(x, y)$, a continuously differentiable function on R;
ϵ, a bound on the absolute error for the final approximation to the lowest f-value on R;
b, an upper bound on the lowest f-value on R, obtained perhaps by a preliminary search;
$i_{\max}$, a bound on the number of subdivisions.

Output: Interval approximations to the location of the minimum and the f-value there, the latter being an interval of size less than ϵ (or a warning that the maximum number of subdivisions has been reached). If the global minimum occurs more than once, then more than one solution will be returned.

Notation: $\mathcal{R}$ is a set of rectangles that might contain the global minimum; a_0 and a_1 are lower and upper bounds, respectively, on the f-value sought, and an *interior* rectangle is one that lies in the interior of R.

Step 1: Initialize: Let $\mathcal{R} = \{R\}$, $i = 0$; $a_0 = -\infty$, $a_1 = b$.
Step 2: The main loop:
While $a_1 - a_0 > \epsilon$ and $i < i_{\max}$:
Let $i = i + 1$;
Let $\mathcal{R}$ = the set of all rectangles that arise by uniformly dividing each rectangle in $\mathcal{R}$ into 4 rectangles;
Let $a_1 = \min(a_1, \min_{T\in\mathcal{R}}(f[T]))$;
Check size: Delete from $\mathcal{R}$ any rectangle T for which the left end of $f[T]$ is not less than a_1;
Check the gradient: Delete from $\mathcal{R}$ any interior rectangle T for which $f_x[T]$ does not contain 0 or $f_y[T]$ does not contain 0;
Let $a_0 = \min_{T\in\mathcal{R}}(f[T])$.
Step 3: Return the centers of the rectangles in $\mathcal{R}$ and of the f-interval for the rectangles in $\mathcal{R}$.

Appendix C.5.3 contains a bare-bones implementation in *Mathematica*, in the simplest form necessary to determine 10 digits of the minimum (both the checking of the number of subdivisions and the distinction involving interior squares are suppressed). Appendix C.4.3 also has code that does the same thing using the INTLAB package for MATLAB. Moreover, there are self-contained packages available that are devoted to the use of interval algorithms in global optimization; two prominent ones are COCONUT[30] and GlobSol,[31] both of which solve this problem with no difficulty. The *Mathematica* version of the algorithm takes under two seconds to solve Problem 4 completely as follows.

A *Mathematica* Session

```
LowestCriticalPoint[f[x, y], {x, -1, 1}, {y, -1, 1}, -3.24, 10^-9]
{{-3.306868647912808, -3.3068686470376663},
  {{-0.024403079696639973, 0.21061242715950385}}}
```

A more comprehensive routine, with many bells and whistles, is available at the web page for this book. That program includes an option for monitoring the progress of the algorithm (including the generation of graphics showing the candidate rectangles) and for getting high accuracy. A run with the tolerance ϵ set to 10^{-12} takes only a few seconds (all timings in this chapter are on a Macintosh G4 laptop with a 1 GHz CPU) and uses

[30] http://www.mat.univie.ac.at/~neum/glopt/coconut/branch.html
[31] http://www.mscs.mu.edu/~globsol/

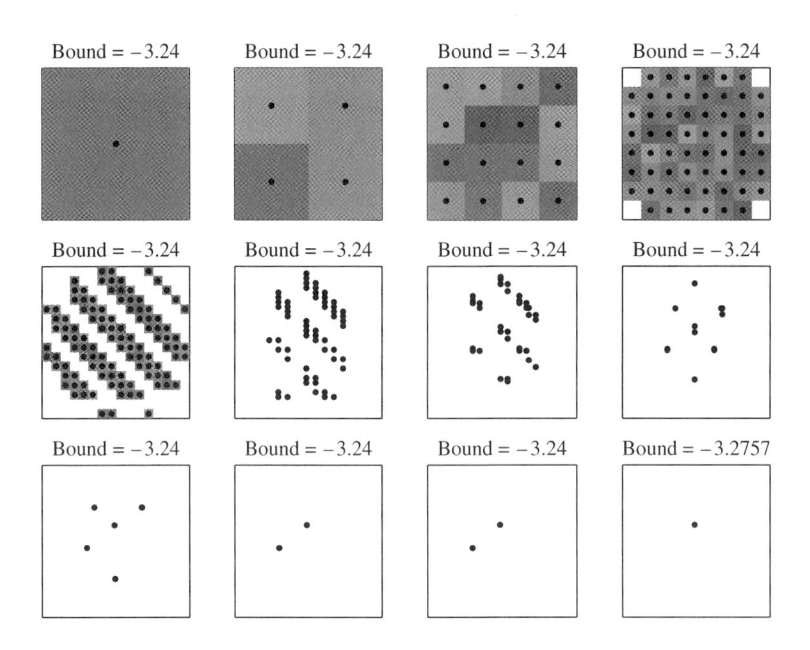

Figure 4.5. *The first 12 iterations of the interval subdivision algorithm. After 47 rounds, 12 digits of the answer are known.*

47 subdivision rounds (iterations of the main While loop). The total number of rectangles examined is 1372, the total number of interval function evaluations is 2210, and the numbers of candidate rectangles after each subdivision step are:

$$4, 16, 64, 240, 440, 232, 136, 48, 24, 12, 12, 4, 4, 4, 4, 4, 4, 4, 4, 4, 4, 4,$$
$$4, 4, 4, 4, 4, 4, 4, 4, 4, 4, 4, 4, 4, 4, 4, 4, 4, 4, 4, 4, 4, 4, 4, 4, 4$$

Thus we see that at the third and later subdivisions, some subrectangles were discarded, and after 11 rounds only a single rectangle remained: it was then subdivided 35 times to obtain more accuracy. The interval returned by this computation is -3.30687^{49}_{56}, which determines 12 digits of the answer (its midpoint is correct to 16 digits). Figure 4.5 shows the candidate rectangles at the end of each of the first 12 subdivision steps.

As with the random search of Algorithm 4.1, an obvious way to improve this basic algorithm is to switch to a rootfinder on the gradient once one has the minimum narrowed down to a small region. But there are many other ways to improve and extend this basic interval subdivision process, and we shall discuss those later in §§4.5 and 4.6.

4.4 Calculus

A natural idea for this problem is to use the most elementary ideas of calculus and try to find all the critical points—the roots of $f_x = f_y = 0$—in the rectangle R; then the smallest f-value among them is the global minimum. While this is not the most efficient way to attack an optimization problem—most of the critical points are irrelevant, and the interval method of Algorithm 4.2 is more efficient at focusing on a region of interest—we present it here because it works, and it is a good general method for finding the zeros of a pair of functions in a rectangle. Moreover, if the objective function were one for which an interval implementation was not easily available, then the method of this section would be a useful alternative.

The partial derivatives are simple enough, but finding all the zeros of two nonlinear functions in a rectangle is a nontrivial matter (see the quote starting this chapter). Yet it can be done for the problem at hand by a very simple method (§4.6 will address the issue of verifying correctness of the list of critical points). The main idea, if one is seeking the roots of $f = g = 0$, is to generate a plot of the zero-set of f and use the data in the plot to help find the zero [SW97]. If a good contour-generating program is available, one can use it to approximate the zero-set of f and then one can design an algorithm to find all zeros in a rectangle as follows.

Algorithm 4.3. Using Contours to Approximate the Roots of $f = g = 0$.
Assumptions: The availability of a contour-drawing program that allows access to the points of the curves.

Inputs: f and g, continuous functions from $\mathbb{R}^2$ to $\mathbb{R}$;
R, the rectangular domain of interest;
r, the grid-resolution of the contour plot that will be used in Step 1.

Output: Numerical approximations to each pair (x, y) that is a root of $f = g = 0$ and lies in R.

Notation: s will contain points that serve as seeds to a root-finder.

Step 1: Generate the curves corresponding to $f = 0$. This is most easily done by a contour plot, set to compute just the zero-level curves, and using r for the resolution in each dimension.
Step 2: For each contour curve, evaluate g on the points that approximate the curve; whenever there is a sign change, put the point just scanned into s.
Step 3: For each point in s, use Newton's method (or a secant method in the case of nondifferentiable functions) to determine a root. Discard the root if it lies outside R.
Step 4: If the number of roots differs from the number of seeds, restart the algorithm with a finer resolution.
Step 5: Return the roots.

The use of a contour routine was given in Step 1 because that is quite simple and is adequate to the task at hand. However, there are more geometric (and faster) methods to compute an approximation to a contour curve that could be used. Such path-following algorithms have seen use in the study of bifurcation of differential equations. See [DKK91] and references therein to the AUTO program.

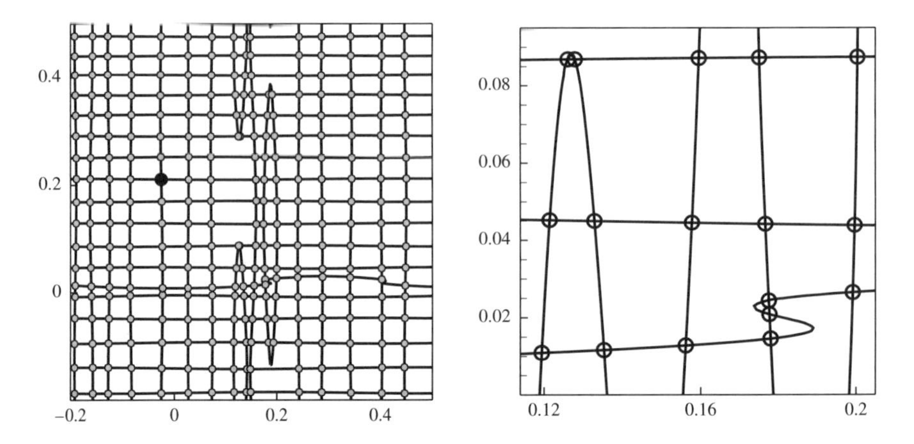

Figure 4.6. *A view of 290 solutions to $f_x = 0$ (vertical curves) and $f_y = 0$ (horizontal), where $f(x, y)$ is the objective function of Problem 4. The black point is the location of the global minimum. The close-up shows that the resolution was adequate for these curves.*

Implementing Algorithm 4.3 is quite simple provided one can get access to the data approximating the curves $f(x, y) = 0$. One might leave Step 4 for a manual check, as opposed to making it part of the program. Figure 4.6, which includes the zero-sets for both functions, shows the result for f_x and f_y on a small section of the plane, where a resolution of 450 was used (this refers to the grid size used by the contour algorithm); five seconds were required to compute all 290 roots. The close-up on the right shows that the resolution was good enough to catch all the intersections.

To solve the problem over all of the square $R = [-1, 1]^2$ it makes sense to work on subsquares to avoid the memory consumption of a very finely resolved contour plot. Therefore, we break R into 64 subsquares, compute all the critical points in each, and keep track of the one that gives the smallest value of f. Doing this takes under a minute and yields 2720 critical points. The one of interest is near $(-0.02, 0.21)$, where the f-value is near -3.3, the answer to the question. Newton's method can then be used to quickly obtain the function value at this critical point to more digits.

Manual examination of the contour plots on the subsquares to confirm that the resolution setting is adequate is not an efficient method for checking the results. A better way to gain confidence in the answer is to run the search multiple times, with increasing resolution, checking for stability; results of such work are in Table 4.1. While even the lowest resolution was enough to find the global minimum, quite high resolution is needed to catch all 2720 critical points in S. Note how, in some cases, 2720 seeds were found but they did not lead to the intended roots, and only 2719 roots were found. The high resolution needed to get 2720 roots (160×160 on each of 64 small squares) justifies the use of the subsquare approach. It is also a simple matter to use the second-derivative test to classify the critical points: 667 are maxima, 693 are minima, and 1360 are saddle points.

Note that the number of extrema ($693 + 667$) coincides with the number of saddles (1360). In some sense this is a coincidence, as it fails for, say, a rectangle that contains but

Table 4.1. *A resolution of* 160 × 160 *on each of 64 subsquares is needed to find all 2720 critical points in* $[-1, 1]^2$.

Resolution	Number of seeds	Number of roots
20	2642	2286
40	2700	2699
60	2716	2712
80	2714	2711
100	2718	2716
120	2720	2719
140	2720	2719
160	2720	2720
180	2720	2720
200	2720	2719
220	2720	2720
240	2720	2720
260	2720	2719
280	2720	2720
300	2720	2720
320	2720	2720
340	2720	2720
360	2720	2720
380	2720	2720
400	2720	2720

a single critical point; it fails also for the rectangle $[-0.999, 0.999]^2$, which has $692 + 667$ extrema and 1357 saddles. Yet this phenomenon is related to the interesting subject known as Morse theory, which provides formulas for

$$\text{number of maxima} + \text{number of minima} - \text{number of saddles}$$

on a closed surface, such as a sphere or torus. These formulas are related to the Euler characteristic; on a torus, this number equals 0 [Mil63]. So let's make our function toroidal by taking f on $R = [-1, 1]^2$ and reflecting the function in each of the four sides, and continuing this reflection process throughout the plane. This gives us a function—call it g—that is doubly periodic on the whole plane, with a fundamental region that is $[-1, 3]^2$; thus g can be viewed as a function on a torus. Each critical point of f in R generates four copies of itself for g. But we have the complication that we have introduced new critical points on the boundary.

To understand the boundary, consider marching along the four edges of R and looking at each one-dimensional extremum: there will be the same number of maxima as minima (we assume none occur at a corner, and no inflection points occur). Moreover, because of the symmetry of the reflection, each such maximum will be either a saddle or a maximum for g, and each minimum will be either a minimum or a saddle for g. If we can enumerate all these occurrences, then we can use the toroidal formula for g to deduce the situation for f in R. But a couple of surprising things happen because of the nature of $f(x, y)$. First, f_x is positive on the left and right edges of R, and f_y is positive on the top and negative on the bottom. This means that all the maxima on the left edge become saddles and all the minima stay minima, with similar results holding for the other edges. We also need to know the numbers of extrema on the border. These are easily computed and turn out to be: top: 24 maxima, 24 minima; bottom: same; left: 29 maxima, 30 minima; right: same. These facts about f combine to show that the toroidal formula applies verbatim to f restricted to R, thus explaining the coincidence. More important, this analysis relies only on a one-dimensional computation and so provides supporting evidence that the counts obtained by the contour method (693, 667, and 1360) are correct.

Figure 4.7 shows another example that arose from a system of differential equations [Col95]. Finding the equilibrium points of the autonomous system

$$x' = 2y\cos\left(y^2\right)\cos(2x) - \cos y, \quad y' = 2\sin(y^2)\sin(2x) - \sin x,$$

is the same as finding all zeros of the right-hand sides. A contour resolution of 60 is sufficient, and the computation of all 73 zeros takes under a second.

The total amount of programming needed to implement Algorithm 4.3 is modest, provided one has a good way of accessing the data in the contour plot. In *Mathematica*, `Cases[ContourPlot[***],_Line,∞]` does the job.

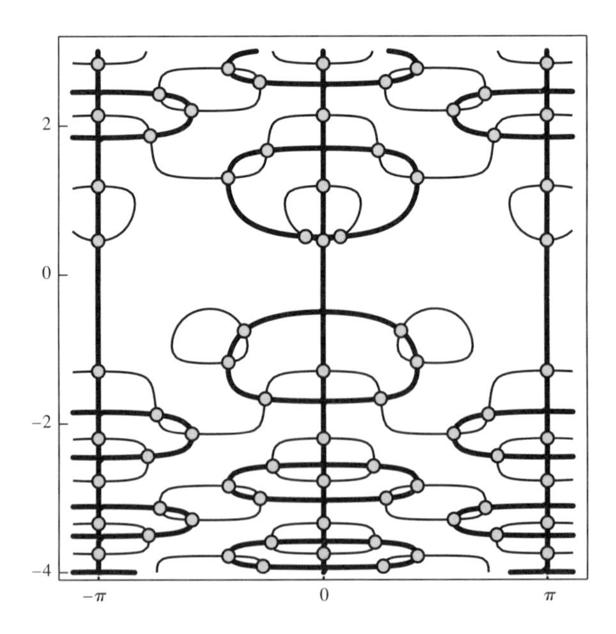

Figure 4.7. *This complicated example of simultaneous equations has 73 solutions in the rectangle shown.*

A *Mathematica* Session

The following is complete code for generating the zeros for the example in Figure 4.7. For a complete routine there are, as always, other issues to deal with: one should eliminate duplicates in the final result, since it is possible that different seeds will converge to the same zero.

```
g1[x_, y_] := 2 y Cos[y^2] Cos[2 x] - Cos[y];

g2[x_, y_] := 2 Sin[y^2] Sin[2x] - Sin[x];

{a, b, c, d} = {-3.45, 3.45, -4, 3};

contourdata = Map[First, Cases[Graphics[
        ContourPlot[g1[x, y], {x, a, b}, {y, c, d}, Contours → {0}, PlotPoints → 60]],
      _Line, ∞]];
seeds =
  Flatten[
    Map[
      #[[1 + Flatten[Position[Rest[ss = Sign[Apply[g2, #, 2]]] * Rest[RotateRight[ss]],
              -1]]]]&, contourdata], 1];
roots =
  Select[Map[{x, y}/.FindRoot[{g1[x, y] == 0, g2[x, y] == 0}, {x, #[[1]]}, {y, #[[2]]}]&,
      seeds], a < #[[1]] < b ∧ c < #[[2]] < d&];

Length[roots]
73
```

This technique also applies to finding all zeros of a complex function $f(z)$ in a rectangle by just applying the method to the system $\text{Re}f = \text{Im}f = 0$. The contour technique can fail badly if the zero-curves for f and g are tangent to each other (multiple zeros), since the approximations to the contours probably fail to have the necessary crossings. The technique is also limited to the plane. In §4.6 we will discuss a slower but more reliable method that addresses both of these problems.

4.5 Newton's Method for Intervals

The interval algorithm of §4.3 solves the problem nicely, but we can improve Algorithm 4.2 in several ways. Extensions of the algorithm so that it is faster and applies to functions in higher dimensions is an active area of research. Here we will only outline two ideas, one easy, one subtle, that can be used to improve it. The first, which might be called *opportunistic evaluation,* is quite simple: right after the subdivision, evaluate the objective function at the rectangle centers. This pointwise functional evaluation might pick up a value that can be used to improve the current best. Then the new current best is used in the size-checking step. This might not lead to great improvement in cases where we start with a good approximation to the answer (such as the -3.24 we had found), but if such an approximation is not available, these extra evaluations will help.

The second improvement relies on a root-finding algorithm called the interval Newton method, a beautiful and important idea originated by R. E. Moore in 1966 [Moo66]. We will use the term *boxes* for intervals in $\mathbb{R}^n$. The starting point is the Newton operator on boxes, defined as follows. Suppose $F:\mathbb{R}^n \to \mathbb{R}^n$ is continuously differentiable; then for a box X in n-space, $N(X)$ is defined to be $m - J^{-1} \cdot F(m)$, where m is the center of X and J is the interval Jacobian $F'[X] = (\partial F_i/\partial x_j[X])_{ij}$. Here J^{-1} is obtained by the standard arithmetic interval operations that arise in the computation of a matrix inverse. If $n = 1$, then $N(X)$ is simply $m - F(m)/F'[X]$, where m is the midpoint of X; note that $N(X) = [-\infty, \infty]$ if $0 \in F'[X]$. This definition leads to several important properties that can be used to design an algorithm to find roots. Here are the main points in the case of one dimension, which is much simpler than the general case.

Theorem 4.1 *Suppose $F:\mathbb{R} \to \mathbb{R}$ is a continuously differentiable function, and $X = [a, b]$ is a finite interval. Then:*

(a) *If r is a zero of F in X, then r lies in $N(X)$.*

(b) *If $N(X) \subseteq X$, then F has a zero in $N(X)$.*

(c) *If $N(X) \subseteq X$, then F has at most one zero in X.*

Proof. (a) If $F'(r) = 0$, then $N(X) = [-\infty, \infty]$; otherwise, the mean-value theorem yields $c \in X$ so that $F(r) - F(m) = F'(c)(r - m)$, and this implies that $r = m - F(m)/F'(c) \in N(X)$.

(b) The hypothesis implies that $F'(x) \neq 0$ on X. Then the mean-value theorem yields values c_i in X so that $m - F(m)/F'(c_1) - a = -F(a)/F'(c_1)$ and $b - (m - F(m)/F'(c_2)) = F(b)/F'(c_2)$; because $N(X) \subseteq [a, b]$, the product of the left sides is positive. But $F'(c_1)$ and $F'(c_2)$ have the same sign, and so the product $F(a)F(b)$ must be negative and F therefore has a zero in $[a, b]$.

(c) If there were two zeros, then the derivative would vanish between them, causing $N(X)$ to become infinite. □

This theorem is very powerful since one can use it to design a simple subdivision algorithm to find zeros. Just subdivide the given interval into subintervals and check, for each subinterval, whether the Newton condition—$N(X) \subseteq X$—holds. If it does, then we know from (c) that there is one and only one zero in X. If it fails in the form $N(X) \cap X = \emptyset$, that is also good, for (a) tells us that there are no zeros in X. If neither situation applies, subdivide and try again. When the Newton condition does hold, we can just iterate the N operator. If we are close enough to the zero, the algorithm converges (see [Kea96, Thm. 1.14]), and in fact will converge quadratically, as does the traditional Newton root-finding method. But some bad things can happen: we might not be close enough for convergence (the exact condition for this depends on, among other things, the tightness of the interval approximation J to the inverse of F'); or there are zeros for which the Newton condition will never hold. Think of the process as a queue: intervals are removed from the queue while, sometimes, their subdivisions are added to the queue, or the interval is added to a list of answers. If the queue becomes empty, we are done. If a max-iteration counter is exceeded and the queue is not empty, then there are some unresolved intervals. This will happen with multiple roots, such as occurs with $x^2 = 0$, for in such cases $N(X) = [-\infty, \infty]$.

For an application of the interval Newton method to a one-dimensional root-finding problem that arises in Problem 8 of the SIAM 100-Digit Challenge, see §8.3.2.

In higher dimensions, parts (a) and (c) remain true, but one must use a variation of the Newton operator to get (b), which is important to guarantee that a zero exists. One approach is by a preconditioning matrix (see [Neu90, Thm. 5.1.7]). Another approach, which we shall follow here, uses the Krawczyk operator. For more information on this important variation to Newton's method, see [Kea96, Neu90]. Let F, m, and J be as defined for the Newton operator. Consider $P(x) = x - YF(x)$, where Y is some type of approximation to J^{-1}, the interval matrix that is $F'[X]^{-1}$. Two natural choices for Y are $F'(m)^{-1}$, or the inverse of the matrix of midpoints of the intervals in the matrix J. The latter is faster, since J has to be computed anyway, and that is what we shall use. Then the Krawczyk operator is defined to be $K(X) = m - YF(m) + (I - YJ)(X - m)$, where I is the $n \times n$ identity matrix (the numeric part of this computation should be done in an interval environment, with m replaced by a small interval around m). To understand the rationale behind the K operator, first recall the mean-value theorem. For a continuously differentiable $F : \mathbb{R}^n \to \mathbb{R}^n$, the mean-value theorem takes the following form: Given x and y in a box X, there are points $c_1, c_2, \ldots, c_n \in X$ so that $F(y) - F(x) = (\nabla F_i(c_i))(y - x)$, where $(\nabla F_i(c_i))$ denotes an $n \times n$ matrix with i indexing the rows.

Now $K(X)$ can be viewed as a "mean-value extension" of P in the following sense: if $P'[X]$ is an interval enclosure of P' on X, the mean-value theorem implies that the box $Q = P(m) + P'[X](X - m)$ contains $P(X)$, which here denotes the exact image, $\{P(x) : x \in X\}$. But because $P'(x) = I - YF'(x)$, we may take $I - YJ$ to be the enclosure. Then Q becomes precisely $K(X)$ and we have proved that $K(X)$ contains $P(X)$. This smooths out the n-dimensional theory nicely as follows.

Theorem 4.2 *Suppose* $F : \mathbb{R}^n \to \mathbb{R}^n$ *is a continuously differentiable function,* X *is a finite box in* $\mathbb{R}^n$, $J = F'[X]$ *is a componentwise interval enclosure, and* Y *is the inverse of the matrix of midpoints of the intervals in* J*;* Y *is assumed to be nonsingular. Let* K *be the corresponding Krawczyk operator. Then:*

(a) *If* r *is a root of* $F = 0$ *in* X*, then* r *lies in* $K(X)$.

(b) *If* $K(X) \subseteq X$*, then* F *has a zero in* $K(X)$.

(c) *If* $K(X) \subset X$*, then* F *has at most one zero in* X.

Proof. (a) $K(X)$ contains $P(X)$, and so $P(r) \in K(X)$. But $P(r) = r - YF(r) = r$.

(b) Since $P(X) \subseteq K(X) \subseteq X$, P is a contraction mapping, and so the Brouwer fixed-point theorem for continuous functions implies that P has a fixed point in X, which means that F has a zero in X.

(c) This one is subtle. Suppose x and y are distinct roots of $F = 0$ in X. Then the mean-value theorem tells us that $F(x) - F(y) = (\nabla F_i(c_i))(x - y)$ for some points $c_i \in X$. It follows that the matrix is singular, and therefore J contains a singular matrix. However, it is not immediately obvious that this contradicts the Krawczyk condition. Yet it does. For a complete proof, see [Neu90, Thm 5.1.8] and note the hypothesis of strict containment here. □

This theorem immediately gives a local root-finding algorithm, one that will come into play in §4.6. If the Krawczyk condition—$K(X) \subset X$—holds, then we know there is one and only one root in X, and that root lies in $K(X)$. Simply iterate K. When we have a small enough interval, we know the root to the desired accuracy.

Here is how root-finding can help in optimization. Once we have the problem reduced to a single box, as happens after 11 subdivisions in the problem at hand, we can check the Krawczyk condition. If it holds, we can use the Newton–Krawczyk iteration to zoom into the unique critical point in the box, and that will give us the answer more quickly than repeated subdivisions (analogous to the advantage Newton's method provides over bisection in one dimension). For the problem at hand and a tolerance of 10^{-6} for the location of the minimum, using this idea (but not opportunistic evaluation) gives a speedup of about 10%.

To be precise, add the following step to the interval method of Algorithm 4.2 right after the gradient check:

> If $\mathcal{R}$ contains only one rectangle X, compute $K(X)$.
>
> If $K(X) \cap X = \emptyset$, then there is no critical point, and the minimum is on the border; do nothing.
>
> If $K(X) \subset X$, then iterate the K operator starting with $K(X)$ until the desired tolerance is reached.
>
> Use the last rectangle to set a_0 and a_1 (the loop will then terminate).

A *Mathematica* Session

An implementation of this extension is available at the web page for this book. Here is how that code would be used to get 100 digits of the answer, using interval arithmetic throughout. The switch to root-finding occurs after the 12th round, and then convergence is very fast. The output shown is the center of an interval of length less than 10^{-102}.

```
IntervalMinimize[f[x, y], {x, -1, 1}, {y, -1, 1}, -3.24,
  AccuracyGoal → 102, WorkingPrecision → 110]
-3.306868647475237280076113770898515657166482361476288217501293085503
   0919983788829503582548807528349918619 3
```

The use of interval methods in global optimization is a well-developed area, with techniques that go well beyond the brief introduction given here (see [Han92, Kea96]; Hansen reports success on a wide variety of problems, including a 10-dimensional one). Nevertheless, the fact that the very simplest ideas give a verified answer to Problem 4 in a few seconds and with only a few lines of code shows how powerful these ideas are. And it is noteworthy that interval analysis has played an important role in diverse areas of modern mathematics. For example, W. Tucker won the Moore prize for his recent work using interval analysis to prove that the Lorenz equations really do have a strange attractor; this solved one of Steven Smale's Problems for the 21st Century (see [Tuc02]); Hales and Ferguson [FH98] used interval methods in their resolution of the Kepler conjecture on sphere-packing, and Lanford [Lan82] used intervals to prove the existence of a universal limit—the Feigenbaum constant—in certain sequences of bifurcations.

The most important points of the interval approach to Problem 4 are:

- The algorithm is very general and will find the lowest critical point in the rectangle (provided interval arithmetic is available in a form that applies to the objective function and its partial derivatives).
- The results are verifiably correct if one uses intervals throughout.
- Getting the results to very high precision is not a problem.
- Having an interval root-finding method can lead to improvements in the optimization algorithm.
- And most important: The interval algorithm is a reasonable way to solve the problem whether or not one wants proved results. In short, interval thinking yields both a good algorithm and proved results.
- The basic ideas apply to functions of more variables, but life becomes more difficult in higher dimensions. See [Han92, Kea96] for discussions of various enhancements that can be used to improve the basic algorithm (one example: using the Hessian to eliminate n-dimensional intervals on which the function is not concave up).

4.6 A Validation Method for Roots

The pessimistic quote at the start of this chapter leads to the question, How can we be certain that the collection of 2720 critical points found in §4.4 is correct and complete? There are several ways to do this. One can use intervals to design a root-finding algorithm and check that it finds the same set of critical points. That can be done by a simple subdivision process where we keep only rectangles that have a chance of containing a zero, and constantly check and use the Krawczyk condition, $K(X) \subset X$. Here is a formal description.

Algorithm 4.4. Using Intervals to Find the Zeros of F: $\mathbb{R}^n \to \mathbb{R}^n$ in a Box.
Assumptions: Interval arithmetic is available for F.

Inputs: F, a continuously differentiable function from $\mathbb{R}^n$ to $\mathbb{R}^n$;
R, the box in which we want all zeros of F;
ϵ, an upper bound on the absolute error of the zero in terms of Euclidean distance;
`mtol`, a tolerance for combining boxes;
$i_{\max}$, a bound on the number of subdivision steps.
Output: A set of intervals with each one trapping a zero, and a set of intervals that are unresolved (these might contain none or one or more zeros).

Notation: For an n-dimensional box X, let m be the center of X and let M be a small box containing m. Then $K(X) = M - YF[M] + (I - YJ)(X - M)$, where J is the Jacobian interval matrix $F'[X]$ and Y is the inverse of the matrix obtained from J by replacing each box by its center. "Subdividing a box" here means dividing it into 2^n pieces by bisecting each side. "Combining" a set of boxes means replacing any two that have nonempty intersection with the smallest box that contains both of them and repeating the process until no intersections remain. The Krawczyk condition for a box X is $K(X) \subset X$.

Step 1: Initialize: Let $\mathcal{R} = \{R\}$, $i = 0$, $a = \emptyset$.
Step 2: The main loop:
While $\mathcal{R} \neq \emptyset$ and $i < i_{\max}$:
    Let $i = i + 1$;
    Let $\mathcal{R}$ be the set of all boxes that arise by uniformly dividing each box in $\mathcal{R}$ into 2^n boxes;
    If all boxes in $\mathcal{R}$ have their maximum side-dimension less than `mtol`, let $\mathcal{R}$ be the result of combining boxes in $\mathcal{R}$ until no further combining is possible;
    For each $X \in \mathcal{R}$ compute $K(X)$, the Krawczyk image of X:
        if $K(X) \cap X = \emptyset$, delete X from $\mathcal{R}$;
        if $K(X) \subset X$,
            iterate K starting from $K(X)$ until the tolerance (the size of the box, or the size of its F-image) is as desired;
            add the resulting box to a and delete X from $\mathcal{R}$.
Step 3: Return a and $\mathcal{R}$, the latter being the unresolved boxes.

The combining step using `mtol` in the main loop requires some explanation. If a zero is near the edge of a box, then it can happen that the subdivision and Krawczyk contraction process will never succeed in isolating it. This can happen even in one dimension if the initial interval is $[-1, 1]$ and the zero is at 0, for then the subdivision process will produce two intervals with a very small overlap, thus placing the zero very near the edge. The combining step checks whether all remaining boxes are so small that we ought to have isolated the zeros (the merging tolerance is provided by the user); the fact that we have not done so ($\mathcal{R}$ is not empty) means that it would be wise to combine small boxes into larger ones. This will likely place the zero nearer the center of a box (but not at the exact center!), and the iterative process then succeeds. Of course, there is the chance of an infinite loop here. Thus it would be reasonable to run this algorithm first with `mtol` $= 0$ so that no combining takes place. If it fails to validate, it can be tried with a setting of, say, `mtol` $= 10^{-6}$.

For zeros x of F at which $F'(x)$ is singular, the Krawczyk condition is never satisfied, and the algorithm does not find those zeros. However, they are not lost, as they will be trapped within the unresolved intervals that are returned, and the user could investigate those further to try to determine if a zero lives in them.

Algorithm 4.4 succeeds in obtaining all 2720 zeros (in eight minutes). It is an important technique, especially in higher dimensions, where we might not have another way to get at the roots (see [Kea96]). But in cases where we think we already have the zeros, we should use an interval algorithm for *validation*; this is a central theme in the field of interval analysis: it is often more appropriate to use traditional numerical algorithms and heuristics to get results that are believed to be complete, and then use interval analysis to validate the results.

Here is a second approach, where we use interval analysis as a validation tool. It is a two-step approach based on the theorem about the Krawczyk operator in §4.5, and it works well for the problem at hand. Suppose the set of approximate zeros is r and has size m.

First we check that each zero in r is roughly correct: for each zero, let X be a small box centered at the zero and check that the Krawczyk condition $K(X) \subset X$ holds; this technique is called *ϵ-inflation* (originally due to Rump; see [Rum98] and references therein). Once this is done, we know that r contains approximations to a set of m true zeros. Then move to verify completeness by carrying out a subdivision process on the given domain, doing two things only with each box that shows up: if interval arithmetic or the Krawczyk operator shows that the enclosure for the F-values on the box does not contain the zero vector, it is discarded; and if the Krawczyk containment holds, so that the box does contain a zero, then it is again discarded, but a running count is incremented by 1. Boxes that remain are subdivided. When no boxes remain, we know that the count equals the number of zeros. If this count equals m, we know the set r was indeed a complete set of approximations to the zeros. Here is a formal description.

Algorithm 4.5. Using Intervals to Validate a Set of Zeros.
Assumptions: Interval arithmetic is available for F and its partial derivatives.

Inputs: F, a continuously differentiable function from $\mathbb{R}^n$ to $\mathbb{R}^n$;
R, the box that is the validation domain;
r, a list, believed to be complete and correct, of the set of zeros of F in R;
ϵ, an upper bound of the absolute error on the zero in terms of Euclidean distance;
`mtol`, a tolerance for combining boxes.

Output: True, if the true zeros in R coincide with the points in r, with the absolute error in each case less than ϵ; False otherwise.

Notation: As in Algorithm 4.4.

Step 1: ϵ-inflation:
For each point in r,
let X be the surrounding box of side-length $2\epsilon/\sqrt{n}$;
check that the Krawczyk condition holds for each X.
If there is any failure, stop and return False.
Step 2: Initialize the subdivision process. Let $\mathcal{R} = \{R\}$; $c = 0$.
Step 3: The main loop:
While $\mathcal{R} \neq \emptyset$:
Let $\mathcal{R}$ be the set of boxes obtained by subdividing each box in $\mathcal{R}$;
Delete from $\mathcal{R}$ any box for which the F-interval does not straddle 0;
Delete from $\mathcal{R}$ any box X for which the Krawczyk condition holds,
increase c by 1 for every such box;
Delete from $\mathcal{R}$ any box X for which $K(X) \cap X = \emptyset$;
If all boxes in $\mathcal{R}$ have their maximum side-dimension less than `mtol`,
let $\mathcal{R}$ be the result of combining boxes in $\mathcal{R}$
until no further combining is possible.
Step 4: If $c =$ the number of points in r, return True, otherwise False.

For the gradient of the function f of Problem 4, the contour method locates the 2720 zeros in 30 seconds. The validation method then takes 30 seconds to check that the zeros are correct and another 5.5 minutes to verify that the set is complete. While the speedup over the method of using intervals from the beginning to find all the zeros is modest, the validation method is a more elegant algorithm and illustrates the important idea that, when possible, one should put off the interval work to the last stage of a computational project.

Algorithms 4.4 and 4.5 can be combined into a single algorithm that finds and validates the zeros. The key observation is that there is no need to iterate the Krawczyk operator when one can just use the traditional Newton method.

Algorithm 4.6. Using Intervals to Find and Validate a Set of Zeros.
Assumptions: Interval arithmetic is available for F and its partial derivatives.

Inputs: F, a continuously differentiable function from $\mathbb{R}^n$ to $\mathbb{R}^n$;
R, the box that is the validation domain;
ϵ, an upper bound on the absolute error of the zeros in terms of Euclidean distance;
`mtol`, a tolerance for combining boxes.

Output: Validated approximations to the zeros, with the absolute error in each case less than ϵ.

Notation: As before, with s being a set of rough approximations to the zeros, and r the set of final approximations.

Step 1: Let $s = \emptyset$; follow Steps 1 and 2 of Algorithm 4.4, except that when $K(X) \subset X$ add the center of X to s.
Step 2: Apply Newton's method to each point in s, putting the result in r.
Step 3: Use ϵ-inflation as in Step 1 of Algorithm 4.5 to verify that r is a complete set of zeros to the desired tolerance.
Step 4: Return r.

Algorithm 4.6 finds and validates all the critical points to the challenge problem in 5.5 minutes, a 9% time savings over the combination of Algorithms 4.4 and 4.5. The material on the web page for this book includes the *Mathematica* programs `ValidateRoots` and `FindAndValidateRoots`.

4.7 Harder Problems

What sort of problem could have been asked instead of Problem 4 that would have been a little, or a lot, harder? There certainly would have been additional difficulties if there were more dimensions or if the objective function did not have interval arithmetic easily available. Indeed, the optimization needed to solve Problem 5 is exactly of this sort (four dimensions, complicated objective), and that is quite a difficult problem for general-purpose minimization algorithms. Yet another sort of problem would arise if the objective function had not a single minimum, but a continuous set, such as a line or a plane (see [Han92]; there are some examples there). But let us look only at the dimension issue for a moment. Here is a slight variation of Problem 4, but in three dimensions: What is the global minimum of

$$g(x, y, z) = e^{\sin(50x)} + \sin(60e^y)\sin(60z) + \sin(70\sin x)\cos(10z) + \sin\sin(80y) - \sin(10(x+z)) + (x^2+y^2+z^2)/4\ ?$$

The methods of this chapter work well on this problem, except for the techniques that tried to find all the critical points in a box: there are probably over 100,000 such points. An approach that combines various methods to advantage would proceed as follows:

1. Use the genetic minimization routine with 200 points per generation and a scale factor of 0.9 to discover that g gets as small as -3.327. Differential evolution works too.

2. Use traditional root-finding (Newton) on the gradient to get the more accurate value of -3.32834 (or hundreds of digits if desired).

3. Use interval arithmetic as in §4.3 to then prove that the global minimum is inside the cube $[-0.77, 0.77]^3$.

4. Use the basic interval algorithm of §4.3, with upper bound -3.328 and Krawczyk iteration taking over after 11 rounds, to prove that the minimum is -3.32834 and occurs near $(-0.15, 0.29, -0.28)$. This takes almost a minute, and the total number of boxes examined is 9408.

A *Mathematica* Session

```
g[x_, y_, z_] := e^Sin[50 x] + Sin[60 e^y] Sin[60 z] + Sin[70 Sin[x]] Cos[10 z] +
    Sin[Sin[80 y]] - Sin[10 (x + z)] + 1/4 (x^2 + y^2 + z^2);

IntervalMinimize[g[x, y, z], {x, -0.77, 0.77}, {y, -0.77, 0.77},
  {z, -0.77, 0.77}, -3.328]
```

```
{Interval[{-3.328338345663281, -3.328338345663262}],
  {{x → Interval[{-0.1580368204689058, -0.1580368204689057}],
    y → Interval[{0.2910230486091526, 0.2910230486091528}],
    z → Interval[{-0.2892977987325703, -0.2892977987325701}]}}}
```

So for at least one complicated example, the algorithm works well in three dimensions. If the dimension is increased in such a way that the number of local minima grows with the space (that is, exponentially), one would expect a slowdown in the algorithm, though it will be sensitive to the location of the minima and still might work well.

As a final example, we mention that the Krawczyk validation method given at the end of §4.6 can solve the related problem of finding the critical points of $g(x, y, z)$. Because there are many critical points we work in a small box only ($[0, 0.1] \times [0, 0.05] \times [0, 0.1]$) and show some of them, together with the three surfaces $g_x = 0$, $g_y = 0$, $g_z = 0$, in Figure 4.8. There are six zeros in this box.

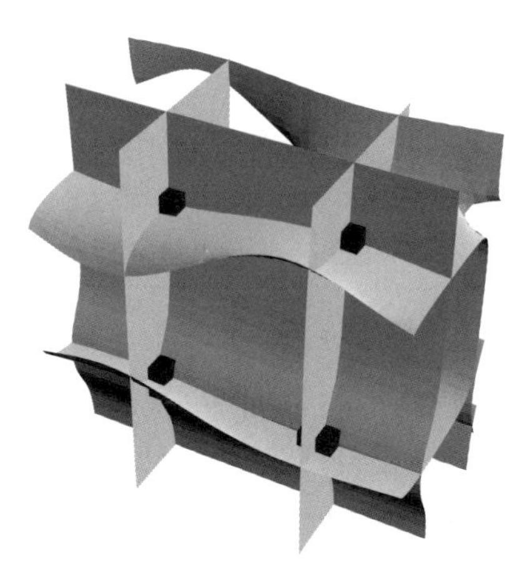

Figure 4.8. *A very small portion—within* $[0, 0.1] \times [0, 0.05] \times [0, 0.1]$*—of the three surfaces* $g_x = 0$*,* $g_y = 0$*,* $g_z = 0$ *whose intersections form the critical points of* g*, with the roots obtained by the interval method shown as small cubes.*

4.8 Summary

Interval arithmetic is powerful and can obtain certifiably correct answers to optimization and root-finding problems. The technique is useful not only as a validation tool, but also as a complete algorithm. For many problems it makes sense to use other techniques first, and then use interval arithmetic to validate the results, if that is desired. Random search methods, especially evolutionary algorithms, can be very efficient at getting approximate results. The contour approach is somewhat specialized, but it too is very efficient at solving this, and other, two-dimensional problems.

Chapter 5
A Complex Optimization

Dirk Laurie

Die Lekkerland se pad is 'n lang, lang pad wat in die rondte loop.
Hy slinger deur die bosse en hy kronkel om die rante tot daar waar sy end moet wees.
En dáár begin hy weer!
Hy dwaal deur die vleie en hy boggel oor die bulte en op een plek raak hy weg.
Maar onder die braambos begin hy weer en hy drentel al om en om en om die diep kuil vol soet water wat nooit opdroog nie.
O, dit is 'n lang, lang pad![32]

—W. O. Kühne [Küh82]

If ζ is a real function of a real variable z, then the relation between ζ and z, which may be written $\zeta = f(z)$, can be visualized by a curve in the plane, namely the locus of a point whose coordinates referred to rectangular axes in the plane are (z, ζ). No such simple and convenient geometrical method can be found for visualizing an equation $\zeta = f(z)$, considered as defining the dependence of one complex number $\zeta = \xi + i\eta$ on another complex number $z = x + iy$.

—E. T. Whittaker and G. N. Watson [WW96, p. 41]

Problem 5

Let $f(z) = 1/\Gamma(z)$, where $\Gamma(z)$ is the gamma function, and let $p(z)$ be the cubic polynomial that best approximates $f(z)$ on the unit disk in the supremum norm $\|\cdot\|_\infty$. What is $\|f - p\|_\infty$?

[32]The Lekkerland road is a long, long road that runs in a circle. It winds through the bushes and it coils round the ridges to there where its end should be.// And there it starts again!// It wanders through the wetlands and it hunches over the hillocks and in one place it dwindles away.// But under the brambles it starts again and it saunters all round and round and round the deep pool of sweet water that never dries up.// Oh, it is a long, long road! —W. O. Kühne, translated by Dirk Laurie

5.1 A First Look

We change Trefethen's notation so that p is a generic cubic polynomial

$$p(z) = az^3 + bz^2 + cz + d,$$

and the optimal polynomial[33] is

$$p_{\text{opt}}(z) = a_{\text{opt}}z^3 + b_{\text{opt}}z^2 + c_{\text{opt}}z + d_{\text{opt}}.$$

We need to compute

$$\epsilon_{\text{opt}} = \|f - p_{\text{opt}}\|_\infty.$$

Since $f - p$ is an entire function, by the maximum principle the maximum of $|f - p|$ over the unit disk occurs on the unit circle. So we can revise the definition of the supremum norm in this case to read

$$\|f - p\|_\infty = \max_{\theta \in [0,2\pi]} |f(e^{i\theta}) - p(e^{i\theta})|.$$

Minimax problems are notoriously hard for general-purpose optimization algorithms, so we will try to get as far as we can before invoking such an algorithm; see §5.2.

Any p trivially gives us an upper bound

$$\epsilon_{\text{opt}} \leqslant \|f - p\|_\infty.$$

One candidate for p that is easy to find comes from the first few terms of the Maclaurin series [AS84, form. (6.1.34)] of f, which gives, to four decimals,

$$p_0(z) = -0.6559z^3 + 0.5772z^2 + z.$$

The usual objections to a Maclaurin polynomial, when seen as a way of approximating a real function on a finite interval, do not apply in the case of approximation on the unit circle. On the contrary, the Maclaurin polynomial is optimal in the L_2 norm, defined by

$$\|g\|_2 = \left(\frac{1}{2\pi}\int_0^{2\pi} |g(e^{i\theta})|^2 \, d\theta\right)^{1/2}.$$

The reason for this optimality is that by setting $z = e^{i\theta}$ the Maclaurin series of $g(z)$ passes into the Fourier series of $g(e^{i\theta})$ on the interval $0 < \theta \leqslant 2\pi$.

Another easy way of getting a first approximation would be to discretize the unit circle using n equidistant points θ_j and then to solve the discrete least squares problem, that is, to minimize

$$E(p, n) = \frac{1}{n}\sum_{j=1}^{n} |f(e^{i\theta_j}) - p(e^{i\theta_j})|^2$$

over all cubic polynomials p. Clearly, $E(p, n)$ is merely a discretization of $\|f - p\|_2^2$, and the polynomial thus found therefore tends rapidly to the Maclaurin polynomial as n increases. By calculating $E(p_0, n)$ for increasing values of n, we find that $\|f - p_0\|_2 \doteq 0.177$.

[33]Existence and uniqueness of the optimal polynomial follows from general results on complex Chebyshev approximation; for references, see §5.8.

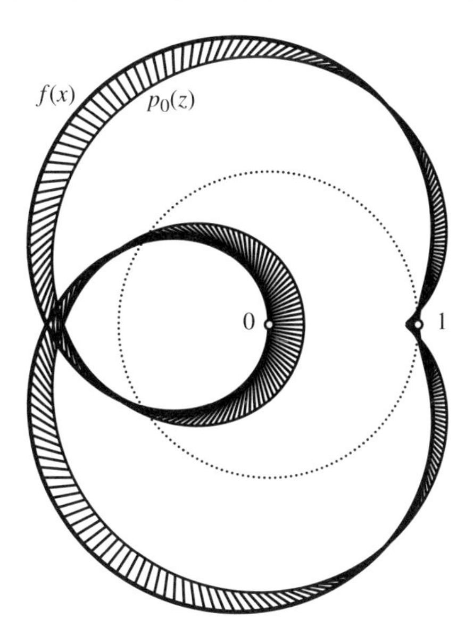

Figure 5.1. *The plots of z, $f(z)$, and $p_0(z)$ in the complex plane when $|z| = 1$.*

To get a visual impression of what we are trying to do, see Figure 5.1, which is a graph of $f(e^{i\theta})$ and of $p_0(e^{i\theta})$ when $\theta = 0^\circ, 1^\circ, 2^\circ, \ldots, 360^\circ$, superimposed on the unit circle. Corresponding points have been connected to make it clear that the quantity $f(z) - p_0(z)$ keeps changing its angle all the time and that $p_0(z)$ is almost never the closest point to $f(z)$ on the graph of p_0.

When θ is restricted to this set of points, we find that $\max |f(\theta) - p_0(\theta)| \doteq 0.235$, so, allowing for a small discretization error, we assert that

$$0.17 < \epsilon_{\text{opt}} < 0.24$$

since

$$\|f - p_0\|_2 \leqslant \|f - p_{\text{opt}}\|_2 \leqslant \|f - p_{\text{opt}}\|_\infty \leqslant \|f - p_0\|_\infty.$$

In a sense, we have one digit: both bounds round to 0.2.

The coefficients of p_0 are real. Can we assume that the coefficients of p_{opt} are also real? The cubic polynomial

$$\hat{p}(z) = \frac{1}{2}(p_{\text{opt}}(z) + \overline{p_{\text{opt}}(\overline{z})})$$

has real coefficients, and for all z, by the symmetry $f(\overline{z}) = \overline{f(z)}$ [AS84, form. (6.1.23)],

$$\begin{aligned} |f(z) - \hat{p}(z)| &\leqslant \frac{1}{2}(|f(z) - p_{\text{opt}}(z)| + |f(z) - \overline{p_{\text{opt}}(\overline{z})}|) \\ &= \frac{1}{2}(|f(z) - p_{\text{opt}}(z)| + |\overline{f(\overline{z})} - \overline{p_{\text{opt}}(\overline{z})}|) \end{aligned}$$

$$= \frac{1}{2}(|f(z) - p_{\text{opt}}(z)| + |f(\bar{z}) - p_{\text{opt}}(\bar{z})|) \leqslant \frac{1}{2}(\epsilon_{\text{opt}} + \epsilon_{\text{opt}}).$$

Thus, $\hat{p}$ is at least as good as p_{opt}, and therefore *it is sufficient to search only among polynomials with real coefficients.*

5.2 Optimization by General-Purpose Methods

General-purpose optimization software does not fare too well on Problem 5. Typically it will halt well before reaching the answer. Yet there is a relatively new technique that does remarkably well here and with a variety of other optimization problems in $\mathbb{R}^n$; it can get a dozen digits in only 40 seconds (*Mathematica* on a 1.25 GHz Macintosh). The method is a type of evolutionary algorithm called *differential evolution*, and it succeeds on many problems where traditional random search methods such as simple evolutionary methods or simulated annealing fail. For an example of success with a simple evolutionary algorithm see §4.2. While no search method can match the analytic methods presented elsewhere in this chapter, the fact that they work at all on this problem is noteworthy and indicates that differential evolution would be a good method to try in situations where there is no known analytic approach.

The idea underlying the algorithm is similar to basic evolutionary methods, with the important twist that the children at each generation are formed by taking a linear combination of parents (as opposed to being simply mutations of a single parent). Each member of the next generation (the new child) has the form $p_1 + r(p_2 - p_3)$, where the p_i are members of the current generation and r is an amplification factor. The formula for children relies heavily on the difference between two of the parents, and this is where the method's name comes from. Here is a formal description.

Algorithm 5.1. Minimization by Differential Evolution.

Input: F, a continuously differentiable function from $\mathbb{R}^d$ to $\mathbb{R}$;
R, a box that serves as the starting seed for the first generation;
N, the population size (number of d-vectors in each generation);
n_{max}, a bound on the number of generations;
r, the amplification factor.

Output: An approximation to the global minimum of F.

Step 1: Initialize the first generation to be a list of N d-vectors by choosing randomly within R. Set up a vector to consist of the F-values at these vectors.

Step 2: The main loop:
For n from 1 to $n_{\text{max}} - 1$, form the $(n+1)$th generation as follows:
For each parent p_i in the nth generation,
construct a child c_i
by randomly choosing indices j, k, and m and letting
$c_i = p_j + r(p_k - p_m)$.

If $F(c_i) < F(p_i)$
let c_i be the new ith member of the $(n+1)$st generation;
else p_i becomes this member.
After all the members of the next generation are determined,
update the value array to contain their F-values.
Step 3: Return the vector in the final generation having the smallest F-value.

Note that a generation is not updated until the entire generation is formed. That is, if a child is to count, it is placed in temporary storage. After the loop is complete, the current generation is replaced by the new generation. Algorithm 5.1 is the barest-possible outline of differential evolution, but it is good enough to solve the problem at hand. One usually selects the indices j, k, and m in such a way that they are distinct from each other and also from i, but ignoring this point speeds up the random integer generation and has no serious consequences. It is also customary to add a crossover step, where some of the entries in the child c_i are replaced by the corresponding entries in the parent p_i, using a probabilistic rule of some sort. But the problem at hand runs faster and more efficiently with no crossover. An amplification factor of 0.4 seems to work.

We need an objective function, of course. Given a cubic $az^3 + bz^2 + cz + d$, where $z = e^{i\theta}$, we can, by symmetry, focus on the domain $\theta \in [0, \pi]$ and use a standard optimization technique starting from a number of seeds, such as is used by *Mathematica*'s `FindMaximum` function, for example. Experiments show that it is sufficient to take seeds near 1.40 and 2.26, and also to throw in the local extremum that always arises (by symmetry) at $\theta = \pi$; that is simply $|a - b + c - d|$. The largest of the three values returned is taken as the value of the objective function.

A *Mathematica* Session

```
maxerror[{a_, b_, c_, d_}] := Max[Abs[a - b + c - d],
   FindMaximum[t = e^(i θ); Abs[a t^3 + b t^2 + c t + d - 1 / Gamma[t]],
      {θ, # - 0.03, # + 0.03}, PrecisionGoal → 12][[1]] & /@ {1.40, 2.26}];
```

Differential evolution is used by *Mathematica*'s `NMinimize` function, but we obtained greater control and understanding by writing our own code. Now, to solve Problem 5 one can let the population size be 60, the amplification factor be 0.4, and the iteration bound be 300. It makes sense to check for convergence, but comparing the best values at successive generations will lead to premature stoppage, since the best might well not change over just one generation. Checking for agreement every 10 generations is more reasonable. When we do that we see convergence to within 10^{-12} in about 180 generations, and therefore about 10,000 evaluations of the objective function. Here is a bare-bones *Mathematica* implementation.[34] One enhancement would be to gradually increase the precision in the objective function as the generations evolve.

[34]MATLAB code can be found on the web page for this book.

A *Mathematica* Session

```
DifferentialEvolution[n_, seeds_] :=
  Module[{best, current, children, vals, childval},
   current = Table[Random[Real, #] & /@ seeds, {n}];
   vals = maxerror /@ current;
   oldbest = Min[vals];
   Do[children = Table[{1, 0.4, -0.4}.
       current[[Table[Random[Integer, {1, n}], {3}]]], {n}];
    Do[If[(childval = maxerror[children[[j]]]) < vals[[j]],
      {current[[j]], vals[[j]]} = {children[[j]], childval}], {j, n}];
    If[Mod[i, 10] == 0, best = Min[vals];
     If[Abs[best - oldbest] < 10^-14, Break[], oldbest = best]], {i, 300}];
   First[Sort[Transpose[{vals, current}]]]];
```

And here is a typical run:

```
SeedRandom[1];
DifferentialEvolution[60, {{-2, 2}, {-2, 2}, {-2, 2}, {-2, 2}}]
```

```
{0.2143352345904104, {-0.6033432200279722,
  0.6252119165808774, 1.019761853131395, 0.005541951112956156}}
```

The numbers in the bottom line are, respectively, ϵ_{opt}, a_{opt}, b_{opt}, c_{opt}, and d_{opt}.

Because of randomness, not every run with these parameters will converge to the right answer, but with $N = 60$ it generally does (353 successes in 400 trials). Figure 5.2 shows the convergence. A comparison of the results of several random runs gives evidence that 12 digits are correct:

$$\epsilon_{\text{opt}} \doteq 0.21433\,52345\,90.$$

For a more comprehensive treatment of the method and further applications, see [CDG99].

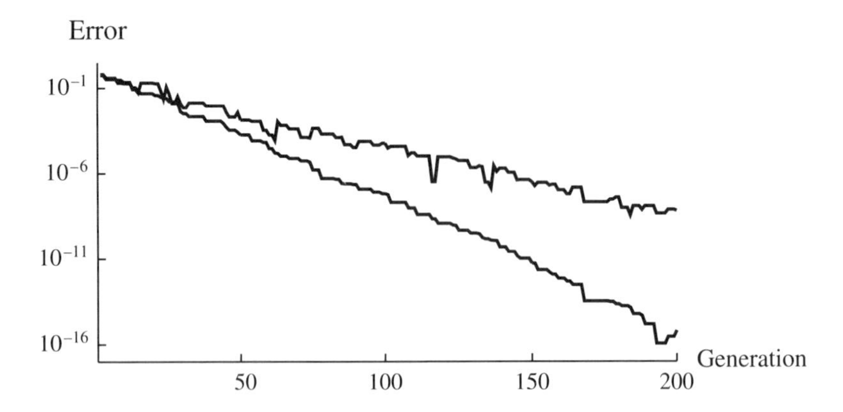

Figure 5.2. *The lower data set shows the convergence to the true answer for a run of 200 generations; the convergence is remarkably steady at about one new digit every 13 generations. The upper data set is the distance from the best cubic at each generation to the truly best cubic (in the Euclidean norm in 4-space).*

5.3 Improving the Lower Bound

Having seen what a sophisticated search technique can do, we return to the basics. Consider the related problems where we minimize the absolute value not of $f - p$, but of its real and imaginary parts separately; that is, we find

$$\epsilon_R = \min_p \epsilon_R(p), \qquad \epsilon_R(p) = \max_{0 \leqslant \theta \leqslant 2\pi} |\mathrm{Re}(f(e^{i\theta}) - p(e^{i\theta}))|;$$

$$\epsilon_I = \min_p \epsilon_I(p), \qquad \epsilon_I(p) = \max_{0 \leqslant \theta \leqslant 2\pi} |\mathrm{Im}(f(e^{i\theta}) - p(e^{i\theta}))|;$$

where in both cases p runs over all cubic polynomials with real coefficients. Clearly ϵ_R and ϵ_I are lower bounds for ϵ_{opt}. These problems involve the Chebyshev approximation of a real-valued function and are therefore easier to solve than the original problem. Let us see what they bring us.

A standard theorem on real Chebyshev approximation (a proof in the case of approximation by polynomials is given in [Rut90, Thms. 7.4 and 7.5] is as follows.

Theorem 5.1 (Alternation Theorem). *Let the continuous functions $f_1, f_2, \ldots, f_n$ be such that the interpolation problem*

$$y_j = \sum_{k=1}^{n} a_k f_k(x_j), \qquad j = 1, 2, \ldots, n,$$

always has a unique solution $a_1, a_2, \ldots, a_n$ when the x_j are distinct points in an interval $[a, b]$. Then the function

$$q(x) = \sum_{k=1}^{n} c_k f_k(x)$$

is the best approximation on $[a, b]$ in the supremum norm to a given continuous function $y(x)$ if and only if $n+1$ points $a \leqslant x_1 < x_2 < \cdots < x_{n+1} \leqslant b$ can be found where $|y(x) - q(x)|$ assumes its maximum value, such that the signs of $y(x_j) - q(x_j)$, $j = 1, 2, \ldots, n+1$, alternate.

The theorem immediately suggests an iterative method, known as the multiple exchange algorithm or the second Remes[35] algorithm [Rem34b, Rem34a].

Algorithm R

0. Find a starting value of q good enough so that step 1, below, is possible. Set ϵ_{old} to some impossibly large value.

1. Find $n+1$ points $x_1 < x_2 < \cdots < x_{n+1}$, where $y - q$ has a local extremum, such that the signs of $y(x_j) - q(x_j)$ alternate.

---

[35] This name is sometimes transliterated as "Remez," based on the current standard for transliteration from Russian into English. However, the papers for which the author is best known were published in French, and presumably he knew how to spell his own name.

2. Solve the linear equations

$$y(x_j) = \sum_{k=1}^{n} c_k f_k(x_j) + \operatorname{sgn}(y(x_j) - q(x_j))\epsilon, \qquad j = 1, 2, \ldots, n+1,$$

for the unknowns $c_1, c_2, \ldots, c_n, \epsilon$.

3. If $|\epsilon| \geqslant \epsilon_{\text{old}}$, exit. Otherwise replace q by $\sum_{k=1}^{n} c_k f_k$, ϵ_{old} by $|\epsilon|$, and return to step 1.

The algorithm becomes slightly more complicated when there are several ways to choose the alternating extrema of $y - q$; fortunately that is not the case here. We can trivially guarantee that the algorithm terminates in a finite number of steps since, on a computer, there are only finitely many values that ϵ can take, and we stop as soon as no progress is made. Of course, if the initial values are bad, we may stop at a point which is not close to a solution to the problem.

If the functions f_k are smooth enough, then the multiple exchange algorithm can be expected to be quadratically convergent; that is, each iteration approximately doubles the number of correct digits. Moreover, as is more fully explained in Chapter 9, when we consider ϵ as a function of $c_1, c_2, \ldots, c_n$ around the optimum, then the value of ϵ will be correct to twice as many digits as the coefficients are.

Let us try this on the real part of f. The approximating functions are $\operatorname{Re}(e^{ik\theta}) = \cos k\theta$, $k = 0, 1, 2, 3$. These functions satisfy the unique interpolation condition (also called the Haar condition) over $[0, \pi]$. Our initial approximation is the real part of p_0, namely

$$q_0(\theta) = -0.6559 \cos 3\theta + 0.5772 \cos 2\theta + \cos\theta.$$

The five extrema occur at approximately 37°, 73°, 108°, 142°, and 180°. Algorithm R gives

$$q_1(\theta) = -0.589085 \cos 3\theta + 0.657586 \cos 2\theta + 1.067908 \cos\theta + 0.032616$$

with $\epsilon_R(q_1) \doteq 0.211379$. We now have

$$0.211 < \epsilon_{\text{opt}} < 0.236;$$

the polynomial p_1 corresponding to q_1 does not improve on the upper bound obtained from p_0.

In the case of the imaginary part, the approximating functions are $\operatorname{Im}(e^{ik\theta}) = \sin k\theta$, $k = 1, 2, 3$, which satisfy the Haar condition over $[0, \pi]$. There are only three parameters, and the optimum is at

$$q_2(\theta) = -0.596767 \sin 3\theta + 0.636408 \sin 2\theta + 1.028102 \sin\theta,$$

with $\epsilon_I(q_2) \doteq 0.212559$. Somewhat surprisingly, this trigonometric polynomial not only gives a better lower bound than q_1 did (which had one more parameter available) but it turns out that the corresponding polynomial

$$p_2(z) = -0.596767z^3 + 0.636408z^2 + 1.028102z$$

does surprisingly well on the real part, so much so that the least-squares upper bound is also improved, and we have

$$0.2125 < \epsilon_{\text{opt}} < \|f - p_2\|_\infty \doteq 0.2319.$$

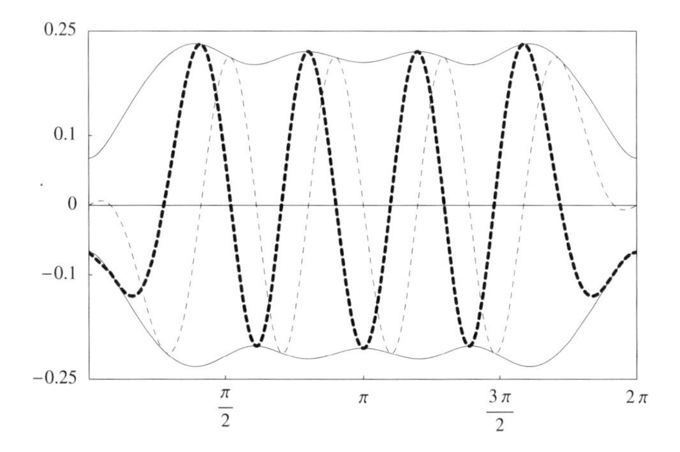

Figure 5.3. *The error in p_2. The solid line gives $|f - p_2|$ and $-|f - p_2|$; the heavy and light dashed lines give the real and imaginary parts of $f - p_2$, respectively.*

(Figure 5.3 shows a graph of the error in p_2.) In retrospect this is perhaps not so surprising, since in principle if we know either the real part or the imaginary part of an analytic function, the other can be found up to an additive constant by applying the Cauchy–Riemann equations.

This picture suggests an easy way to improve the approximation: the constant term does not affect the imaginary part, but it does affect the real part. Note that $\max \operatorname{Re}(f - p_2) > \max \operatorname{Re}(p_2 - f)$. Therefore, for small values of $d > 0$, we should find that $p_2(z) + d$ is a better approximation to f than p_2.

The optimal d satisfies

$$\|f - p_2 - d\|_\infty = |f(-1) - p_2(-1) - d| = p_2(-1) + d,$$

which can be solved by fixed-point iteration from the equivalent equation

$$d = \tfrac{1}{2}(|f - p_2 - d|_\infty - p_2(-1) + d).$$

This gives us the improved approximation

$$p_3(z) = p_2(z) + d = -0.596767z^3 + 0.636408z^2 + 1.028102z + 0.014142.$$

The upper bound is reduced quite a bit by doing this, giving

$$0.2125 < \epsilon_{\text{opt}} < \|f - p_3\|_\infty < 0.2193.$$

We now have two digits: $\epsilon_{\text{opt}} \doteq 0.21$.

The graph of $|f - p_3|$ in Figure 5.4 shows five peaks, one more than the number of parameters, but the two inner peaks are slightly lower than the central and outer peaks. We suspect that the optimal error curve will look like this one, but with all five peaks at the same height.

5.4 Discrete Complex Approximation

There is another way of finding lower bounds: suppose that we somehow are able to solve the minimax problem on a subset S of the unit circle; that is, we can find p_S such that $\epsilon_S(p_S)$

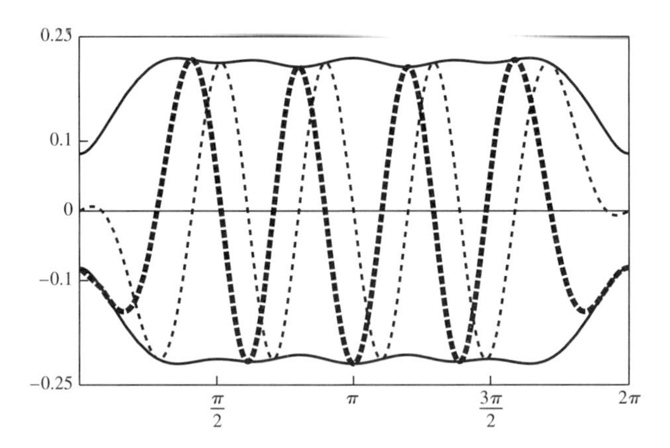

Figure 5.4. *The error in p_3. The solid line gives $|f - p_3|$ and $-|f - p_3|$; the heavy and light dashed lines give the real and imaginary parts of $f - p_3$, respectively.*

is a minimum, where $\epsilon_S(q) = \max_{z \in S} |f(z) - q(z)|$. Then $\epsilon_S(p_S)$ is a lower bound for ϵ_{opt}. This is because

$$\epsilon_S(p_S) \leqslant \epsilon_S(p_{\text{opt}}) \leqslant \max_{|z|=1} |f(z) - p_{\text{opt}}(z)| = \epsilon_{\text{opt}}.$$

Now suppose that S consists only of points where $|f(z) - p_S(z)| = \|f(z) - p_S(z)\|_\infty$; that is, there is no point z outside of S where $|f(z) - p_S(z)|$ is greater than its maximum on S. Then $\epsilon_S(p_S)$ is also an upper bound for ϵ_{opt}, which means that $\epsilon_S(p_S) = \epsilon_{\text{opt}}$. Since we expect the equation $|f(z) - p_S(z)| = \epsilon_S(p_S)$ to hold for only a finite number of values of z, the case of particular interest occurs when S is finite.

We are thus led to the following analogue of Algorithm R for minimizing $\|f - p\|_\infty$ when f is a given complex-valued function and p comes from some n-dimensional linear space $\mathcal{P}$.

Algorithm C

0. Find a starting value of p good enough so that step 1, below, is possible. Set ϵ_{old} to some impossibly large value.

1. Find the set $S = \{z_1, z_2, \ldots, z_m\}$ of points where $|f - p|$ has a local maximum.

2. Solve the restricted problem of finding p that minimizes $\epsilon = \max_j |f(z_j) - p(z_j)|$.

3. If $\epsilon \geqslant \epsilon_{\text{old}}$, exit. Otherwise replace ϵ_{old} by ϵ, and return to step 1.

If this algorithm converges, the answer is the solution we want; of course, it is possible that it might diverge.

The crunch is in step 2. Let $p(z) = \sum_{k=1}^{n} x_k f_k(z)$, where the functions $f_1, f_2, \ldots, f_n$ form a basis for the space $\mathcal{P}$. Then step 2 is the special case where $a_{j,k} = f_k(z_j)$, and $b_j = f(z_j)$ of the discrete linear complex Chebyshev approximation problem:

Given an $m \times n$ matrix A and an m-vector b with complex entries, find a complex n-vector x such that $\|Ax - b\|_\infty$ is a minimum.

Although we happen to know that the x_k are real, there is no theoretical advantage in making that assumption. Watson [Wat88] gave the following characterization theorem.

Theorem 5.2 *The vector x minimizes $\|r\|_\infty$, where $r = b - Ax$, if and only if there exists a set $\mathcal{I}$ containing d indices, where $d \leqslant 2n+1$, and a real m-vector w, such that:*

(a) $|r_j| = \|r\|_\infty$, $j \in \mathcal{I}$;

(b) $w_j > 0$ *if* $j \in \mathcal{I}$, *and* $w_j = 0$ *for* $j \notin \mathcal{I}$;

(c) $A^* W r = 0$ *with* $W = \mathrm{diag}(w_j)$, *where* A^* *is the Hermitian transpose of* A.

The set $\mathcal{I}$ is called an active set *and the vector w a* dual solution. *In addition, if all $n \times n$ submatrices of A are nonsingular (the discrete Haar condition), the Chebyshev solution x (but not necessarily the active set $\mathcal{I}$ or the dual solution w) is unique. In that case, $d \geqslant n+1$.*

The practical implications of this theorem are that, once the set $\mathcal{I}$ is fixed, one obtains d real equations of the form $|r_j| = \epsilon$, $j \in I$, from (a) and n complex equations from (c) for the $d+1$ real unknowns $\epsilon, w_1, w_2, \ldots, w_d$ and the n complex unknowns x_k. However, the system of equations is not underdetermined, since it is homogeneous in the w_j; we could pick any of the w_j, set it to 1, and solve for the others. If any w_j is then nonpositive, we would know that $\mathcal{I}$ is wrong.

It is in general a combinatorial problem, not at all easy when m is large, to find the active set $\mathcal{I}$ (see, for example, [LV94]). In the present case we are lucky: the error graph for p_3, currently our best available approximation, gives reason to think that S contains only five points, so $m = 5 = n+1$; and since the Haar condition holds in the space of cubic polynomials, all five points are active.

By symmetry, we can write S in terms of two unknowns as

$$S = \{e^{i\theta_1}, e^{i\theta_2}, -1, e^{-i\theta_2}, e^{-i\theta_1}\},$$

where θ_1 is near 65° and θ_2 is near 113°. The characterization equations then reduce to

$$|f(z_j) - (az_j^3 + bz_j^2 + cz_j + d)|^2 = \epsilon^2, \qquad j = 1, 2, 3, 4, 5;$$

$$\sum_{j=1}^{5} \overline{z_j}^k w_j (f(z_j) - (az_j^3 + bz_j^2 + cz_j + d)) = 0, \qquad k = 0, 1, 2, 3.$$

We have squared the first set of equations in order to obtain expressions that are analytic in terms of the coefficients. Now let $w_3 = 1$; note that, by symmetry, we can discard the equations with $j = 4, 5$, and also take $w_5 = \overline{w_1}$, $w_4 = \overline{w_2}$; then we are left with seven equations in seven unknowns $a, b, c, d, w_1, w_2, \epsilon$, since for the inner iteration the z_j values are fixed. So Algorithm C involves the solution of a system of nonlinear equations *at every iteration*. Also, the question of initial values for the w_j arises.

Surely one can simplify things a little more? Indeed one can, and there are two ways. One of these does away with the outer iteration by differentiating the square of the residual (treated in §5.5), so that the inner iteration converges to the correct answer; the other does away with the inner iteration by explicitly solving the nonlinear system (treated in §5.6). These two sections are independent of each other.

5.5 A Necessary Condition for Optimality

Consider the second set of equations in Theorem 5.2:

$$\sum_{j=1}^{5} \overline{z_j}^k w_j r_j = 0, \qquad k = 0, 1, 2, 3.$$

If we knew the r_j's, these equations (taking $w_3 = 1$) would form a system of four equations in four unknowns w_1, w_2, w_4, w_5. By symmetry, $r_5 = \overline{r_1}, r_4 = \overline{r_2}, w_5 = \overline{w_1}, w_4 = \overline{w_2}$; moreover, we make the assumption that $r_3 < 0$, leading to $r_3 = -\epsilon$. This assumption is based on what we know of p_3, which is already a very good approximant. What is left of the characterization constraints? Well, nothing so far forces the w_j values to be real; this yields the two equations $\operatorname{Im} w_1(z_1, z_2) = 0$ and $\operatorname{Im} w_2(z_1, z_2) = 0$, where the functions w_1 and w_2 are implicitly defined by the system of linear equations. Analytical expressions for $r_1 w_1$ and $r_2 w_2$ are easy to obtain using Cramer's rule and Vandermonde determinants: let

$$V(a, b, c, d) = \begin{vmatrix} 1 & 1 & 1 & 1 \\ a & b & c & d \\ a^2 & b^2 & c^2 & d^2 \\ a^3 & b^3 & c^3 & d^3 \end{vmatrix} = (a - b)(a - c)(a - d)(b - c)(b - d)(c - d),$$

then

$$r_1 w_1 = \frac{V(-1, \overline{z_2}, z_2, z_1)}{V(\overline{z_1}, \overline{z_2}, z_2, z_1)} \epsilon, \qquad r_2 w_2 = \frac{V(\overline{z_1}, -1, z_2, z_1)}{V(\overline{z_1}, \overline{z_2}, z_2, z_1)} \epsilon. \tag{5.1}$$

Some further simplifications are possible; for example, the denominators are clearly real and therefore irrelevant when testing whether w_1 and w_2 are real.

If Algorithm C converges, the optimal points z_j are such that

$$\left(\frac{d}{d\theta} |f(e^{i\theta}) - p(e^{i\theta})|^2 \right)_{\theta=\theta_j} = 0.$$

Since for f and p, differentiation with respect to $z = e^{i\theta}$ is natural, we use $z'(\theta) = iz$ and the chain rule, giving

$$\begin{aligned} 0 &= \left(\frac{d}{d\theta} |f(e^{i\theta}) - p(e^{i\theta})|^2 \right)_{\theta=\theta_j} \\ &= \overline{(f(e^{i\theta_j}) - p(e^{i\theta_j}))}\, iz_j (f'(e^{i\theta_j}) - p'(e^{i\theta_j})) \\ &\quad + (f(e^{i\theta_j}) - p(e^{i\theta_j}))\, \overline{iz_j (f'(e^{i\theta_j}) - p'(e^{i\theta_j}))} \\ &= -2\operatorname{Im} \left(\overline{r}_j z_j (f'(e^{i\theta_j}) - p'(e^{i\theta_j})) \right). \end{aligned}$$

Collecting all our information, we see that the following six equations in the six unknowns a, b, c, d, θ_1, θ_2 are necessary optimality conditions:

$$|r_j| - \epsilon = 0, \quad \operatorname{Im} w_j(z_1, z_2) = 0, \quad \operatorname{Im}\left(\overline{r_j} z_j (f'(z_j) - p'(z_j))\right) = 0, \quad j = 1, 2, \tag{5.2}$$

where the functions w_1 and w_2 are defined in (5.1) and the auxiliary quantities are

$$\epsilon = d - c + b - a, \qquad z_j = e^{i\theta_j}, \qquad r_j = f(z_j) - p(z_j).$$

The equation for ϵ comes from $1/\Gamma(-1) = 0$, plus the assumption that $r_3 = -\epsilon$. The derivative of f is given by $f'(z) = -\psi(z)f(z)$, where $\psi(z) = \Gamma'(z)/\Gamma(z)$ is the digamma function.[36]

Instead of the approach with which the previous section ends, whereby a system of seven nonlinear equations must be solved at each step of an iteration involving the z_j's, we now have the alternative of solving the system (5.2) of only six nonlinear equations and obtaining p_{opt} and the critical values z_j without any outer iteration. It is true that our derivation does not guarantee that the solution of (5.2) is unique. To check that the solution thus found is indeed optimal, one could simply plot $|f(z) - p(z)|$ over the whole interval of interest. But it is easier simply to check that $w_1(z_1, z_2)$ and $w_2(z_1, z_2)$ are positive: in that case, a result due to Vidensky (see [Sin70b, Thm. 1.4, p. 182] or [SL68, Lemma 1, p. 450]) guarantees optimality.

Six equations in six unknowns, with no outer iteration: that is surely a viable approach, given the good initial values that we already have. Newton's method,

$$u_{m+1} = u_m - J(F, u_m)^{-1} F(u_m),$$

for improving the approximate solution u_m of a nonlinear system $F(u) = 0$ when the Jacobian $J(F, u)$ can be calculated is likely to work.

An Octave Session

The functions `p5func` (which evaluates the six functions in (5.2) given the six unknowns as a vector) and `jac` (which calculates a numerical Jacobian of a given function) have been predefined and can be found on the web page for this book. I have deleted some padding from the actual output.

```
>> u=[0.014142; 1.028102; 0.636408; -0.596767; [65;113]*pi/180];
>> while true,
>>     f=p5func(u); [u(1)-u(2)+u(3)-u(4), max(abs(f))]
>>     if abs(f)<5e-15, break, end;
>>     u=u-jac('p5func',u)\p5func(u);
>> end
>> u

   ans =
      2.19215000000000e-01   4.38761683759301e+00
```

[36]The complex gamma and digamma functions are available for Octave and MATLAB as contributed implementations by Paul Godfrey:
`http://www.mathworks.com/matlabcentral/fileexchange/loadFile.do?objectId=978.`

```
      2.09343333405806e-01   7.48097169131265e-01
      2.12244608568497e-01   3.48454464754578e-01
      2.14102287307571e-01   4.63264212148370e-03
      2.14335804172636e-01   4.31572906936413e-05
      2.14335234577871e-01   2.74815371456513e-09
      2.14335234590463e-01   5.82867087928207e-15
      2.14335234590459e-01   2.27453094590858e-15
  u =
      5.54195073311451e-03
      1.01976185298384e+00
      6.25211916433389e-01
     -6.03343220407797e-01
      1.40319917081600e+00
      2.26237744961289e+00
```

The convergence criterion has not been plucked out of thin air: since $\epsilon = d-c+b-a \doteq 0.2$ but $|d|+|c|+|b|+|a| \doteq 2$, we expect to lose one digit to roundoff in the calculation of ϵ, and therefore in IEEE double precision (almost 16 digits)[37] one expects that the result will be off by a few units in the 15th digit. We obtain

$$p_4(z) = -0.60334322040780z^3 + 0.62521191643339z^2 + 1.01976185298384z + .00554195073311$$

(the error in p_4 is shown in Figure 5.5) and

$$\theta_1 \doteq 1.4031992, \qquad \theta_2 \doteq 2.2623774, \qquad \epsilon_{\text{opt}} \doteq 0.21433523459046.$$

The angles are near 80.4° and 129.6°, quite some distance from the starting angles; if our initial values were less good, Newton's method might well have failed to converge.

Note that although ϵ_{opt} agrees to 13 digits with the result found in §5.2, the coefficients of p_4 agree to only 8 digits with those given there. As a general rule, in an optimization problem one can find the value of a smooth function at the optimum much more accurately than one can find the location of the optimum. The reason for this is explained more fully in §9.3 in a univariate setting. Thus, when working to 16 digits, we can expect only about 8 correct digits in the coefficients even though the function value itself is correct to nearly the full working precision.

This does not mean that we can afford to display the coefficients of p_4 to less than full accuracy, only that there are certain perturbations of the coefficients (e.g., to change (a, b, c, d) by a multiple of $(1, 1, 1, 1)$) to which $\|f-p\|_\infty$ is very insensitive when p is near p_4. There are other perturbations (e.g., to change (a, b, c, d) by a multiple of $(1, -1, 1, -1)$) to which $\|f-p\|_\infty$ is not insensitive. Thus, the quantity $\|f-p\|_\infty$, when seen as a function of four variables, is smooth in some directions and nonsmooth in others. This property may account for the difficulty that general-purpose optimization methods have on this problem.

It is astonishing just how flat the whole graph is, barely discernible from a horizontal line in the range $72° < \theta < 288°$, which represents three-fifths of the whole. In fact,

[37]The methods of this and the next section also work for 10,000 digits; see Appendix B.

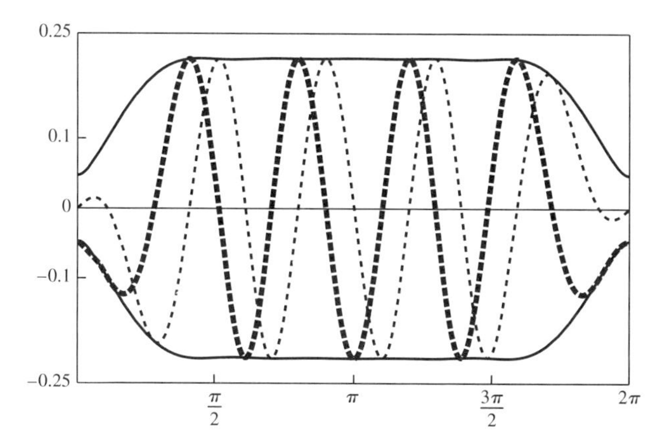

Figure 5.5. *The error in p_4. The solid line gives $|f - p_4|$ and $-|f - p_4|$; the heavy and light dashed lines give the real and imaginary parts of $f - p_4$, respectively.*

$0.21280 < |f(z) - p_4(z)| < 0.21434$ for every $z = e^{i\theta}$ in that region. This property[38] is also one that could cause difficulties for general-purpose optimization methods.

5.6 Implementation of Algorithm C

The discrete linear complex Chebyshev approximation problem has an explicit solution [LV94] when $m = n + 1$:

> Note that the left nullspace of A is one-dimensional. There is thus a unique vector $y = (y_1, y_2, y_3, y_4, y_5)$ in that space satisfying $y_3 = -1$; therefore $\epsilon y = Wr$, so that $w_j = |y_j|$ and $r_j = \epsilon \operatorname{sgn} y_j$. Find that vector y, and then solve the $m \times m$ linear system
>
> $$\begin{pmatrix} 1 & z_1 & z_1^2 & z_1^3 \\ 1 & z_2 & z_2^2 & z_2^3 \\ 1 & z_3 & z_3^2 & z_3^3 \\ 1 & z_4 & z_4^2 & z_4^3 \\ 1 & z_5 & z_5^2 & z_5^3 \end{pmatrix} \begin{pmatrix} d \\ c \\ b \\ a \end{pmatrix} + \epsilon \begin{pmatrix} \operatorname{sgn} y_1 \\ \operatorname{sgn} y_2 \\ \operatorname{sgn} y_3 \\ \operatorname{sgn} y_4 \\ \operatorname{sgn} y_5 \end{pmatrix} = \begin{pmatrix} f(z_1) \\ f(z_2) \\ f(z_3) \\ f(z_4) \\ f(z_5) \end{pmatrix} \tag{5.3}$$
>
> for the unknowns a, b, c, d, and ϵ.

The system can be simplified to take advantage of conjugate symmetry so that only real quantities appear. So each iteration of Algorithm C involves finding two extrema of $|f(e^{i\theta}) - p(e^{i\theta})|$ near their previous values, followed by the solution of a 5×5 linear system to update p.

It is also possible to take a purely dual view of the problem: all the quantities in (5.3) depend only on the two critical angles θ_1 and θ_2. We can think of (5.3) as defining a function $\epsilon(\theta_1, \theta_2)$. This value of ϵ, as we have seen, is a lower bound for ϵ_{opt}. The dual point of view is simply this: find a local *maximum* of ϵ as a function of θ_1 and θ_2.

[38] The flatness of the error in the best approximation is a well-known feature (called "near-circularity of the error curve") of complex Chebyshev approximation, and in fact is less pronounced for $1/\Gamma$ than for more well-behaved functions [Tre81].

An Octave Session *(Solving the dual problem by Newton's method)*

The routines `hessian` and `cjac` are prewritten, giving numerical approximations to, respectively, the Hessian matrix and the Jacobian of a function of several variables. Their code can be found on the web page for this book.

```
>> function [c,err]=p5solve(z)
>>   for p=1:4, A(:,p)=z.^(p-1); end
>>   b=isgamma(z); d=sign(b-A*(A\b)); c=[A d]\b; err=c(5); c(5)=[];
>> endfunction
>> function y=epsilon(th)
>>   z=z=exp(i*th); z=[z;-1;conj(z)]; [c,e]=p5solve(z); y=real(e);
>> endfunction
>> th=[80;130]*pi/180; f0=epsilon(th);
>> while true,
>>    dth=hessian('epsilon',th)\cjac('epsilon',th)';
>>    new=th-dth; f1=epsilon(new);
>>    [new' f1], if abs(f1-f0)<1e-15, break; end
>>    f0=f1; th=new;
>> end

   ans =
     1.403354225245182   2.262356557886901   0.214335228788773
     1.403199207182680   2.262377441623623   0.214335234590459
     1.403199170789188   2.262377449611380   0.214335234590460
```

For the dual algorithm, the initial values from p_3 are not good enough for Newton's method to converge to the correct values, which is why the very good initial values of 80° and 130° were used here for the demo version of the code. The version of the code on the web page for this book uses a slightly more sophisticated variation of Newton's method in which monotonic convergence of ϵ is enforced, which does converge from the p_3 initial values.

5.7 Evaluation of the Gamma Function

Some people feel that one can trust developers who produce software to evaluate a mathematical function: surely they will have made such a deep study of the function in question that the casual user need not be concerned with the details. This is probably true for the elementary functions as implemented in hardware on standard IEEE-compliant processors, but in the case of the gamma function, there are at least two reasons why one must know something about the implementation. One reason is that only a handful of languages provide the gamma function over the whole complex plane; another is that the gamma function takes much longer to evaluate than do the elementary functions. We may have to write our own routine, and we certainly need to understand the behavior of other people's routines.

In the Octave programs above, use was made of a handwritten routine `isgamma`, since neither Octave nor MATLAB comes with a built-in routine that can evaluate `Gamma(z)` when z is not real. This routine is a straightforward implementation of [AS84, form. (6.1.34)], which gives $1/\Gamma(z)$ as a power series around $z = 0$, taking into account all the coefficients that are significant when working to 16 decimal places. The series seems to be Taylor-made for our application, where $1/\Gamma(z)$ is required in IEEE double precision for $|z| = 1$; it is not suitable when $|z|$ is much larger than that.

The same series was given to 20 decimal places by Luke [Luk75, pp. 1,2] who additionally warmed the hearts of do-it-yourself enthusiasts by supplying a recursion formula for the coefficients in the series. That formula requires the values of Euler's constant γ and of $\zeta(k)$, $k = 2, 3, 4, \ldots$, where ζ is the Riemann zeta function, and is therefore not yet the ultimate answer.

A popular way of evaluating $\Gamma(z)$ is via Stirling's formula [AS84, form. (6.1.42)]

$$\log \Gamma(z) = (z - \tfrac{1}{2}) \log z - z + \tfrac{1}{2} \log(2\pi) + \sum_{m=1}^{n} \frac{B_{2m} z^{-2m+1}}{2m(2m-1)} + R_n, \tag{5.4}$$

$$|R_n| = \frac{|B_{2n+2} z^{-2n-1}| K_n}{(2n+1)(2n+2)}, \tag{5.5}$$

where B_n is the nth Bernoulli number, defined recursively by

$$B_n = \sum_{k=0}^{n} \binom{n}{k} B_k, \qquad n = 2, 3, \ldots;$$

starting from $B_0 = 1$ (note that the recursion formula says nothing about B_n, but can be thought of as defining B_{n-1} in terms of its predecessors). Several estimates for K_n are given in [Luk75, pp. 8, 9], but we will only need the simplest: $K_n \leqslant 1$ when $\arg z \leqslant \pi/4$. For those values of z, the error term in (5.5) is smaller in magnitude than the first neglected term.

For $n \geqslant 4$, the estimate (see [AS84, forms. (6.1.38) and (23.1.15)])

$$4\sqrt{\pi n} \left(\frac{n}{\pi e}\right)^{2n} \leqslant |B_{2n}| \leqslant 4.08\sqrt{\pi n} \left(\frac{n}{\pi e}\right)^{2n} \tag{5.6}$$

holds. Thus for $K_n \leqslant 1$ and $n \geqslant 4$,

$$|R_n| \leqslant \frac{2.04|z|\sqrt{\pi(n+1)}}{(n+1)(2n+1)} \left(\frac{n+1}{\pi e|z|}\right)^{2n+2}.$$

A reasonable (not the best) place to stop is when $n + 1 = \lfloor \frac{1}{2}\pi e|z| \rfloor$. In that case, for $\arg z \leqslant \pi/4$ and $|z| > 1$,

$$|R_n| < 2^{-\pi e|z|}. \tag{5.7}$$

We have no way, for fixed z, of substantially decreasing this bound, but we can increase z to make the bound small enough and then recurse back, using the relation

$$\log \Gamma(z + N) - \log \Gamma(z) = \log \prod_{k=0}^{N-1} (z + k).$$

This process will cause some cancellation of significant figures, but nothing catastrophic: for example, for z on the unit circle, if we want 10,000 significant digits, we will need to take N near 4000. Then $\log \Gamma(z + N)$ will be near 30,000, so we can expect to lose about five digits to cancellation.

This, in essence, is the way almost all multiprecision languages with a routine for Γ compute it. By making pessimistic decisions all along the way, a careful implementation can come close to delivering a guaranteed precision.

Why have we gone to such lengths to explain what every implementor knows? This is the reason: *Not only is the gamma function a very expensive function to compute, it also has some rather counterintuitive timing behavior.*

The normal paradigm when comparing optimization methods for speed is to base the comparison on how many function evaluations are required. Underlying this way of thinking is the tacit assumption that not only are the function values far and away the most expensive part of the computation, but also that all function values take approximately the same time to compute. *This assumption is not true in the case of the gamma function.*

A PARI/GP Session

```
? \p1000
?  realprecision = 1001 significant digits (1000 digits displayed)
? #
   timer = 1 (on)
? g=gamma(Pi/4);
   time = 11,121 ms.
? g2=gamma(Pi/4+0.1*I);
   time = 703 ms.
? z3=zeta(3);
   time = 78 ms.
? b2500=bernreal(2500);
   time = 0 ms.
```

Note that it took over 11 seconds to compute the first value of the gamma function, but less than 1 second to compute the next one, even though the second argument is complex. The clue lies in that Bernoulli number that we got instantly. In a fresh PARI/GP session, we get:

```
? b2500=bernreal(2500);
   time = 8,940 ms.
```

So almost all the time goes into the one-off computation of the necessary Bernoulli numbers. Once they are known, the second and later evaluations of the gamma function go quickly. Other multiprecision packages like Maple and *Mathematica* show similar behavior.

To summarize: When going for 10,000 digits on *this* problem (see Appendix B), there is little point in trying to optimize the solution method itself, since that first value of the gamma function totally swamps the computing time.

5.8 More Theory and Other Methods

We have no more than touched upon the very rich theory of approximation in the Chebyshev norm, and in particular we have barely scratched the surface of the results available for approximation by elements of subspaces of a complex vector space. Much more has been proved; for example, there is a general theorem due to Kolmogorov [Kol48] that gives existence and uniqueness of the solution, containing Theorems 5.1 and 5.2 as simple corollaries. The interested reader is referred to Watson [Wat00] for a historical survey of theory and computational methods for approximation in real vector spaces, and to Singer [Sin70b] for a self-contained treatise in a functional analysis framework, which includes complex vector spaces but omits numerical methods. Both authors have a strong sense of historical responsibility and give numerous references to original sources. Another useful work is [SL68], also a self-contained treatise, but less dauntingly abstract than [Sin70b].

A Google™ search on January 22, 2004, for the exact phrase "complex Chebyshev approximation" returned 181 hits. One of these, an algorithm by Tang [Tan88], was successfully used by at least one winning team.

The special case of which this problem is an example, namely, Chebyshev approximation by a polynomial on the unit circle, has been the subject of some recent papers [Tse96, BT99].

Further, there is the MATLAB package COCA,[39] written by Fischer and Modersitzki, that calculates linear Chebyshev approximations in the complex plane based on techniques similar to those in §5.6. Using it, Problem 5 can successfully be solved with a few lines of code, which can be found on the web page for this book.

5.9 A Harder Problem

Problem 5 can be made substantially harder by introducing one tiny change in the wording:

> *Let $f(z) = 1/\Gamma(z)$, where $\Gamma(z)$ is the gamma function, and let $p(z)$ be the cubic polynomial that best approximates $f(z)$ on the unit disk in the L_1 norm $\|\cdot\|_1$. What is $\|f - p\|_1$?*

The definition of the L_1 norm is

$$\|g\|_1 = \frac{1}{\pi} \int_0^{2\pi} \int_0^1 |g(re^{i\theta})| \, r \, dr \, d\theta.$$

[39] http://www.math.mu-luebeck.de/workers/modersitzki/COCA/coca5.html

Chapter 6

Biasing for a Fair Return

Folkmar Bornemann

It was often claimed that [direct and "exact" numerical solution of the equations of physics] would make the special functions redundant .... The persistence of special functions is puzzling as well as surprising. What are they, other than just names for mathematical objects that are useful only in situations of contrived simplicity? Why are we so pleased when a complicated calculation "comes out" as a Bessel function, or a Laguerre polynomial? What determines which functions are "special"?

—Sir Michael Berry [Ber01]

People who like this sort of thing will find this the sort of thing they like.

—Abraham Lincoln. Quoted by Barry Hughes in his appendix "Special Functions for Random Walk Problems" [Hug95, p. 569]

Problem 6

A flea starts at $(0, 0)$ *on the infinite two-dimensional integer lattice and executes a biased random walk: At each step it hops north or south with probability* $1/4$*, east with probability* $1/4 + \epsilon$*, and west with probability* $1/4 - \epsilon$*. The probability that the flea returns to* $(0, 0)$ *sometime during its wanderings is* $1/2$*. What is* ϵ*?*

Asking for the ϵ that gives a certain probability p of return yields a problem hardly any more difficult than calculating the probability for a given ϵ: it just adds the need to use a numerical root-finder. But the problem looks more interesting the way it is stated. In §6.1 we give a short argument on why the problem is solvable.

We will discuss several methods for calculating the probability of return. In §6.2, using virtually no probability theory, we transform the problem to one of linear algebra. Solving a sparse linear system of dimension 25,920 gives us 15 correct digits. The main story, told in §§6.3–6.5, is based on the relation between the probability p of return and the expected number E of visits to the starting site, namely, $E = 1/(1-p)$. We represent E as an infinite series and, by stepwise decreasing the computational effort while increasing the level of analytic sophistication, go from a brute force numerical approximation to a symbolic evaluation using special functions. The latter results in a closed formula involving the arithmetic-geometric mean M,

$$p = 1 - M\left(\sqrt{1-(1+\eta)^2/4}, \sqrt{1-(1-\eta)^2/4}\right), \quad \eta = \sqrt{1-16\epsilon^2}. \tag{6.1}$$

To answer Berry's question from the quote at the beginning of this chapter: *we* are so pleased that it "comes out" as such an expression because there is an exceedingly fast algorithm known for its evaluation. It allows us to solve Problem 6 to 10,000 digits in less than a second. And even more, we can validate these digits using interval arithmetic. Finally, in §6.6 we use the technique of lattice Green functions and Fourier analysis to establish a further expression for E, a double integral. Using adaptive numerical quadrature we can thus solve Problem 6 with just three lines of MATLAB code.

6.1 A First Look: Is It Solvable?

Before we start developing methods for calculating the probability of return, let us convince ourselves that root-finding for Problem 6 will yield a result. To this end, we look at the two extreme cases of the bias $0 \leqslant \epsilon \leqslant 1/4$.[40] In one of the earliest papers on random walks [Pól21], dating back to 1921, Pólya established the classic result that for the unbiased random walk in two dimensions the probability of return is 1, that is, $p|_{\epsilon=0} = 1$. We will come back to this point with a proof in §6.5. On the other hand, in the case of maximum bias, $\epsilon = 1/4$, at each step the walker (aka the flea) hops to the east with probability $1/2$, but never to the west. In going a step to the east he would thus prevent himself from ever returning. Hence, the probability of return is at most that of not going east in the first step, $p|_{\epsilon=1/4} \leqslant 1/2$. By continuity we can conclude that there is a bias $0 < \epsilon_* \leqslant 1/4$ such that $p_{\epsilon=\epsilon_*} = 1/2$.

Notation and Terminology. Throughout the chapter we adopt the following notation. The (nonnegative) probabilities for a step to the east, west, north, or south, are denoted by p_{E}, p_{W}, p_{N}, or p_{S}, respectively. This way we are more general than the problem demands, but we will benefit from much cleaner formulas. We also do not assume that the possibilities of a transition to either direction are exhaustive; that is, we do not assume that $p_{\mathrm{E}} + p_{\mathrm{W}} + p_{\mathrm{N}} + p_{\mathrm{S}} = 1$, but only that

$$p_{\mathrm{E}} + p_{\mathrm{W}} + p_{\mathrm{N}} + p_{\mathrm{S}} \leqslant 1.$$

The excess probability $p_{\mathrm{kill}} = 1 - p_{\mathrm{E}} - p_{\mathrm{W}} - p_{\mathrm{N}} - p_{\mathrm{S}}$ will be interpreted as the probability that the walker vanishes altogether, that is, that the walk stops at the current lattice point. Sometimes p_{kill} is called the *killing rate* [Hug95, §3.2.4, p. 123], and the walker is said to be *mortal* if $p_{\mathrm{kill}} > 0$.

We introduce some further useful terminology, common in the literature on random walks [Hug95, p. 122]. If the eventual return to the starting site is certain, that is, if $p = 1$,

[40] By the symmetries of the problem we can restrict ourselves to nonnegative ϵ.

the random walk is called *recurrent*. Otherwise, when $p < 1$, the random walk is called *transient*.

6.2 Using Numerical Linear Algebra

Sometimes in mathematics a problem becomes much easier if we try to solve for more. Instead of just asking for the probability p that the walker starting at $(0, 0)$ reaches $(0, 0)$ *again*, we consider the probability $q(x, y)$ that the walker, starting at a lattice point (x, y), *ever* reaches $(0, 0)$. Of course, we have

$$q(0, 0) = 1. \tag{6.2}$$

So, how can $q : \mathbb{Z}^2 \to [0, 1]$ help us in calculating p? The point is that a returning walker has left $(0, 0)$ in his first step to a nearest neighbor and has reached $(0, 0)$ from there. Thus, we obtain

$$p = p_{\mathrm{E}}\, q(1, 0) + p_{\mathrm{W}}\, q(-1, 0) + p_{\mathrm{N}}\, q(0, 1) + p_{\mathrm{S}}\, q(0, -1). \tag{6.3}$$

A similar argument links the values of q themselves: the probability that the walker reaches $(0, 0)$ from the lattice point $(x, y) \neq (0, 0)$ can be expressed in terms of the probability that he moves to any of its nearest neighbors and reaches $(0, 0)$ from there. In this way we obtain the *partial difference equation*

$$q(x, y) = p_{\mathrm{E}}\, q(x+1, y) + p_{\mathrm{W}}\, q(x-1, y) + p_{\mathrm{N}}\, q(x, y+1) + p_{\mathrm{S}}\, q(x, y-1), \tag{6.4}$$

with $(x, y) \neq (0, 0)$, subject to the boundary condition (6.2). We will encounter a similar argument in §10.2, where we will approximate a Brownian motion by a random walk.

The partial difference equation (6.4) constitutes an infinite-dimensional linear system of equations for the unknowns $q(x, y)$, $(x, y) \in \mathbb{Z}^2 \setminus (0, 0)$. If we confine ourselves to a finite spatial region of the lattice, say

$$\Omega_n = \{(x, y) \in \mathbb{Z}^2 : |x|, |y| \leqslant n\},$$

we have to supply the neighboring values of q as boundary values. Now, it is reasonable to expect that reaching $(0, 0)$ becomes increasingly unlikely for a walker starting at points further and further away from the origin:

$$\lim_{n\to\infty} \sup_{(x,y)\notin\Omega_n} q(x, y) = 0.$$

Mimicking this, we approximate q by the solution q_n of the linear system

$$q_n(x, y) = p_{\mathrm{E}}\, q_n(x+1, y) + p_{\mathrm{W}}\, q_n(x-1, y) + p_{\mathrm{N}}\, q_n(x, y+1) + p_{\mathrm{S}}\, q_n(x, y-1), \tag{6.5}$$

with $(x, y) \in \Omega_n \setminus (0, 0)$, subject to the boundary values

$$q_n(0, 0) = 1, \quad q_n(x, y) = 0 \ \text{ for } \ (x, y) \notin \Omega_n.$$

One can prove the convergence $q_n(x, y) \rightarrow q(x, y)$, uniformly in (x, y). In fact, the convergence is exponentially fast in n. A proof can be based on the properties of the lattice Green function that we will determine in §6.6. However, we will not go into details here because the result obtained will be validated by other methods later on, which are easier to analyze.

The difference equation (6.5) forms a linear system of N equations in the unknowns $q_n(x, y)$, $(x, y) \in \Omega_n \setminus (0, 0)$, with the dimension $N = (2n + 1)^2 - 1$. In analogy to the five-point stencil discretization of Poisson's equation (cf. §10.3), this linear system can be brought to the matrix-vector form

$$A_N x_N = b_N,$$

with a given vector $b_N \in \mathbb{R}^N$ and a sparse matrix $A_N \in \mathbb{R}^{N \times N}$ with just five nonzero diagonals. In Appendix C.3.1 the reader will find the very short MATLAB function `ReturnProbability`, which generates A_N using ideas from discrete Poisson's equations (such as the Kronecker tensor product; cf. [Dem97, §6.3.3]), solves the linear system using MATLAB's built-in sparse linear solver, and outputs an approximation to the return probability p using (6.3). A *Mathematica* version is given in Appendix C.5.2.

Having this in hand, we can solve the equation $p|_{\epsilon=\epsilon_*} = 1/2$ for ϵ_* using MATLAB's root-finder. Figure 6.1, obtained with small n for plotting accuracy only, shows that there is just one positive solution $\epsilon_* \approx 0.06$.

A MATLAB Session

```
>> f = inline('ReturnProbability(epsilon,n)-0.5','epsilon','n');
>> for n=10*2.^(0:3)
>>     out=sprintf('n = %3i\t N = %6i\t\t epsilon* = %18.16f',n,...,
>>        (2*n+1)^2-1,fzero(f,[0.06,0.07],optimset('TolX',1e-16),n));
>>     disp(out);
>> end
n = 10   N =    440  epsilon* = 0.0614027632354456
n = 20   N =   1680  epsilon* = 0.0619113908202284
```

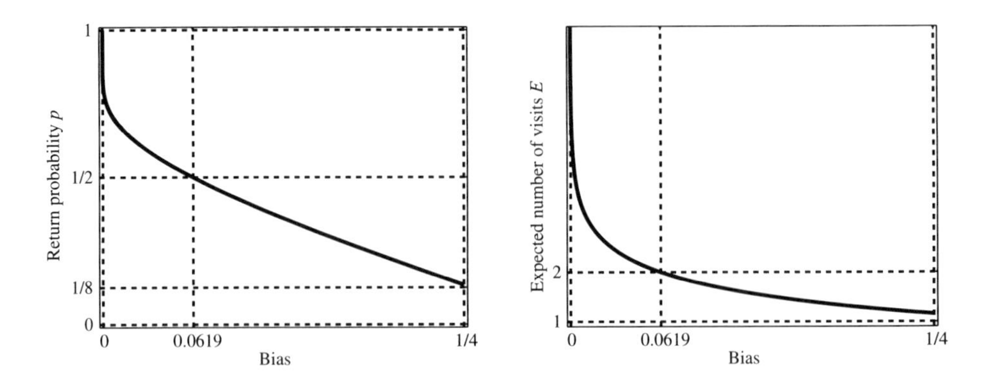

Figure 6.1. *Return probability p and expectation E as a function of the bias ε.*

```
n = 40   N =  6560  epsilon* = 0.0619139543917373
n = 80   N = 25920  epsilon* = 0.0619139544739909
```

The run with $n = 80$ takes 11 seconds on a 2 GHz PC.

How do we assess the accuracy of these four approximations? First, we have to be sure that the sparse linear solver itself does not distort the result. This can be checked using techniques from a posteriori error control that we will discuss in §7.4.1. Given that the linear solver is accurate to at least 15 digits in IEEE arithmetic, we observe that the first two approximations agree to 2 digits, the second and third to 4 digits, the third and fourth to 8 digits. It appears that doubling n doubles the number of correct digits, which is experimental evidence for the claimed exponentially fast convergence rate. Assuming this, we would expect about 15 correct digits at $n = 80$ (which means solving a linear system of $N = 25{,}920$ equations), that is,

$$\epsilon_* \doteq 0.0619139544739909.$$

The following sections will show that these digits are indeed correct.

6.3 Expectations

From now on we utilize the relation of the probability p of return to the expected number E of visits to the starting site, including the initial visit to this spot. To establish this relation we observe that the probability of exactly k visits is given by $p^{k-1}(1-p)$: the walker has to return $k-1$ times in succession and then to hop away forever. Thus, the expected number of visits is simply [Fel50, Thm. 2, §12.3]

$$E = \sum_{k=1}^{\infty} k\, p^{k-1}(1-p) = \frac{1}{1-p}, \tag{6.6}$$

with the understanding that in the case of a *recurrent* walk ($p = 1$) we have $E = \infty$. Alternatively, we could derive (6.6) by arguing that after the second visit to the starting site the walker's future is, in probability, just the same as initially. That is, the expected number of visits satisfies the simple equation $E = 1 + pE$.

The point is that the quantity E can be expressed in various other ways that are computationally accessible. In fact, we will give several such expressions in the course of this chapter. A first useful expression is the series

$$E = \sum_{k=0}^{\infty} p_k,$$

where p_k denotes the probability of occupying the starting site at step $2k$. If the series diverges, the walk is recurrent ($E = \infty$) by [Fel50, Thm. 2, §12.3]. We will approximate E by the partial sums

$$E_K = \sum_{k=0}^{K-1} p_k$$

for K sufficiently large. As it turns out, there will be a trade-off between the complexity of the algorithm for calculating the terms $p_0, \dots, p_{K-1}$ and the mathematical sophistication of its derivation: the faster the algorithm, the more theory we have to invest.

6.3.1 Using Brute Force

Here we will start with a straightforward simple algorithm of complexity $O(K^3)$. Convergence acceleration will help us to keep K reasonably small. Later, in §6.4, we will increase the efficiency of calculating E_K in two steps. Some combinatorics reduces the complexity to $O(K^2)$. Zeilberger's creative telescoping algorithm yields a three-term recurrence for p_k that allows us to calculate $p_0, \dots, p_{K-1}$ with optimal complexity $O(K)$. Then, calculating partial sums for sufficiently large K is so affordable that acceleration of the convergence is no longer necessary. Finally, in §6.5, we will use special functions to evaluate the series symbolically.

We consider the probabilities $P_k(0,0)$ of visiting, at step k, the starting site $(0,0) \in \mathbb{Z}^2$. Then the expected number of visits to $(0,0)$ is [Fel50, Thm. 2, §12.3]

$$E = \sum_{k=0}^{\infty} P_k(0,0). \tag{6.7}$$

Now, we can calculate $P_k(0,0)$ directly from the rules of the random walk, if we know the probabilities of occupying the neighbors of the starting site at the step $k-1$. For those neighbors we do the same, recursively up to the initial step $k=0$ where all is known. Thus, if we introduce the lattice function $P_k : \mathbb{Z}^2 \to [0,1]$ that assigns to each lattice point the probability of being occupied at step k, we obtain the partial difference equation

$$\begin{aligned} P_k(x,y) = {} & p_{\mathrm{E}}\, P_{k-1}(x-1,y) + p_{\mathrm{W}}\, P_{k-1}(x+1,y) \\ & + p_{\mathrm{N}}\, P_{k-1}(x,y-1) + p_{\mathrm{S}}\, P_{k-1}(x,y+1), \end{aligned} \tag{6.8}$$

subject to the initial condition

$$P_0(0,0) = 1, \quad P_0(x,y) = 0 \quad \text{for } (x,y) \neq (0,0). \tag{6.9}$$

With a little thought we can arrange the difference equation in a form that avoids the handling and storage of zero probabilities. Because in step k the only sites that can be occupied are those no farther than k steps away from the starting site, we observe that

$$P_k(x,y) = 0 \quad \text{for} \quad |x| + |y| > k.$$

In fact, there are further lattice points that cannot be occupied at step k. Taking a checkerboard coloring of the lattice, the walker has to switch colors, that is, the parity of $x+y$, at each step:

$$P_{2k}(x,y) = 0 \quad \text{for } x+y \text{ odd}, \qquad P_{2k+1}(x,y) = 0 \quad \text{for } x+y \text{ even}.$$

Now, we arrange all the nontrivial probabilities at step k into the $(k+1) \times (k+1)$ matrix Π_k defined by

$$\Pi_k = \begin{pmatrix} P_k(-k,0) & P_k(-k+1,1) & \cdots & P_k(0,k) \\ P_k(-k+1,-1) & P_k(-k+2,0) & \cdots & P_k(1,k-1) \\ \vdots & \vdots & \ddots & \vdots \\ P_k(0,-k) & P_k(1,-k+1) & \cdots & P_k(k,0) \end{pmatrix}.$$

The partial difference equation (6.8) can be rewritten as the matrix recurrence

$$\Pi_{k+1} = p_{\mathrm{E}} \left(\begin{array}{c|ccc} 0 & 0 & \cdots & 0 \\ \hline 0 & & & \\ \vdots & & \Pi_k & \\ 0 & & & \end{array}\right) + p_{\mathrm{W}} \left(\begin{array}{ccc|c} & & & 0 \\ & \Pi_k & & \vdots \\ & & & 0 \\ \hline 0 & \cdots & 0 & 0 \end{array}\right)$$
$$+ p_{\mathrm{N}} \left(\begin{array}{c|ccc} 0 & & & \\ \vdots & & \Pi_k & \\ 0 & & & \\ \hline 0 & 0 & \cdots & 0 \end{array}\right) + p_{\mathrm{S}} \left(\begin{array}{ccc|c} 0 & \cdots & 0 & 0 \\ \hline & & & 0 \\ & \Pi_k & & \vdots \\ & & & 0 \end{array}\right), \tag{6.10}$$

subject to the initial condition $\Pi_0 = \big(1\big)$. The probabilities for the starting site are obtained as

$$P_{2k+1}(0,0) = 0, \quad P_{2k}(0,0) = \text{center entry of } \Pi_{2k}.$$

Since the starting site can be occupied only at every second step, we simplify our notation and use

$$p_k = P_{2k}(0,0), \quad E = \sum_{k=0}^{\infty} p_k. \tag{6.11}$$

The matrix recurrence (6.10) can easily be coded; the reader will find the quite compact MATLAB function `OccupationProbability` in Appendix C.3.1. Using this algorithm, the cost of calculating $p_0, \ldots, p_{K-1}$, and hence that of calculating the partial sum

$$E_K = \sum_{k=0}^{K-1} p_k,$$

grows as $O(K^3)$. Let us have faith and solve Problem 6. That is, with

$$p_{\mathrm{E}} = 1/4 + \epsilon, \quad p_{\mathrm{W}} = 1/4 - \epsilon, \quad p_{\mathrm{N}} = 1/4, \quad p_{\mathrm{S}} = 1/4,$$

we look for the bias ϵ_* that solves the equation $E|_{\epsilon=\epsilon_*} = 2$, which by (6.6) is equivalent to biasing for a fair return, $p|_{\epsilon=\epsilon_*} = 1/2$.

A MATLAB Session

```
>> f=inline('sum(OccupationProbability(epsilon,K))-2','epsilon','K');
>> for K=125*2.^(0:3)
>>    out=sprintf('K = %4i \t\t epsilon* = %17.15f',K,...
>>      fzero(f,[0.06,0.07],optimset('TolX',1e-14),K));
>>    disp(out);
>> end

K =  125       epsilon* = 0.061778241155115
K =  250       epsilon* = 0.061912524106289
K =  500       epsilon* = 0.061913954180807
K = 1000       epsilon* = 0.061913954473991
```

The run with $K = 125$ takes 9 seconds, the one with $K = 1000$, about 2.5 hours on a 2 GHz PC.

How do we assess the accuracy of these four approximations? We observe that the first two agree to 2 digits, the second and third to 4 digits, the third and fourth to 8 digits. It appears that doubling K doubles the number of correct digits; that is, the convergence of the series is roughly exponential.[41] Taking this for granted we would expect 16 correct digits at $K = 1000$. However, the absolute error of the root-finder was put to 10^{-14}, which restricts the accuracy of the fourth approximation to about 12 digits. Because the 11th and 12th digits read `39`, we have to be careful with them: they could be `40` as well. In any case, we have no reason to question the correctness of the first 10 digits,

$$\epsilon_* \doteq 0.06191395447.$$

In fact, we will see later that the $K = 1000$ run is correct to 13 digits.

6.3.2 Using Convergence Acceleration

The run time of 2.5 hours is a good reason for trying convergence acceleration. We use E_K and estimate the tail $E - E_K$ by Wynn's epsilon algorithm (see Appendix A, p. 245), which is particularly well suited for the near-exponential convergence of the series. A good choice of parameters requires some experimentation. A reasonable compromise between run time and accuracy turns out to be $K \approx 400$, for which E_K itself gives just 6 correct digits. To determine the number of extrapolation steps, and hence the number of extra terms, we follow the recommendations of Appendix A (see p. 253) and look, for the particular choice $\epsilon = 0.06$, at the differences of the first row of the extrapolation table; cf. Figure 6.2. We observe that round-off error becomes significant at about $j \geqslant 3$, where the magnitude of the differences settles at about 10^{-12}. This corresponds to 12 correct digits of E, a gain of 6 digits over just using E_K with $K = 400$. The reader will find the MATLAB function `ExpectedVisitsExtrapolated` in Appendix C.3.1. We fix $K = 393$ and $j = 3$, using 7 extra terms of the series for extrapolation.

[41] Indeed, we will prove this later; cf. Lemma 6.1.

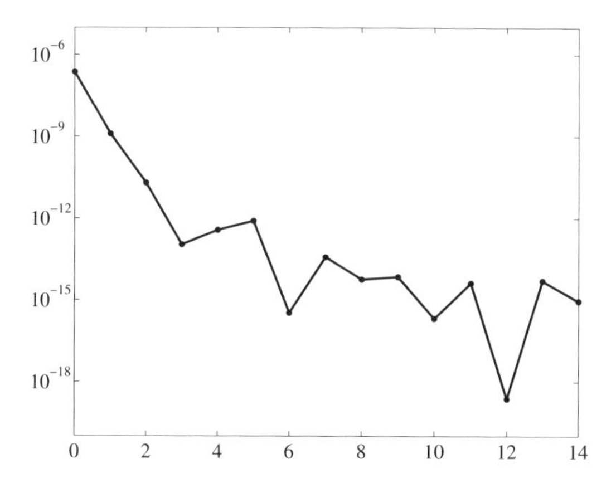

Figure 6.2. *Differences* $|s_{1,2j+2} - s_{1,2j}|$ *of the first row of the extrapolation table vs. j, using Wynn's epsilon algorithm* ($K = 393$, $\epsilon = 0.06$).

A MATLAB Session

```
>> f=inline('ExpectedVisitsExtrapolated(epsilon,K,extraTerms)-2',...
>>       'epsilon','K','extraTerms');
>> epsilon=fzero(f,[0.06,0.07],optimset('TolX',1e-14),393,7)

epsilon = 6.191395447397203e-002
```

The run takes less than eight minutes on a 2 GHz PC, a considerable speedup. The absolute tolerance of the root-finder was set to 10^{-14}, which matches the accuracy of the extrapolation given by our preparatory experiments. As in the $K = 1000$ run, there remains some uncertainty about the 11th and 12th digits. In any case, we have now collected enough evidence to be confident about the correctness of 10 digits.

6.4 Reducing the Complexity by Increasing the Level of Sophistication

In this section we show that simple combinatorics and symbolic calculations reduce the complexity of generating $p_0, \ldots, p_{K-1}$ from $O(K^3)$ to the optimal $O(K)$. This enables us to efficiently use the partial sum E_K to solve the problem.

6.4.1 Using Combinatorics

To begin with, we observe that calculating all the probabilities stored in Π_{2k} is overkill if we are only interested in $p_k = P_{2k}(0, 0)$. This is because a walker visiting $(0, 0)$ at step $2k$ has followed a very particular type of walk: a combination of two one-dimensional walks. In fact, he has spent $2j$ steps in the east–west direction and $2k - 2j$ steps in the north–south direction. The number of eastbound steps has to be equal to the number of westbound ones, and likewise for north–south. Simple combinatorics yields

$$p_k = \sum_{j=0}^{k} \binom{2k}{2j} \cdot \underbrace{\binom{2j}{j} p_E^j p_W^j}_{\text{east–west}} \cdot \underbrace{\binom{2k-2j}{k-j} p_N^{k-j} p_S^{k-j}}_{\text{north–south}}.$$

The binomial coefficients reflect the possibilities that the walker has $2j$ steps in the east–west direction out of $2k$ steps altogether, j eastbound steps out of $2j$ steps in the east–west direction, and $k-j$ northbound steps out of $2k-2j$ steps in the north–south direction.

The expression for p_k can be simplified further by a little fiddling with binomial coefficients,

$$p_k = \sum_{j=0}^{k} \binom{2k}{k} \binom{k}{j}^2 (p_E p_W)^j (p_N p_S)^{k-j},$$

a formula that can be found, though without much explanation, in the work of Barnett [Bar63, form. (7)]. We observe that p_k depends on p_E, p_W only via their product, and the same for p_N, p_S. Therefore, it makes sense to consider their geometric means

$$p_{EW} = \sqrt{p_E p_W}, \qquad p_{NS} = \sqrt{p_N p_S}.$$

We obtain[42]

$$p_k = \sum_{j=0}^{k} \binom{2k}{k} \binom{k}{j}^2 p_{EW}^{2j} p_{NS}^{2k-2j}. \tag{6.12}$$

Because of the size of the binomial coefficients, using this expression requires some care with machine arithmetic. For instance, at $k = 1000$ the term $\binom{2k}{k}$ would become as large as 10^{600}, which far exceeds the range of IEEE double-precision numbers. We have to balance this enormous size with the powers of p_{EW} and p_{NS}. For $p_{EW} \leqslant p_{NS}$,[43] which is true for Problem 6, we therefore rewrite (6.12) as

$$p_k = \underbrace{\binom{2k}{k} p_{NS}^{2k}}_{=a_k} \sum_{j=0}^{k} \underbrace{\binom{k}{j}^2 \left(\frac{p_{EW}}{p_{NS}}\right)^{2j}}_{=b_j}.$$

[42] If $p_{EW} = 0$, this expression simplifies to $p_k = \binom{2k}{k} p_{NS}^{2k}$, yielding the return probability

$$p = 1 - 1/E = 1 - \sqrt{1 - 4p_{NS}^2}.$$

Here, the random walk is effectively a biased one-dimensional one in the north–south direction; cf. [Hug95, p. 123]: the nonzero transition probability out of p_E or p_W just adds to the killing rate. In the setting of Problem 6 we get $p_{EW} = 0$ for the maximum bias $\epsilon = 1/4$. Hence, because of $p_{NS} = 1/4$, we obtain $p|_{\epsilon=1/4} = 1 - \sqrt{3}/2 \doteq 0.1339745962$, which is slightly larger than $1/8$; cf. Figure 6.1.

[43] Without loss of generality we can generally assume this to be the case. By the symmetry of the problem we would otherwise just change the roles of p_{NS} and p_{EW}.

To be efficient we calculate the coefficients a_k and b_j recursively:

$$a_k = \frac{2k(2k-1)}{k^2} p_{\mathrm{NS}}^2 \cdot a_{k-1}, \quad a_0 = 1;$$

$$b_j = \left(\frac{k-j+1}{j}\right)^2 \frac{p_{\mathrm{EW}}^2}{p_{\mathrm{NS}}^2} \cdot b_{j-1}, \quad b_0 = 1.$$

The resulting algorithm for calculating $p_0, \ldots, p_{K-1}$ is of complexity $O(K^2)$.

A *Mathematica* Session

```
ExpectedVisits[ε_Real, K_] :=
  (pE = 0.25 + ε; pW = 0.25 - ε; pN = pS = 0.25; w = pE pW; z = pN pS;
   (a = 1) +
    Sum[(a *= (2k (2k - 1))/k^2 z)
     ((b = 1) + Sum[b *= ((k - j + 1)/j)^2 w/z, {j, 1, k}]), {k, 1, K - 1}])
```

```
{K == #,
    ε* ==
     (ε/.FindRoot[ExpectedVisits[ε, #] == 2, {ε, 0.06, 0.07},
        WorkingPrecision → MachinePrecision,
        PrecisionGoal → 12])}&/@{125, 250, 500, 1000}//TableForm
```

```
K == 125    ε* == 0.06177824115511528
K == 250    ε* == 0.06191252410628889
K == 500    ε* == 0.06191395418080741
K == 1000   ε* == 0.06191395447399104
```

Mathematica's output confirms the result of the corresponding run using the $O(K^3)$ algorithm in §6.3.1. However, for $K = 1000$ we now need, using a 2 GHz PC, just 45 seconds instead of 2.5 hours. We recall that $K = 1000$ gives us at least 10 correct digits.

6.4.2 Using Symbolic Computation

For the initiated, the form of the expression (6.12) shows that there is a holonomic[44] M-term recurrence for p_k and that its polynomial coefficients can be calculated by Zeilberger's *creative telescoping algorithm.* This algorithm falls into the realm of symbolic hypergeometric summation [Koe98] and allows for the computerized proofs [PWZ96] of many if not most identities of sums involving binomial coefficients. For decades, the derivation and verification of such identities was considered to be a challenging problem. Indeed,

[44]A linear, homogeneous recurrence with polynomial coefficients is called *holonomic*.

a paper coauthored by Zeilberger and presenting his algorithm was entitled "How to Do *Monthly* Problems with Your Computer" [NPWZ97].

Let us briefly explain the framework of Zeilberger's algorithm. It applies to sums of the form

$$f_k = \sum_{j \in \mathbb{Z}} F(k, j),$$

where $F(k, j)$ is a *proper hypergeometric term* (cf. [PWZ96, p. 64] or [Koe98, p. 110]), which is essentially a term of the form

$$\begin{aligned} F(k, j) = {} & (\text{polynomial in } k, j) \\ & \times (\text{product of binomial coefficients of arguments integer-linearly in } k, j) \\ & \times w^k z^j . \end{aligned} \tag{6.13}$$

Then a theorem of Zeilberger [PWZ96, Thm. 6.2.1] states that F satisfies a recurrence of the form

$$\sum_{m=0}^{M} a_m(k) F(k+m, j) = G(k, j+1) - G(k, j),$$

where $a_0, \ldots, a_M$ are polynomials and $G(k, j)/F(k, j)$ is rational in k, j. Moreover, there is an algorithm (cf. [PWZ96, §6.3] or [Koe98, Chap. 7]) to construct all these functions. Now, we can sum over all integer values of j and obtain by telescoping the recurrence

$$\sum_{m=0}^{M} a_m(k) f_{k+m} = 0.$$

In the problem at hand, since $\binom{k}{j} = 0$ for $j > k$ or $j < 0$, the probability (6.12) can be written as a sum over the integers $j \in \mathbb{Z}$ of terms that have the required proper hypergeometric form (6.13). Thus, we should be able to calculate such an M-term recurrence for p_k. In fact, Maple ships with the package `sumtools`, which includes the command `sumrecursion` that performs exactly this task. Details and source code of a similar command can be found in the book by the package's author Koepf [Koe98, p. 100].

A Maple Session

We use the abbreviations $w = p_{\text{EW}}^2$, $z = p_{\text{NS}}^2$.

```
> with(sumtools):
> sumrecursion(binomial(2*k,k)*binomial(k,j)^2*w^j*z^(k-j),j,p(k));
```

$$4\,(w - z)^2\,(2\,k - 1)\,(2\,k - 3)\,p\,(k - 2) - 2\,(2\,k - 1)^2\,(w + z)\,p\,(k - 1) + p\,(k)\,k^2$$

Written in full, the Maple output shows that p_k satisfies the three-term recurrence

$$k^2 p_k = 2(2k-1)^2\,(p_{\text{EW}}^2 + p_{\text{NS}}^2)\,p_{k-1} - 4(2k-1)(2k-3)\,(p_{\text{EW}}^2 - p_{\text{NS}}^2)^2\,p_{k-2}, \tag{6.14}$$

subject to the initial conditions

$$p_0 = 1, \quad p_1 = 2(p_{\mathrm{EW}}^2 + p_{\mathrm{NS}}^2).$$

Hence, we have obtained an algorithm that allows us to calculate $p_0, \ldots, p_{K-1}$ in optimal complexity $O(K)$.

Now, we are ready to solve the equation $E|_{\epsilon=\epsilon_*} = 2$ for ϵ_* using a software package root-finder. From §6.3.1 we know that the convergence of the series for E is essentially exponential. Therefore, it is reasonable to sum the terms of the series until the value of the sum no longer changes in finite-precision arithmetic.

A *Mathematica* Session

```
ExpectedVisits[ϵ_Real] :=
 (pE = 1/4 + ϵ; pW = 1/4 - ϵ; pN = pS = 1/4; w = pE pW; z = pN pS;
  s = pold = 1; pnew = 2 (w + z); k = 2;
  While[s =! = (s+ = pnew),
    pnew = (-4 (2 k - 1) (2 k - 3) (w - z)^2 pold + 2 (2 k - 1)^2 (w + z) (pold = pnew)) / k++^2];
  s);
```

```
{Precision == #,
    ϵ* ==
     (ϵ/.FindRoot[ExpectedVisits[ϵ] == 2, {ϵ, 0.06, 0.07},
        WorkingPrecision → #, AccuracyGoal → #])}&/@
  {13, MachinePrecision, 19, 22, 25}//TableForm
```

```
Precision == 13                  ϵ* == 0.06191395447253
Precision == MachinePrecision    ϵ* == 0.0619139544739901
Precision == 19                  ϵ* == 0.06191395447399094135
Precision == 22                  ϵ* == 0.06191395447399094284664
Precision == 25                  ϵ* == 0.06191395447399094284817374
```

We observe that we lose about the last three digits. In machine precision the run takes under a second on a 2 GHz PC and uses 884 terms of the series. With a working precision of 103 digits, and using 7184 terms of the series, it takes less than a minute to get

$$\begin{aligned}\epsilon_* \doteq 0.&06191395447\,39909\,42848\,17521\,64732\,12176\,99963\,87749\,98362\\ &07606\,14672\,58859\,93101\,02975\,96158\,45907\,10564\,57520\,87861,\end{aligned}$$

which is correct to the printed 100 digits.

6.5 The Joy of Special Functions

In the last section we learned that the occupation probabilities p_k are homogeneous polynomials of degree k in p_{EW}^2 and p_{NS}^2 satisfying a three-term recurrence. The odds are good that they are related to one of the canonized families of orthogonal polynomials. Let us see what happens if we ask Maple to evaluate (6.12) symbolically.

A Maple Session
We use the abbreviations $w = p_{\mathrm{EW}}^2$, $z = p_{\mathrm{NS}}^2$.

```
> simplify(sum(binomial(2*k,k)*binomial(k,j)^2*z^j*w^(k-j),j=0..k))
    assuming k::integer;
```

$$\binom{2k}{k} LegendreP\left(k, \frac{w+z}{w-z}\right)(w-z)^k$$

We encounter the *Legendre polynomials* P_k. Written in full, the Maple output reads as

$$p_k = \binom{2k}{k} P_k \left(\frac{p_{\mathrm{EW}}^2 + p_{\mathrm{NS}}^2}{p_{\mathrm{EW}}^2 - p_{\mathrm{NS}}^2}\right)(p_{\mathrm{EW}}^2 - p_{\mathrm{NS}}^2)^k.$$

Certainly, using this expression instead of (6.14) is, computationally, no improvement. We already had arrived at the optimal complexity of $O(K)$ for calculating the probabilities $p_0, \ldots, p_{K-1}$.

However, the Legendre polynomials are useful in deriving precise asymptotic formulas for p_k and, further below, in deriving a closed expression for the expected value E. We recall our observation that doubling K doubles the number of digits to which E_K correctly approximates E; the convergence appears to be roughly exponential. In fact, we are now able to prove this to be true.

Lemma 6.1

(a) *If* $\sigma = 2(p_{\mathrm{EW}} + p_{\mathrm{NS}})$, *then* $\sigma \leqslant 1$, *with equality if and only if*

$$p_{\mathrm{E}} = p_{\mathrm{W}}, \qquad p_{\mathrm{N}} = p_{\mathrm{S}}, \qquad p_{\mathrm{E}} + p_{\mathrm{W}} + p_{\mathrm{N}} + p_{\mathrm{S}} = 1.$$

(b) *If* $p_{\mathrm{EW}} \cdot p_{\mathrm{NS}} \neq 0$, *there is the asymptotic formula*

$$p_k \simeq \frac{\sigma^{2k+1}}{4\pi k \sqrt{p_{\mathrm{EW}}\, p_{\mathrm{NS}}}}; \quad \textit{otherwise} \quad p_k \simeq \frac{\sigma^{2k}}{\sqrt{\pi k}}.$$

(c) *The random walk is recurrent, that is,* $p = 1$, *if and only if* $\sigma = 1$.

Proof. (a) Using the inequality between geometric and arithmetic means we obtain

$$\sigma = 2p_{\mathrm{EW}} + 2p_{\mathrm{NS}} \leqslant (p_{\mathrm{E}} + p_{\mathrm{W}}) + (p_{\mathrm{N}} + p_{\mathrm{S}}) \leqslant 1.$$

Since equality of the arithmetic and geometric means of two quantities holds only if these quantities are equal, $\sigma = 1$ is characterized as asserted.

(b) Suppose that $p_{\mathrm{EW}} \cdot p_{\mathrm{NS}} \neq 0$. The Laplace–Heine asymptotic formula for the Legendre polynomials [Sze75, Thm. 8.21.1] yields after a short calculation

$$P_k\left(\frac{p_{\mathrm{EW}}^2 + p_{\mathrm{NS}}^2}{p_{\mathrm{EW}}^2 - p_{\mathrm{NS}}^2}\right)(p_{\mathrm{EW}}^2 - p_{\mathrm{NS}}^2)^k \simeq \frac{(p_{\mathrm{EW}} + p_{\mathrm{NS}})^{2k+1}}{2\sqrt{\pi k}\,\sqrt{p_{\mathrm{EW}}\,p_{\mathrm{NS}}}}.$$

A multiplication with Stirling's formula $\binom{2k}{k} \simeq 4^k/\sqrt{\pi k}$ proves the assertion.

On the other hand, if $p_{\mathrm{EW}} \cdot p_{\mathrm{NS}} = 0$, we may assume without loss of generality that $p_{\mathrm{EW}} = 0$. By (6.12) we obtain, using Stirling's formula again,

$$p_k = \binom{2k}{k} p_{\mathrm{NS}}^{2k} \simeq \frac{4^k p_{\mathrm{NS}}^{2k}}{\sqrt{\pi k}} = \frac{\sigma^{2k}}{\sqrt{\pi k}}.$$

(c) By d'Alembert's ratio test for the convergence of series with positive terms, the asymptotic formulas in (b) prove that $E = \sum_{k=0}^{\infty} p_k < \infty$ if $\sigma < 1$. On the other hand, if $\sigma = 1$, by the comparison test [Kno56, p. 56] the series $\sum_{k=0}^{\infty} p_k$ inherits the divergence of $\sum_{k=1}^{\infty} 1/k$ or $\sum_{k=1}^{\infty} 1/\sqrt{k}$, respectively.

Summarizing, the series converges—that is, the walk is transient—if and only if $\sigma < 1$. Therefore, by (a), recurrence is equivalent to $\sigma = 1$. □

For the particular biasing of Problem 6 we have $\sigma = (1 + \sqrt{1 - 16\epsilon^2})/2$. Here, the walk is recurrent if and only if it is unbiased ($\epsilon = 0$). As promised at the beginning of this chapter, we have thus proven the recurrence of the unbiased random walk in two dimensions, a classic result of Pólya [Pól21].

The Legendre polynomials are also useful if we aim directly at the expected value

$$E = \sum_{k=0}^{\infty} p_k = \sum_{k=0}^{\infty} \binom{2k}{k} P_k\left(\frac{p_{\mathrm{EW}}^2 + p_{\mathrm{NS}}^2}{p_{\mathrm{EW}}^2 - p_{\mathrm{NS}}^2}\right)(p_{\mathrm{EW}}^2 - p_{\mathrm{NS}}^2)^k.$$

The point is that there is some hope of finding a closed expression for E in terms of special functions, which might be computationally advantageous. For instance, in many compilations of formulas, such as [EMOT53, Vol. 2, §10.10] or [AS84, Table 22.9], one finds the generating functions

$$\sum_{k=0}^{\infty} P_k(x)\,z^k = \frac{1}{\sqrt{1 - 2xz + z^2}}, \qquad \sum_{k=0}^{\infty} \frac{1}{k!} P_k(x)\,z^k = e^{xz} J_0(z\sqrt{1 - x^2}) \tag{6.15}$$

with J_0 being the Bessel function of the first kind of zero order. A systematic way to derive such expressions was communicated to us by Herbert Wilf: one simply plugs Laplace's first integral for the Legendre polynomials (cf. [WW96, §15.23] or [Rai60, Chap. 10, §97])

$$P_k(x) = \frac{1}{\pi}\int_0^{\pi} (x + \sqrt{x^2 - 1}\cos\theta)^k\,d\theta, \quad x \in \mathbb{C},$$

into the series at hand and changes the order of summation and integration. If the power series $\sum_{k=0}^{\infty} a_k z^k$ is a known function $f(z)$, one gets

$$\begin{aligned}\sum_{k=0}^{\infty} a_k P_k(x) z^k &= \frac{1}{\pi}\int_0^{\pi} \sum_{k=0}^{\infty} a_k (x+\sqrt{x^2-1}\cos\theta)^k z^k \, d\theta \\ &= \frac{1}{\pi}\int_0^{\pi} f\left((x+\sqrt{x^2-1}\cos\theta)z\right) d\theta.\end{aligned}$$

The integral form might be advantageous because there are more methods known for the closed evaluation of integrals than of sums. We leave the actual calculations to a computer algebra system such as *Mathematica* and just assist the symbolic integration by restricting the possible values of p_{EW} and p_{NS}:

$$0 \leqslant p_{\mathrm{EW}} \leqslant p_{\mathrm{NS}}, \quad \sigma = 2(p_{\mathrm{EW}} + p_{\mathrm{NS}}) < 1.$$

Because of the symmetry of the problem, the first inequality is no loss of generality. By Lemma 6.1 we know that the second inequality is equivalent to $E < \infty$.

A *Mathematica* Session

```
LegendreKernel[x_, k_] := 1/π (x + Sqrt[x^2 - 1] Cos[θ])^k ;
```

```
Integrate[
  Sum[Binomial[2k, k] LegendreKernel[(p_EW^2 + p_NS^2)/(p_EW^2 - p_NS^2), k] (p_EW^2 - p_NS^2)^k,
   {k, 0, ∞}], {θ, 0, π}, Assumptions → {0 ≤ p_EW ≤ p_NS, 2 (p_EW + p_NS) < 1}]
```

```
2 EllipticK[ - (16 p_EW p_NS)/(-1+4 (p_EW-p_NS)^2) ]
------------------------------------------------
        π Sqrt[1 - 4 (p_EW - p_NS)^2]
```

The reader must be alert to the fact that *Mathematica* uses the complete elliptic integral of the first kind,

$$K(k) = \int_0^1 \frac{dt}{\sqrt{1-t^2}\,\sqrt{1-k^2t^2}},$$

not as a function of the modulus k, but of the parameter $m = k^2$ instead. Summarizing, we have obtained the closed-form expression

$$E = \frac{2}{\pi\sqrt{1-4(p_{\mathrm{EW}}-p_{\mathrm{NS}})^2}}\, K\left(\frac{4\sqrt{p_{\mathrm{EW}}p_{\mathrm{NS}}}}{\sqrt{1-4(p_{\mathrm{EW}}-p_{\mathrm{NS}})^2}}\right), \tag{6.16}$$

a result that was discovered in the early 1960s independently by Henze [Hen61, form. (3.3)] and Barnett [Bar63, form. (43)].[45] The advantage of K being involved is that it can be evaluated [BB87, Alg. 1.2(a)] exceedingly fast with the help of the *arithmetic-geometric mean* $M(a, b)$ of Gauss,

$$K\left(\sqrt{1-k^2}\right) = \frac{\pi}{2\,M(1, k)}. \tag{6.17}$$

The reader will find more details on M in §6.5.1. From (6.17) we infer the final formula of this section,

$$E = 1/M\left(\sqrt{1-4(p_{\text{EW}}+p_{\text{NS}})^2}, \sqrt{1-4(p_{\text{EW}}-p_{\text{NS}})^2}\right), \tag{6.18}$$

which we now use to solve Problem 6.

A *Mathematica* Session

```
ExpectedVisits[ϵ_] :=
  1/ArithmeticGeometricMean[√(1 - 4 (p_EW + p_NS)^2), √(1 - 4 (p_EW - p_NS)^2)] /.
   {p_EW → √((1/4 + ϵ) (1/4 - ϵ)), p_NS → 1/4}

{Precision == #,
    ϵ_* ==
     (ϵ /. FindRoot[ExpectedVisits[ϵ] == 2, {ϵ, 0.06, 0.07},
         WorkingPrecision → #, AccuracyGoal → #])}&/@
  {13, MachinePrecision, 19, 22, 25}//TableForm
```

```
Precision == 13                  ϵ_* == 0.06191395447402
Precision == MachinePrecision    ϵ_* == 0.06191395447399095
Precision == 19                  ϵ_* == 0.06191395447399094287
Precision == 22                  ϵ_* == 0.06191395447399094284820
Precision == 25                  ϵ_* == 0.06191395447399094284817519
```

All but about the last two digits are correct. The code is so fast that using it with a working precision of 10,010 digits takes less than a second (on a 2 GHz PC) to get 10,000 correct digits.

---

[45] One motivation of Barnett was to characterize recurrent biased random walks, that is, $E = \infty$. From (6.16) he could infer that $E = \infty$ if and only if the argument of K equals 1,

$$\frac{4\sqrt{p_{\text{EW}}\,p_{\text{NS}}}}{\sqrt{1-4(p_{\text{EW}}-p_{\text{NS}})^2}} = \frac{4\sqrt{p_{\text{EW}}\,p_{\text{NS}}}}{\sqrt{1-\sigma^2+16p_{\text{EW}}\,p_{\text{NS}}}} = 1.$$

This is the case if and only if $\sigma = 1$, which is consistent with Lemma 6.1.

6.5.1 Using Interval Arithmetic

The arithmetic-geometric mean $M(a, b)$ of nonnegative numbers a and b is defined as the common limit of the two quadratically convergent sequences that are built by taking, successively, the geometric and arithmetic means,

$$M(a, b) = \lim_{n\to\infty} a_n = \lim_{n\to\infty} b_n; \quad a_0 = a, \ b_0 = b, \ a_{n+1} = \sqrt{a_n b_n}, \ b_{n+1} = \frac{a_n + b_n}{2}.$$

It enjoys many monotonicity properties that make it particularly well suited for interval calculations. First, it is easy to check for $0 \leqslant a_* \leqslant a \leqslant a^*, 0 \leqslant b_* \leqslant b \leqslant b^*$ that

$$M(a_*, b_*) \leqslant M(a, b) \leqslant M(a^*, b^*).$$

It follows that

$$M([a, b], [c, d]) = [M(a, c), M(b, d)], \qquad 0 \leqslant a \leqslant b, \ 0 \leqslant c \leqslant d.$$

Next, the above iteration is always enclosing [BB87, §1.1],

$$a_n \leqslant a_{n+1} \leqslant M(a, b) \leqslant b_{n+1} \leqslant b_n, \qquad n \geqslant 1.$$

Thus, using downward rounding for a_n and upward rounding for b_n yields a straightforward, tight, and rigorous implementation using an interval arithmetic package such as INTLAB. The reader will find the INTLAB function `AGM` in Appendix C.4.3.

For Problem 6, appropriate interval root-finding can be based on the interval-bisection method. The INTLAB procedure `IntervalBisection` of Appendix C.4.3 is coded essentially in analogy to the interval-minimization algorithm of §4.3.

A MATLAB/INTLAB Session

```
>> E=inline('1/AGM(sqrt(1-4*(pEW+pNS)^2),sqrt(1-4*(pEW-pNS)^2))',...
>>      'pEW','pNS');
>> f=inline('E(sqrt((0.25+epsilon)*(0.25-epsilon)),0.25)-2',...
>>      'epsilon','E');
>> epsilon=IntervalBisection(f,infsup(0.06,0.07),1e-15,E)

intval epsilon = [6.191395447399049e-002, 6.191395447399144e-002]
```

Just using IEEE machine arithmetic, the run takes less than four seconds on a 2 GHz PC and validates the correctness of 13 digits,

$$\epsilon_* \doteq 0.06191395447399.$$

For higher precision we use *Mathematica*. Here, no directive for directional rounding is accessible to the user, and a rigorous implementation of the arithmetic-geometric mean has to use interval arithmetic for the iteration itself. In machine arithmetic the resulting code is slightly less efficient and less tight than the INTLAB implementation. The reader will find an overload[46] of the command `ArithmeticGeometricMean` in Appendix C.5.3, and the code `IntervalBisection` is given in Appendix C.5.3.

[46] This way we can reuse the function `ExpectedVisits` from p. 137 for interval arguments.

As discussed in §4.5, there is no need to use the interval method from scratch. Instead we start by using our favorite method to obtain an approximation of ϵ_* that is very likely correct to the requested number of digits. After inflation of this approximation to a small interval, we apply the interval-bisection method and obtain a validated enclosure.

A *Mathematica* Session *(cont. of session on p. 137)*[47]

```
prec = 10010;
ϵTry = ϵ/.FindRoot[ExpectedVisits[ϵ] - 2, {ϵ, 0.06, 0.07},
    WorkingPrecision → prec, AccuracyGoal → prec];
ϵInterval = IntervalBisection[ExpectedVisits[#] - 2&, ϵTry + {-1, 1}10^(-prec+5),
   10^-prec];
DigitsAgreeCount[ϵInterval]
10006
```

This technique of approximation first, validation last shows the correctness of 10,006 digits in less than a minute of computing time on a 2 GHz PC.

6.6 Using Fourier Analysis

In this final section we follow a different approach to calculating E. Instead of looking at the expected number of visits to just the starting site, we consider it for *all* sites at once. That is, generalizing (6.7) we introduce the expected number $E(x, y)$ of visits to the site (x, y),

$$E(x, y) = \sum_{k=0}^{\infty} P_k(x, y), \qquad E = E(0, 0).$$

Now, summing the partial difference equation (6.8) over all k yields a partial difference equation for $E(\cdot, \cdot)$,

$$E(x, y) = P_0(x, y) + p_{\mathrm{E}}\, E(x-1, y) + p_{\mathrm{W}}\, E(x+1, y) \\ + p_{\mathrm{N}}\, E(x, y-1) + p_{\mathrm{S}}\, E(x, y+1).$$

Because of the initial condition (6.9), written with the Kronecker δ symbol as $P_0(x, y) = \delta_{x,0}\delta_{y,0}$, we obtain without any further reference to probabilities

$$E(x, y) = \delta_{x,0}\delta_{y,0} + p_{\mathrm{E}}\, E(x-1, y) + p_{\mathrm{W}}\, E(x+1, y) \\ + p_{\mathrm{N}}\, E(x, y-1) + p_{\mathrm{S}}\, E(x, y+1). \quad (6.19)$$

In analogy with partial differential equations, the solution of this difference equation is called the *lattice Green function* of the random walk [Hug95, p. 132].

Concerning the existence and construction of the lattice Green function one encounters some technical difficulties of convergence for an immortal walker, that is,

[47] The command `DigitsAgreeCount` can be found in Appendix C.5.1.

if $p_{\text{kill}} = 1 - p_{\text{E}} - p_{\text{W}} - p_{\text{N}} - p_{\text{S}} = 0$. We will avoid them by temporarily assuming $p_{\text{kill}} > 0$. As soon as it is safe to take the limit $p_{\text{kill}} \to 0$, we will drop this assumption. By continuity, formulas will then be correct for $p_{\text{kill}} = 0$ also.

A convenient way to solve the linear partial difference equation (6.19) is to use Fourier series,

$$\hat{E}(\phi, \theta) = \sum_{(x,y)\in\mathbb{Z}^2} E(x, y)e^{ix\phi}e^{iy\theta}. \tag{6.20}$$

Multiplying (6.19) by $e^{ix\phi}e^{iy\theta}$ and summing over all integer values of x and y yields

$$\hat{E}(\phi, \theta) = 1 + \underbrace{\left(p_{\text{E}}e^{i\phi} + p_{\text{W}}e^{-i\phi} + p_{\text{N}}e^{i\theta} + p_{\text{S}}e^{-i\theta}\right)}_{=\lambda(\phi,\theta)} \hat{E}(\phi, \theta),$$

where $\lambda(\phi, \theta)$ is called the *structure function* or *symbol* of the random walk [Hug95, form. (3.116)]. This equation for $\hat{E}$ is readily solved by

$$\hat{E}(\phi, \theta) = \frac{1}{1 - \lambda(\phi, \theta)} = \frac{1}{1 - p_{\text{E}}e^{i\phi} - p_{\text{W}}e^{-i\phi} - p_{\text{N}}e^{i\theta} - p_{\text{S}}e^{-i\theta}}.$$

Up to here, our arguments were purely formal. Now, by our temporary assumption $p_{\text{kill}} > 0$ we see that

$$\sum_{(x,y)\in\mathbb{Z}^2} |E(x, y)| = \sum_{(x,y)\in\mathbb{Z}^2} E(x, y) = \hat{E}(0, 0) = p_{\text{kill}}^{-1} < \infty.$$

Therefore, the Fourier series (6.20) converges absolutely and the coefficients $E(x, y)$ are established as the solution of the difference equation.

The value of interest, $E = E(0, 0)$, is obtained by an inverse Fourier transformation,

$$E = \frac{1}{4\pi^2} \int_{-\pi}^{\pi} \int_{-\pi}^{\pi} \frac{d\phi \, d\theta}{1 - p_{\text{E}}e^{i\phi} - p_{\text{W}}e^{-i\phi} - p_{\text{N}}e^{i\theta} - p_{\text{S}}e^{-i\theta}}. \tag{6.21}$$

Numerical quadrature for this double integral becomes less involved if we avoid the use of complex numbers. To this end we use a trick that is provided by the following lemma.

Lemma 6.2 *Let $f(z)$ be a function that is analytic in the disk $|z| < R$. For $a, b > 0$ with $a + b < R$, the integral*

$$I(a, b) = \frac{1}{2\pi} \int_{-\pi}^{\pi} f(ae^{ix} + be^{-ix}) \, dx$$

is well defined. Moreover, the following transformation is valid:

$$I(a, b) = I(\sqrt{ab}, \sqrt{ab}) = \frac{1}{2\pi} \int_{-\pi}^{\pi} f(2\sqrt{ab} \cos x) \, dx.$$

Proof. Note that $|ae^{ix} + be^{-ix}| \leqslant a + b < R$ for $x \in [-\pi, \pi]$ and that $2\sqrt{ab} \leqslant a + b < R$. Let C be the positively oriented unit circle. We have

$$I(a, b) = \frac{1}{2\pi i} \int_C f(az + b/z) \, \frac{dz}{z}.$$

The substitution $z = \sqrt{b/a} \cdot w$ and Cauchy's integral theorem show that

$$I(a, b) = \frac{1}{2\pi i} \int_{\sqrt{a/b}\, C} f(\sqrt{ab}\, w + \sqrt{ab}/w)\, \frac{dw}{w}$$
$$= \frac{1}{2\pi i} \int_{C} f(\sqrt{ab}\, w + \sqrt{ab}/w)\, \frac{dw}{w} = I(\sqrt{ab}, \sqrt{ab}).$$

The observation $\sqrt{ab}\, e^{ix} + \sqrt{ab}\, e^{-ix} = 2\sqrt{ab} \cos x$ finishes the proof. □

If we apply this lemma to (6.21) twice, which is possible by our temporary assumption $p_{\text{kill}} > 0$, we obtain by using the symmetry of the cosine

$$E = \frac{1}{4\pi^2} \int_{-\pi}^{\pi} \int_{-\pi}^{\pi} \frac{d\phi\, d\theta}{1 - 2p_{\text{EW}} \cos\phi - 2p_{\text{NS}} \cos\theta}$$
$$= \frac{1}{\pi^2} \int_{0}^{\pi} \int_{0}^{\pi} \frac{d\phi\, d\theta}{1 - 2p_{\text{EW}} \cos\phi - 2p_{\text{NS}} \cos\theta}. \tag{6.22}$$

Now, since

$$|2p_{\text{EW}} \cos\phi + 2p_{\text{NS}} \cos\theta| \leqslant 2p_{\text{EW}} + 2p_{\text{NS}} = \sigma$$

the integral expression (6.22) is well defined as long as $\sigma < 1$. Therefore, from now on we may drop the assumption $p_{\text{kill}} > 0$ and replace it by the transience condition $\sigma < 1$; cf. Lemma 6.1. As we have already noted following that lemma, the specific transition probabilities of Problem 6 yield that $\sigma < 1$ is equivalent to the condition $0 < \epsilon \leqslant 1/4$ on the bias.

Using adaptive numerical quadrature, expression (6.22) can readily be used to solve Problem 6, that is, to solve the equation $E|_{\epsilon=\epsilon_*} = 2$ for ϵ_*.

A MATLAB Session

```
>> g=inline(...
     '1./(1-2*sqrt((1/4+ep)*(1/4-ep))*cos(phi)-cos(theta)/2)/pi^2',...
     'phi','theta','ep');
>> f=inline('dblquad(g,0,pi,0,pi,1e-13,@quadl,ep)-2','ep','g');
>> epsilon=fzero(f,[0.06,0.07],optimset('TolX',1e-16),g)

epsilon = 6.191395447399090e-002
```

The parameters are chosen so that the code should deliver at least 13 correct digits. In fact, 15 digits are correct. The run time on a 2 GHz PC is eight minutes.

On comparing the two expressions (6.16) and (6.22) for E, we have obtained

$$\frac{1}{\pi^2} \int_{0}^{\pi} \int_{0}^{\pi} \frac{d\phi\, d\theta}{1 - a\cos\phi - b\cos\theta} = \frac{2}{\pi\sqrt{1-(a-b)^2}}\, K\left(\frac{2\sqrt{ab}}{\sqrt{1-(a-b)^2}}\right), \tag{6.23}$$

assuming $a \geqslant 0$, $b \geqslant 0$, and $a + b < 1$. Given all the success of symbolic computation in this chapter so far, it is amusing to note that neither *Mathematica* nor Maple is currently able to deliver this formula. We therefore cannot refrain from showing the reader how to get it directly by hand. From the straightforward fact [PBM86, form. (1.5.9.15)]

$$\frac{1}{\pi}\int_0^\pi \frac{d\theta}{c+d\cos\theta} = \frac{1}{\sqrt{c^2-d^2}}, \qquad c > |d|,$$

which is certainly no problem for any of the computer algebra systems, we infer that

$$\frac{1}{\pi^2}\int_0^\pi\int_0^\pi \frac{d\phi\, d\theta}{1-a\cos\phi-b\cos\theta} = \frac{1}{\pi}\int_0^\pi \frac{d\phi}{\sqrt{(1-a\cos\phi)^2-b^2}}$$

$$= \frac{1}{\pi}\int_{-1}^1 \frac{ds}{\sqrt{(1-s^2)((1-as)^2-b^2)}} = \int_{-1}^1 \frac{ds}{\sqrt{\text{quartic polynomial in } s}}.$$

We recognize the last expression as an elliptic integral of the first kind. It can be brought to canonical form by the substitution

$$t = \sqrt{\frac{(1+b-a)(1+s)}{2(1+b-a\,s)}}$$

(cf. [BF71, form. (252.00)]), which finally yields the desired result

$$\frac{1}{\pi}\int_{-1}^1 \frac{ds}{\sqrt{(1-s^2)((1-as)^2-b^2)}}$$

$$= \frac{2}{\pi\sqrt{1-(a-b)^2}}\int_0^1 \frac{dt}{\sqrt{1-t^2}\sqrt{1-\frac{4ab}{1-(a-b)^2}\,t^2}}$$

$$= \frac{2}{\pi\sqrt{1-(a-b)^2}}\,K\!\left(\frac{2\sqrt{ab}}{\sqrt{1-(a-b)^2}}\right).$$

6.7 Harder Problems

For two-dimensional random walks on the square lattice, E is expressible in terms of the arithmetic-geometric mean (6.18), and so there is no conceivable question about return probabilities that would increase the computational difficulty beyond that of Problem 6. This changes if we consider other lattices, say, triangular two-dimensional, cubic three-dimensional, or hypercubic d-dimensional ones.

Cubic Lattices. For cubic lattices the techniques of §§6.2–6.4 and 6.6 are still applicable, though the computational effort will increase considerably. However, we leave it as a challenge to the reader to obtain for the general biased random walk on cubic lattices nice formulas using special functions such as those in §6.5—if there are any at all.

To highlight the difficulty that will be encountered, let us review what is known for the random walk on the simple cubic lattice. In the unbiased case, that is, if all the next-neighbor transition probabilities are equal to $1/6$, one immediately generalizes (6.22) to the triple integral

$$E = \frac{1}{\pi^3}\int_0^\pi\int_0^\pi\int_0^\pi \frac{d\phi_1\, d\phi_2\, d\phi_3}{1-(\cos\phi_1+\cos\phi_2+\cos\phi_3)/3}.$$

Watson proved in 1939, in a famous *tour de force* [Wat39], the result[48]

$$E = \frac{12}{\pi^2}(18 + 12\sqrt{2} - 10\sqrt{3} - 7\sqrt{6})\, K^2(k_6), \qquad k_6 = (2-\sqrt{3})(\sqrt{3}-\sqrt{2}).$$

In 1977 Glasser and Zucker [GZ77] gave the formula[49]

$$E = \frac{\sqrt{6}}{32\pi^3}\,\Gamma\left(\frac{1}{24}\right)\Gamma\left(\frac{5}{24}\right)\Gamma\left(\frac{7}{24}\right)\Gamma\left(\frac{11}{24}\right),$$

which by [BZ92, Table 3(vii)] can be simplified to

$$E = \frac{\sqrt{3}-1}{32\pi^3}\,\Gamma^2\left(\frac{1}{24}\right)\Gamma^2\left(\frac{11}{24}\right).$$

The two formulas involving the gamma function may look more pleasing and elementary than Watson's. However, from a computational perspective the latter is still the best. For instance, to get 1000 digits the formulas involving the gamma function require about 16 seconds of CPU time, while Watson's does it in under a tenth of a second; this is because of the cost of evaluating the gamma function to high precision (see §5.7). In fact, Borwein and Zucker suggested in 1992 to use in turn such relations between the gamma function and the complete elliptic integral of the first kind for a fast evaluation of the gamma function at particular rational arguments; cf. [BZ92].

In the general biased case, let us—in addition to the two-dimensional transition probabilities p_E, p_W, p_N, and p_S—denote the probabilities for a step upwards or downwards in the third dimension by p_U and p_D, respectively. Analogously to p_{EW} and p_{NS} we use the abbreviation $p_{UD} = \sqrt{p_U p_D}$. Now, formula (6.22) is readily generalized to

$$E = \frac{1}{\pi^3}\int_0^\pi\int_0^\pi\int_0^\pi \frac{d\phi_1\, d\phi_2\, d\phi_3}{1 - 2p_{EW}\cos\phi_1 - 2p_{NS}\cos\phi_2 - 2p_{UD}\cos\phi_3}.$$

Using (6.23) we obtain[50]

$$E = \frac{2}{\pi^2}\int_0^\pi \frac{K(k)\, d\phi}{\sqrt{(1-2p_{UD}\cos\phi)^2 - 4(p_{EW}-p_{NS})^2}},$$

$$k = \frac{4\sqrt{p_{EW}\, p_{NS}}}{\sqrt{(1-2p_{UD}\cos\phi)^2 - 4(p_{EW}-p_{NS})^2}},$$

a formula that can be straightforwardly evaluated numerically. Evaluating it symbolically is, however, still an open problem. Recently, the particular case $p_{EW} = p_{NS}$ was solved

[48] In particular we observe that $E < \infty$. Therefore, unlike in one or two dimensions, the unbiased random walk in three dimensions is transient ($p < 1$), a classic fact already known to Pólya in 1921 [Pól21]. Watson evaluated his expression as $E/3 \doteq 0.5054620197$ (cf. [Wat39, p. 267]), which gives the return probability $p = 1 - 1/E \doteq 0.3405373295$—a value correct to the 10 digits given.

[49] Glasser communicated to Hughes [Hug95, p. 614] that "the expression reported in this paper is spoiled by the accidental omission in transcription of a factor of 384π."

[50] This result and the one that follows were brought to our attention by John Boersma, who learned them from John Zucker.

in another *tour de force* by Delves and Joyce [DJ01], later simplified in collaboration with Zucker [JDZ03]. Their result [DJ01, form. (5.13)], respectively, [JDZ03, form. (4.26)], reads in the notation used so far as

$$E|_{p_{\mathrm{EW}}=p_{\mathrm{NS}}} = \frac{8\,K(k_+)\,K(k_-)}{\pi^2\big(\sqrt{1-4(p_{\mathrm{UD}}-2p_{\mathrm{EW}})^2}+\sqrt{1-4(p_{\mathrm{UD}}+2p_{\mathrm{EW}})^2}\,\big)},$$

with the moduli

$$k_\pm^2 = \frac{1}{2}\left(1 - \frac{\sqrt{(1-2p_{\mathrm{UD}})^2-16p_{\mathrm{EW}}^2}+\sqrt{(1+2p_{\mathrm{UD}})^2-16p_{\mathrm{EW}}^2}}{\big(\sqrt{1-4(p_{\mathrm{UD}}-2p_{\mathrm{EW}})^2}+\sqrt{1-4(p_{\mathrm{UD}}+2p_{\mathrm{EW}})^2}\,\big)^3}\right.$$
$$\left.\cdot\left(\sqrt{1-4p_{\mathrm{UD}}^2}\left(\sqrt{(1-4p_{\mathrm{EW}})^2-4p_{\mathrm{UD}}^2}+\sqrt{(1+4p_{\mathrm{EW}})^2-4p_{\mathrm{UD}}^2}\right)^2 \pm 64p_{\mathrm{EW}}^2\right)\right).$$

What a triumph of dedicated men; for such problems current computer algebra systems are of little help.

Hypercubic Lattices. The extension to higher dimensions requires further thinking. None of the methods presented so far directly generalizes to a really efficient numerical method for calculating E in dimensions $d \gg 3$. However, the result of §6.6 can be transformed appropriately to address biased random walks on the *hypercubic lattice* in d dimensions, that is, on the integer lattice $\mathbb{Z}^d$.

We denote the transition probabilities for a step forward or backward in the jth dimension by p_j^+ and p_j^-, respectively, and their geometric mean by

$$p_j^* = \sqrt{p_j^+ p_j^-}.$$

Just like in Lemma 6.1 we can prove that $0 \leqslant \sigma = 2(p_1^* + \cdots + p_d^*) \leqslant 1$. Formula (6.22) is readily generalized to

$$E = \frac{1}{\pi^d}\int_0^\pi \cdots \int_0^\pi \frac{d\phi_1 \ldots d\phi_d}{1-2p_1^*\cos\phi_1-\cdots-2p_d^*\cos\phi_d}. \tag{6.24}$$

In higher dimensions numerical integration of this formula is prohibitively expensive. However, there is now a neat trick (see [Mon56, §2]) that transforms the d-dimensional integral to a single integral of a product of modified Bessel functions: we insert the integral

$$\frac{1}{1-2p_1^*\cos\phi_1-\cdots-2p_d^*\cos\phi_d} = \int_0^\infty e^{-t(1-2p_1^*\cos\phi_1-\cdots-2p_d^*\cos\phi_d)}\,dt$$

into (6.24), interchange orders of integration, use the representation [Olv74, p. 82]

$$I_0(x) = \frac{1}{\pi}\int_0^\pi e^{x\cos\phi}\,d\phi$$

of the modified Bessel function I_0 of the first kind of order zero, and deduce that[51]

$$E = \int_0^\infty e^{-t} I_0(2p_1^* t) \cdots I_0(2p_d^* t) dt. \tag{6.25}$$

Since $I_0(x)$ grows like (see [Olv74., p. 83])

$$I_0(x) \sim \frac{e^x}{\sqrt{2\pi x}} \quad \text{as } x \to \infty,$$

the integrand in (6.25) decays like $O(e^{-(1-\sigma)t} t^{-d/2})$ as $t \to \infty$. In particular, we observe that $E < \infty$ for $d \geqslant 3$;[52] that is, the biased random walk is always *transient* in dimensions higher than 2.

The efficient numerical evaluation of the integral (6.25) can be based on a double-exponential quadrature formula (see §§3.6.1 and 9.4). However, because of the exponential growth of $I_0(x)$ we have to exercise some care to avoid overflow. It is advisable to perform the calculation using the scaled function $\tilde{I}_0(x) = e^{-x} I_0(x)$, $x \geqslant 0$, and the expression

$$E = \int_0^\infty e^{-(1-\sigma)t} \tilde{I}_0(2p_1^* t) \cdots \tilde{I}_0(2p_d^* t) dt.$$

In MATLAB the scaled function $\tilde{I}_0(x)$ can be evaluated with the built-in command `BesselI(0,x,1)`. A *Mathematica* implementation `BesselITilde[x]` can be found on the web page for this book. It is based on an asymptotic expansion of $I_0(x)$ with rigorous bounds [Olv74, p. 269]: for $x > 0$

$$I_0(x) = \frac{e^x}{\sqrt{2\pi x}} \left(\sum_{k=0}^{n-1} \frac{(2k-1)!!^2}{k!(8x)^k} + R_n \right), \quad |R_n| \leqslant 2e^{1/4x} \frac{(2n-1)!!^2}{n!(8x)^n}.$$

We can use this approach to solve Problem 6 once more:

A *Mathematica* Session

```
ExpectedVisits[ϵ_Real] := (pE = 1/4 + ϵ; pW = 1/4 - ϵ;
   pN = pS = 1/4; pEW = Sqrt[pE pW]; pNS = Sqrt[pN pS]; σ = 2 (pEW + pNS);
   NIntegrate[Exp[-(1 - σ) t] BesselITilde[2 pEW t] BesselITilde[2 pNS t],
    {t, 0, ∞}, PrecisionGoal → 15, Method → DoubleExponential]);
FindRoot[ExpectedVisits[ϵ] == 2, {ϵ, 0.06, 0.07}, AccuracyGoal → 15]

{ϵ → 0.0619139544739909}
```

The run, in IEEE arithmetic, takes about one second on a 2GHz PC and yields 15 correct digits.

Finally, we apply the approach to higher-dimensional problems and calculate, for various dimensions d, the probability of return, $p = 1 - 1/E$, of the *unbiased* random walk, that is, for the specific transition probabilities $p_j^* = 1/2d$, $j = 1, \ldots, d$.

[51] This generalizes the result [Mon56, form (2.11)] that Montroll had obtained in 1956 for the particular case $p_j^+ = p_j^-$, $j = 1, \ldots, d$.

[52] And for $d \leqslant 2$, if and only if $\sigma < 1$, in agreement with Lemma 6.1.

Table 6.1. *Probability p of return for the d-dimensional hypercubic lattice.*

d	p	d	p	d	p
3	0.340537329550999	10	0.056197535974268	17	0.031352140397027
4	0.193201673224984	11	0.050455159821331	18	0.029496289133281
5	0.135178609820655	12	0.045789120900621	19	0.027848522338807
6	0.104715495628822	13	0.041919897078975	20	0.026375598694496
7	0.085844934113379	14	0.038657877090674	30	0.017257643569441
8	0.072912649959384	15	0.035869623125357	40	0.012827098305686
9	0.063447749652725	16	0.033458364465789	100	0.005050897571251

A *Mathematica* Session *(generates Table 6.1)*

```
ReturnProbability[d_Integer] :=
 1 - 1 / NIntegrate[BesselITilde[t / d]^d, {t, 0, ∞}, WorkingPrecision → 20,
   PrecisionGoal → 17, Method → DoubleExponential]
{#, NumberForm[ReturnProbability[#], {15, 15}]} & /@
  Join[Range[3, 20], {30, 40, 100}] // TableForm
```

The data of Table 6.1 have previously been calculated by Griffin [Grif90, Table 4] (differing in the last digit for $d = 11$ and $d = 12$, however). He used a dimensional recursion that generalizes the combinatorial method of §6.4.1 and accelerated the convergence of the underlying series by Aitken's Δ^2-method.

Chapter 7

Too Large to Be Easy, Too Small to Be Hard

Folkmar Bornemann

> *My mathematical tastes were formed at a time when the most common adjectives in mathematical conversation were "deep" and "trivial." First-rate mathematics was deep, all the rest was trivial.*
>
> —Freeman Dyson [Dys96, p. 612]

> *If I assert the existence of a number but cannot actually compute its value, then I have not finished the problem.... There are no deep theorems—only theorems that we have not understood very well. That is the constructive impulse.*
>
> —Nicholas Goodman [Goo83, p. 63]

Problem 7

> *Let A be the* $20{,}000 \times 20{,}000$ *matrix whose entries are zero everywhere except for the primes* $2, 3, 5, 7, \ldots, 224737$ *along the main diagonal and the number* 1 *in all the positions* a_{ij} *with* $|i-j| = 1, 2, 4, 8, \ldots, 16384$*. What is the* $(1,1)$ *entry of* A^{-1}*?*

The dimension of the problem is *too small* to make its solution in IEEE double-precision arithmetic really hard using current software and hardware environments: we will solve it with essentially a single MATLAB command in §7.2, and, having a particularly efficient iterative solver in hand, we will be able to increase the dimension by a factor of 100 to 2,000,000 without any difficulty in §7.3.

However, the dimension of the problem is *too large* for us to be naively confident that round-off errors will not compromise some of the precious digits that were so easily obtained. In §7.4 we will discuss two methods, namely a posteriori round-off error analysis and interval arithmetic, which will restore our confidence in these digits.

Finally, the dimension of the problem is *just right* to make the completely unexpected calculation of the exact rational solution only just feasible and therefore absolutely impressive: we will sketch in §7.5 the methods that allowed three team members of the LinBox project to solve the problem as a ratio of two 97,389-digit integers.

7.1 A First Look: A Linear System of Equations

Having the matrix A in hand, how do we calculate the $(1,1)$ entry of A^{-1}? As we learn in a first course on numerical analysis [DH03, p. 2], calculating—and using—the inverse of a matrix is almost never a good idea: it is too slow and bound to be numerically unstable. A better idea is to reformulate the problem as a system of linear equations. For the problem at hand, we observe that the entire first column x of the inverse

$$A^{-1} = \left(\begin{array}{c|ccc} x_1 & * & \cdots & * \\ \vdots & \vdots & \ddots & \vdots \\ x_n & * & \cdots & * \end{array} \right)$$

satisfies the linear system of equations

$$Ax = b, \qquad b = (1, 0, 0, \ldots, 0)^T. \tag{7.1}$$

Problem 7 is now answered by the first entry of x, namely, x_1.

Generating the Sparse Matrix *A*. By using dimension as a parameter we can naturally embed the problem into a family of similar problems. This point of view, that is, not looking at the problem as just a single instance, will help us get a better feeling for the inherent difficulties and complexity of the task and to learn from smaller instances about larger ones. We denote the dimension by n and consider the matrix $A_n \in \mathbb{R}^{n \times n}$ that is defined along the diagonal by

$$(A_n)_{ii} = p_i \qquad \text{with } p_i \text{ the } i\text{th prime number,} \tag{7.2}$$

and along the off-diagonal by

$$(A_n)_{ij} = 1 \qquad \text{if } |i-j| \text{ is a power of 2;}$$

all other entries are zero.

This is a highly sparse matrix with $O(n \log n)$ nonzero elements. For $n = 20{,}000$ only 554,466 of the $n^2 = 4 \cdot 10^8$ entries are nonzero, that is, about 0.14%; cf. Figure 7.1. For $n = 2{,}000{,}000$ the ratio is 0.002%.

MATLAB and *Mathematica*[53] can deal with general sparse matrices; that is, they offer storage schemes and commands for building sparse matrices and operating on them. Here is code that generates A_{20000} and the right-hand side b in MATLAB.

A MATLAB Session

```
n = 20000; p_n = 224737; b = sparse(n,1); b(1) = 1;
A = spdiags(primes(p_n)',0,n,n); e = ones(n,2);
for k = 2.^(0:floor(log2(n))), A = A + spdiags(e,[-k k],n,n); end
```

[53] Version 5.0 and higher.

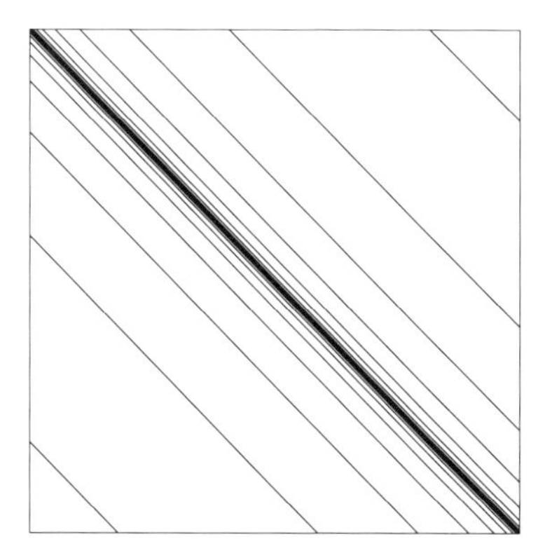

Figure 7.1. *Sparsity pattern of* A_{20000}.

The following code does the same in *Mathematica*.

A *Mathematica* Session

```
n = 20000; b = Table[0, {n}]; b[[1]] = 1;
A = SparseArray[{{i_, i_} → Prime[i]}, n] + (# + Transpose[#])&@
    SparseArray[
      Flatten@Table[{i, i + 2^j} → 1, {i, n - 1}, {j, 0, Log[2., n - i]}], n];
```

Table 7.1 shows how much time[54] is needed to generate A_n for various n using the MATLAB code. Further, the table shows how much memory is consumed for storing A_n. For the *Mathematica* code the timings are nearly identical, however; the amount of space is less since *Mathematica* stores the entries as integers and not as double-precision floats.

Table 7.1. *Time and space for generating* A_n *(MATLAB).*

n	time	space
2,000	0.2s	0.48MB
20,000	2.7s	6.4MB
200,000	45s	79MB
2,000,000	720s	944MB

7.2 Solving It Directly

Let us have faith and try solving the linear system with a direct sparse solver [GL81, Chap. 5] such as the one in MATLAB. The following runs in 6 hours and 11 minutes.

[54] In this chapter all timings are for a Sun™ UltraSPARC®-III+ CPU with 900 MHz.

A MATLAB Session *(cont. of session on p. 148)*

```
>> x = A\b;
>> x1 = x(1)

x1 = 7.250783462684010e-001
```

Let us look under the hood. Because the matrix is symmetric with a positive diagonal, MATLAB makes a guess and assumes that the matrix is also *positive definite*. Based on this assumption it tries to calculate a Cholesky factorization of A_n,

$$A_n = LL^T, \qquad L \text{ lower triangular with positive diagonal.}$$

In general, the factorization succeeds if and only if the symmetric matrix at hand is indeed (numerically) positive definite. Specifically, it does succeed for A_{20000}. The creation of nonzero entries in the process of calculating L, that is, the amount of *fill-in*, depends critically on the ordering of the unknowns of the linear system; cf. [GL81, p. 115]. By default, MATLAB applies the *minimal degree ordering* [GL81, §5.3], which tends to minimize the fill-in. Thus, written in full, the MATLAB line `x = A\b` performs the following ($R = L^T$):

```
x = zeros(size(b));
perm = symmmd(A);                % minimal degree ordering
R = chol(A(perm,perm));
x(perm,:) = R\(R'\b(perm,:));
```

Here, the lower triangle of L is filled to 41.26% (cf. Figure 7.2) and consumes 944 MB of memory. Memory consumption scales as $O(n^2)$, and the time for calculating L, as $O(n^3)$. Obviously, this scaling would prohibit any calculation for much larger dimensions such as $n = 200{,}000$.

There is an alternative ordering of the unknowns, the *reverse Cuthill–McKee ordering* [GL81, §4.3.1], that tends to minimize bandwidth of the sparse matrix instead of the fill-in. In cases such as the problem at hand, with an already relatively large amount of fill-in

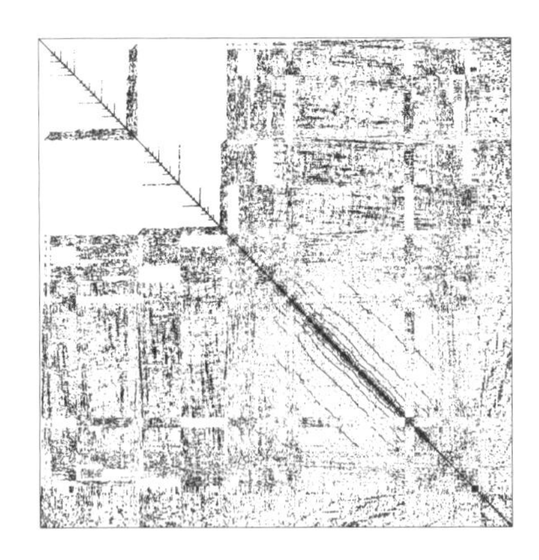 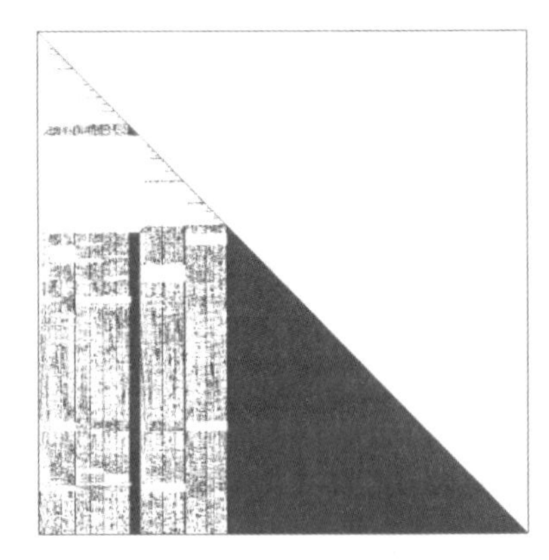

Figure 7.2. *Minimal degree ordering: sparsity pattern of reordered A_{5000} (left) and its Cholesky factor L (right).*

for the minimal degree ordering, this ordering requires comparatively only a little more memory to store the L factor. However, because MATLAB is now able to switch to band methods [GL81, Chap. 4], the run time for calculating the Cholesky factorization reduces substantially.

A Matlab Session *(cont. of session on p. 148)*

```
>> x = zeros(n,1);
>> perm = symrcm(A);          % reverse Cuthill-McKee ordering
>> R = chol(A(perm,perm));
>> x(perm) = R\(R'\b(perm));
>> x1 = x(1)

x1 = 7.250783462684012e-001
```

Now, the lower triangle of L is filled to 50.46% (cf. Figure 7.3) and consumes 1.13 GB of memory. However, the run time reduces by nearly 1 hour, to 5 hours and 18 minutes.

7.2.1 Assessment I: A Priori Round-Off Error Analysis

Round-off is the only possible source of inaccuracy for the solution given by a direct solver. So we ask: How many digits of x_1 given in the MATLAB session on p. 150 could have been distorted this way?

The Cholesky factorization "enjoys perfect normwise [numerical] stability" [Hig96, p. 206]; that is, the resulting error is comparable to the size of the effect on the solution that would result from a perturbation of the linear system on the order of machine precision. No reason to worry, the reader might think. But let us be more specific.

Let $\hat{x}$ denote the solution vector calculated in the realm of machine numbers. The stability of the Cholesky method yields the following a priori error estimate (cf. [Hig96, p. 142]):

$$|x_1 - \hat{x}_1| \leqslant \|x - \hat{x}\| \leqslant c_n \kappa(A_n)\|x\| \cdot u + O(u^2). \tag{7.3}$$

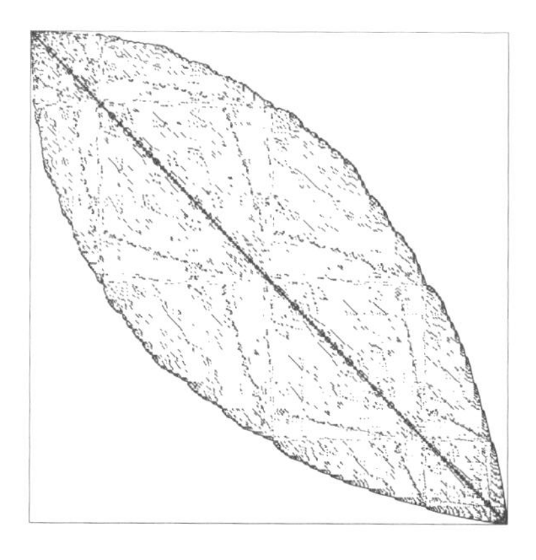

Figure 7.3. *Reverse Cuthill–McKee ordering: sparsity pattern of reordered A_{5000} (left) and its Cholesky factor L (right).*

Here, $\|\cdot\|$ denotes the Euclidean norm, u the unit round-off[55] of the machine arithmetic, $\kappa(A_n)$ the spectral condition number of A_n, and c_n a constant that depends on the dimension n only. In fact, the following bound can be proved (cf. [Hig96, form. (10.7)]):

$$c_n \leqslant 4n(3n+1).$$

To make use of these estimates we need an idea about the size of the condition number $\kappa(A_n) = \lambda_{\max}(A_n)/\lambda_{\min}(A_n)$, that is, the ratio of the largest to the smallest eigenvalue of the symmetric positive definite matrix A_n. Because the diagonal of A_n is dominant, at least at its lower end, a reasonable approximation of the condition number is given by the ratio of the largest to the smallest value of that diagonal,[56]

$$\kappa(A_n) \approx p_n/2, \qquad \text{that is,} \quad \kappa(A_{20000}) \approx 10^5. \tag{7.4}$$

For dimension $n = 20{,}000$, plugging all these data into (7.3) results in the estimate

$$|x_1 - \hat{x}_1| \lessapprox 0.04.$$

Thus these arguments do not guarantee correctness of even the first digit of $\hat{x}_1$. The reason for this pessimistic bound is twofold: first, the relatively large condition number of the linear system, which would indeed be cause for concern, and second, the large dimension that results in a very bad factor c_n. Experience with the a priori bounds of round-off error analysis teaches us that constants such as c_n are overly pessimistic [Hig96, §§2.6/3.2]. If we are unduly optimistic and assume $c_n \approx 1$, the a priori bound results in

$$|x_1 - \hat{x}_1| \lessapprox 8 \cdot 10^{-12}.$$

This would certainly guarantee the correctness of the first 10 digits, but all these considerations eat at our confidence in this approach. Thus, the lesson to be learned is that while a priori round-off error analysis is essential for a proper qualitative understanding of the worst-case effects caused by round-off errors, it is of little practical help as far as sharp quantitative bounds for a concrete case are concerned [Hig96, §3.2].

Later, after we have discussed much more efficient solvers, we will see that in fact 15 digits of the solution delivered by the direct solver were correct. To this end we will use more refined methods of assessing the accuracy, such as using higher precision arithmetic (§7.3.1), a posteriori error estimates (§7.4.1), or interval arithmetic (§7.4.2).

7.2.2 Why Not Extrapolate?

Because extrapolation works so well in many chapters of this book, the reader might ask whether the results of small instances can be extrapolated to approximate the solution to the problem at hand. Figure 7.4 shows the number of digits for which the solutions for A_n agree with that for A_{2000}, respectively A_{20000}. The data for A_{2000} can be obtained with the direct sparse solver on any reasonable machine.

We observe that if n passes a power of 2, there are jumps in the approximation. That is because at those n a new off-diagonal starts in the building of A_n. The size of the jumps

[55] IEEE double-precision arithmetic has $u = 2^{-53} \doteq 1.11 \cdot 10^{-16}$.

[56] A comparison with Table 7.2 shows that this rough approximation underestimates $\kappa(A_n)$ by just a factor of 2.

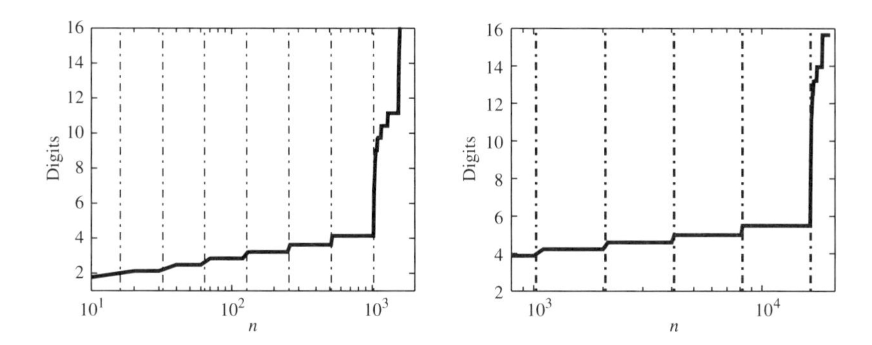

Figure 7.4. *Approximation of x_1 by lower dimensions. The graphs show the number of digits for which the solution for A_n agrees with that for A_{2000} (left), respectively A_{20000} (right). The dashed lines indicate the dimensions that are a power of two.*

appears to be pretty predictable until the very last one: suddenly the accuracy jumps not just by a fraction of a digit but by more than 6 digits at once. Conversely, this particular behavior means that any extrapolation based on data of dimensions smaller than the start of the last off-diagonal is not able to predict the final result to more than about 6 digits. For the particular case $n = 20{,}000$, the last off-diagonal starts at about $2^{14} = 16{,}384$. Thus, with the $O(n^3)$ scaling of the computing time, we could reduce our efforts by at best a factor of only $(20000/16384)^3 \approx 1.8$.

7.3 Solving It Iteratively

The enormous amount of memory and the large computing time needed for the Cholesky factorization in §7.2 naturally lead to the question of how well iterative solvers will do on Problem 7.

Let us remark right at the beginning that using an iterative solver does not require the matrix to be explicitly stored in whatever sparse format. For instance, a characteristic of *Krylov space methods*[57] is that they access the matrix A_n only via a black box for the matrix-vector product mapping $x \mapsto A_n x$. This map can be realized by two nested `for` loops, consuming optimal $O(n)$ memory for storing the primes $2, 3, \dots, p_n$ in a vector variable that we will call `diagonal`. As an example, here is a MATLAB function that implements the matrix-vector product (*Mathematica* code can be found in Appendix C.5.2):

```
function y = ATimes(x,diagonal)
 % realizes: y=A*x; initialize by: diagonal = primes(p_n)'
 y = diagonal.*x; n=length(x);
 for k=1:n-1,for j=2.^(0:floor(log2(n-k))), y(k)=y(k)+x(k+j); end,end
```

[57]The kth Krylov space generated by a matrix A and a vector b is the linear space spanned by the first k applications of A to b, that is, by the vectors $b, Ab, A^2b, \dots, A^kb$.

```
 for k=2:n,   for j=2.^(0:floor(log2(k-1))), y(k)=y(k)+x(k-j), end,end
return
```

However, for efficiency reasons we recommend using the sparse matrix A_n explicitly as long as sufficient memory is available. That way one can use the highly optimized vectorization implemented for the explicit matrix-vector product in MATLAB or *Mathematica*.

For Problem 7, the iterative method of choice is the classic *conjugate gradient* (CG) method of Hestenes and Stiefel from 1952; cf. [DH03, Alg. 8.16] or [TB97, Alg. 38.1]. This is a Krylov space method that is tailored to symmetric positive definite matrices. There are many different derivations of the method to be found in the literature, but in the end they all result in the same algorithm.

Algorithm 7.1. CG Iteration for Solving $Ax = b$ with Initial Vector $x^{(0)}$.

$p^{(1)} = r^{(0)} = b - Ax^{(0)}$;

for $k = 1$ **to** $k_{\max}$ **do**

$\alpha_k = \langle r^{(k-1)}, r^{(k-1)}\rangle / \langle Ap^{(k)}, p^{(k)}\rangle$;

$x^{(k)} = x^{(k-1)} + \alpha_k p^{(k)}$;

$r^{(k)} = r^{(k-1)} - \alpha_k A p^{(k)}$;

if termination criterion satisfied **then exit**;

$\beta_{k+1} = \langle r^{(k)}, r^{(k)}\rangle / \langle r^{(k-1)}, r^{(k-1)}\rangle$;

$p^{(k+1)} = r^{(k)} + \beta_{k+1} p^{(k)}$;

end for

Here $\langle x, y\rangle = x^T y$ denotes the Euclidean inner product. At each round the `for` loop involves one matrix-vector product, namely, $Ap^{(k)}$ (which appears twice in the algorithm but need be computed only once). In exact arithmetic the vector $r^{(k)}$ equals the residual,

$$r^{(k)} = b - Ax^{(k)}.$$

Now, for a working algorithm we have to fill in the missing part, that is, the termination criterion. A judicious choice should be based on a little theory and some rough estimates.

A Termination Criterion. The convergence properties of the CG method are determined by the spectrum of A. Namely, one can prove the error reduction [DH03, Cor. 8.18]

$$\|x^{(k)} - x\|_A \leqslant \epsilon\, \|x\|_A, \qquad \text{if } k \geqslant \left\lceil \frac{\sqrt{\kappa(A)}}{2} \log(2/\epsilon) \right\rceil. \tag{7.5}$$

Here, $\|\cdot\|_A$ denotes the *energy norm* defined by $\|x\|_A^2 = \langle Ax, x\rangle$, and $\kappa(A)$ is the spectral condition number of A, which we already considered in §7.2.1, that is, the ratio $\lambda_{\max}(A)/\lambda_{\min}(A)$ of the largest to the smallest eigenvalue of A. Let us improve on the rough estimate (7.4) for $\kappa(A_n)$ that we had obtained before.

Table 7.2. *Spectral data of A_n.*

n	p_n	$\lambda_{\max}(A_n)$	$\lambda_{\min}(A_n)$	$\kappa(A_n)$	$\kappa(D_n^{-1}A_n)$
20	71	$7.1512 \cdot 10^1$	1.1335	$6.3 \cdot 10^1$	4.36
200	1223	$1.2234 \cdot 10^3$	1.1217	$1.1 \cdot 10^3$	4.45
2000	17389	$1.7390 \cdot 10^4$	1.1207	$1.6 \cdot 10^4$	4.45
20,000	224737	$2.2474 \cdot 10^5$	1.1205	$2.0 \cdot 10^5$	4.45

The maximum eigenvalue is mainly influenced by the large dominant entries of the diagonal of A_n. In fact, with $e_n = (0, 0, \ldots, 0, 1)^T$, the nth canonical basis vector, we get by [HJ85, Thm. 4.2.2/Thm. 5.6.9]

$$p_n = \langle A_n e_n, e_n \rangle \leqslant \lambda_{\max}(A_n) \leqslant \|A_n\|_\infty = p_n + O(\log n), \text{ that is } \lambda_{\max}(A_n) \simeq p_n.$$

Good estimates for the minimum eigenvalue are not that easy to obtain.[58] Table 7.2 displays some numerical eigenvalue calculations[59] that provide numerical evidence for $\lambda_{\min}(A_n) \geqslant 1$. If we assume this simple estimate to be true, we may infer

$$\langle x, x \rangle \leqslant \langle A_n x, x \rangle \leqslant \langle A_n^2 x, x \rangle = \langle A_n x, A_n x \rangle, \qquad \text{that is,} \qquad \|x\| \leqslant \|x\|_A \leqslant \|A_n x\|,$$

which implies the following estimate for the error in the first component:

$$|x_1 - x_1^{(k)}| \leqslant \|x - x^{(k)}\| \leqslant \|x - x^{(k)}\|_A \leqslant \|r^{(k)}\|, \qquad r^{(k)} = b - A_n x^{(k)}. \tag{7.6}$$

Therefore, accuracy of 10 digits is obtained by taking $\epsilon = 10^{-11} < 5 \cdot 10^{-11}|x_1|/\|x\|_A$ in (7.5).[60] That is, 10 digits are guaranteed if we take a number of iterations $k \geqslant 5819$. Even though this a priori estimate predicts the correct order of magnitude, in practice it can be a considerable overestimate. It is therefore advisable to terminate the iteration as soon as a simple a posteriori error estimate indicates sufficient accuracy. Here, because of (7.6) we choose

$$\|r^{(k)}\| \leqslant \texttt{tol} \tag{7.7}$$

with $\texttt{tol} = 10^{-11}$ as the termination criterion.

MATLAB comes with a built-in implementation of the CG method, called by the function `pcg`, that implements the termination criterion (7.7) by default. *Mathematica* also offers CG as an optional method for its linear solver, which we will use later on p. 158.

[58]Later, in Lemma 7.1, we will prove with quite some effort that $1 \leqslant \lambda_{\min}(A_n) \leqslant 2$ for at least the dimensions in the range $1 \leqslant n \leqslant 2^{100}$.

[59]The cases $n = 20$, $n = 200$, and $n = 2000$ were straightforwardly calculated using MATLAB's `eig` command, which is based on the QR-algorithm. The case $n = 20{,}000$ was obtained by the inverse power method, using the iterative method of §7.3.1 as the underlying linear solver.

[60]The fact that $5|x_1| > \|x\|_A$ can be guessed from experiments with smaller dimensions.

A MATLAB Session *(cont. of session on p. 148)*

```
>> tol = 1e-11; kmax = 5819;
>> [x,fail,r,k] = pcg(A,b,tol,kmax);
>> if not(fail), k, x1 = x(1), err = r/x1, end

k   = 1743
x1  = 7.250783462683977e-001
err = 1.328075010788348e-011
```

Hence, after 1743 iterations, in about three minutes, we have obtained a solution with an estimate of the relative error that indicates 10 digits to be correct,

$$x_1 \doteq 0.72507\,83462.$$

However, the discussion in §7.2.1 taught us to be concerned about the possible influence of round-off errors, which might have distorted the calculation of the residuals so that they become worthless for reliable error estimation. We will come back to this point in §§ 7.4.1 and 7.4.2.

7.3.1 Preconditioning

For fixed accuracy ϵ, the number k_ϵ of iterates needed by the CG method grows by (7.5) at worst like

$$k_\epsilon = O\left(\sqrt{\kappa(A_n)}\right) = O\left(p_n^{1/2}\right) = O\left(n^{1/2}\log^{1/2} n\right),$$

since by the prime number theorem $p_n = O(n \log n)$; cf. [CP01, Thm. 1.1.4]. Hence, the complexity of solving the system of equations is $O(n^{3/2}\log^{3/2} n)$; quite a reduction from the $O(n^3)$ scaling of the direct methods. However, for using these ideas for dimensions n much higher than 20,000 the growth rate is still too large.

A remedy is provided by *preconditioning* [TB97, Lect. 40]. That is, with a "simple" symmetric positive definite approximation $B \approx A^{-1}$ in hand, we apply the CG method to the linear system

$$BAx = Bb. \tag{7.8}$$

Note that BA is symmetric with respect to the inner product $\langle x, y\rangle_{B^{-1}} = x^T B^{-1} y$. If we use this inner product instead of $\langle\cdot,\cdot\rangle$ in Algorithm 7.1 as applied to the linear system (7.8), and rewrite everything in terms of the Euclidean inner product, we obtain the *preconditioned conjugate gradient (PCG)* method.

Algorithm 7.2. PCG Iteration for Solving $Ax = b$ with Initial Vector $x^{(0)}$.

$r^{(0)} = b - Ax^{(0)};\ \ p^{(1)} = Br^{(0)};$

for $k = 1$ to $k_{\max}$ do

$\alpha_k = \langle Br^{(k-1)}, r^{(k-1)}\rangle / \langle Ap^{(k)}, p^{(k)}\rangle;$

$x^{(k)} = x^{(k-1)} + \alpha_k p^{(k)};$

$r^{(k)} = r^{(k-1)} - \alpha_k A p^{(k)};$

if termination criterion satisfied **then exit**;

$\beta_{k+1} = \langle Br^{(k)}, r^{(k)} \rangle / \langle Br^{(k-1)}, r^{(k-1)} \rangle$;

$p^{(k+1)} = Br^{(k)} + \beta_{k+1} p^{(k)}$;

end for

The convergence is now determined by the spectral properties of BA; estimate (7.5) generalizes to

$$\|x^{(k)} - x\|_A \leqslant \epsilon \, \|x\|_A \qquad \text{if } k \geqslant \left\lceil \frac{\sqrt{\kappa(BA)}}{2} \log(2/\epsilon) \right\rceil. \tag{7.9}$$

In problems with matrices A_n of varying dimension n, the idea of preconditioning is now to choose the preconditioner B_n in a way that the condition number $\kappa(B_n A_n)$ does not grow as fast as $\kappa(A_n)$. The ultimate goal of preconditioning is to obtain a uniform bound,

$$\kappa(B_n A_n) = O(1).$$

For the sparse matrices arising from finite element discretizations of second-order elliptic boundary value problems, sophisticated methods such as multilevel preconditioning schemes have been invented to establish such a uniform bound; cf. [BY93]. However, for the problem at hand, the simplest possible idea—the *diagonal preconditioner*—actually works: we take

$$B_n = D_n^{-1}, \qquad D_n = \text{diag}(A_n) = \text{diag}(2, 3, 5, \ldots, p_n).$$

Table 7.2 gives numerical evidence that

$$\kappa(D_n^{-1} A_n) \doteq 4.45 \tag{7.10}$$

for large n. That is, independent of the dimension n, the a priori bound (7.9) of the number k of iterations that are necessary to achieve 10 digits of accuracy reduces considerably to $k \geqslant 28$.

A MATLAB Session *(cont. of session on p. 148)*

```
>> tol = 1e-11; kmax = 28;
>> D=spdiags(diag(A),0,n,n);
>> [x,fail,r,k] = pcg(A,b,tol,kmax,D);
>> if not(fail), k, x1 = x(1), err = r/x1, end

k   = 14
x1  = 7.250783462684012e-001
err = 1.218263630055319e-012
```

After just 14 iterations we have obtained a solution with an error estimate that indicates 11 digits to be correct,

$$x_1 \doteq 0.72507\,83462\,6.$$

Table 7.3. *Results of the diagonally PCG method,* `tol` $= 10^{-11}$.

n	# iter.	x_1	Est. of rel. error	Time
20,000	14	0.72507 83462	$1.2 \cdot 10^{-12}$	1.7s
200,000	14	0.72508 09785	$1.4 \cdot 10^{-12}$	29s
2,000,000	14	0.72508 12561	$1.5 \cdot 10^{-12}$	360s

The run time is below 2 seconds, an improvement of more than a factor of 100 over the case without any preconditioning. Now moving to much higher dimension is no problem; see the results in Table 7.3.[61] The fact that we always need just 14 iterations nicely reflects the uniform condition number bound (7.10).

A reliable check of the accuracy that circumvents possible problems with round-off is provided by increasing the precision of the arithmetic. Because we have now found a very efficient algorithm, we should give bigfloats a try. *Mathematica*[62] is especially convenient for that purpose because of its built-in PCG method for sparse matrices.

A *Mathematica* Session *(cont. of session on p. 149)*

```
diagonal = Table[A[[i, i]], {i, n}]; prec = 100; b = SetPrecision[b, prec + 5];
x =
  LinearSolve[A, b,
   Method → {Krylov, Method → ConjugateGradient, Preconditioner → (#/diagonal &),
      Tolerance → 10^(-prec-1)}];
N[x[[1]], 100]
0.7250783462684011674686877192511609688691805944795089578781647692077731
        899945962835735923927864782020
```

In less than a minute, with 64 iterations, we get 100 correct digits. We can play this game to the extreme and obtain 10,000 digits in 5.3 days using 2903 iterations.[63]

7.4 How Many Digits Are Correct?

We will now discuss two methods that allow the reliable assessment (§7.4.1), or the validation (§7.4.2), of the calculated solution including round-off error effects. In contrast to the a priori analysis (§7.2.1), the results will be useful.

[61] The run time shown is without the time needed to generate A_n; cf. Table 7.1.

[62] Version 5.0 or higher.

[63] Dan Lichtblau and Yifan Hu of Wolfram Research communicated to us that the 10,000-digit run can be sped up by a factor of 2 if one uses the machine-precision Cholesky factorization of §7.2 as a preconditioner. This is in accordance with the convergence theory, since the condition number is reduced from about 4.45 to about 1.

7.4.1 Assessment II: A Posteriori Round-Off Error Analysis

We start with the general problem of solving a linear system of equations,

$$Ax = b.$$

Given a calculated vector $\hat{x}$, we want to estimate the error $\|x - \hat{x}\|$. This error comprises eventual approximation errors as caused by an iterative method as well as round-off errors.

Now, the general trick of a posteriori error estimates is to go from the error to an appropriately scaled residual. The reason for this is that backwards-stable direct methods such as the Cholesky factorization tend to produce very small residuals, whereas iterative methods such as the CG iteration can be forced into very small residuals by the termination criterion. A first try of getting the residual r involved is the fairly simple estimate

$$\|x - \hat{x}\| = \|A^{-1} \underbrace{(b - A\hat{x})}_{=r} \| \leqslant \|A^{-1}\| \cdot \|r\|. \tag{7.11}$$

In general, using the submultiplicativity of the norm might already cause us to lose too much information about the error. However, for the problem at hand we infer from the numerical evidence of Table 7.2 that[64]

$$\|A_n^{-1}\| = 1/\lambda_{\min}(A_n) \leqslant 1.$$

Generally speaking, if we can scale the problem to $\|A^{-1}\| \approx 1$, estimate (7.11) turns out to be useful. For instance, this is also the case in §10.3.1.

Now, it looks as if (7.11) is already the perfect bound. In fact, this was what we used in §7.3. However, we have to be careful, because the calculation of r itself can be severely distorted by round-off errors. In the realm of finite-precision arithmetic, instead of r we calculate a perturbed vector $\hat{r} = r + \Delta r$. It has been shown that the perturbation Δr satisfies [Hig96, form. (7.26)][65]

$$|\Delta r| \leqslant \gamma_m(|A|\,|\hat{x}| + |b|), \qquad \gamma_m = \frac{mu}{1 - mu}.$$

Here, u is the unit round-off, and $m - 1$ is the maximum number of nonzero entries in any row of A. The absolute values and the inequality have to be understood componentwise. Thus, we replace (7.11) by the practical bound

$$\|x - \hat{x}\| \leqslant \|A^{-1}\| \Big(\|\hat{r}\| + \gamma_m \big\| \, |A|\,|\hat{x}| + |b| \, \big\| \Big). \tag{7.12}$$

The round-off error made in the evaluation of this formula is of higher order in the unit round-off and can be neglected for all practical purposes.

[64] The spectral norm $\|A\|$ of a general matrix A, that is, the matrix norm induced by the Euclidean vector norm, satisfies (cf. [Hig96, p. 120]) $\|A\| = \sigma_{\max}(A)$ and $\|A^{-1}\| = 1/\sigma_{\min}(A)$, where $\sigma_{\max}(A)$ and $\sigma_{\min}(A)$ denote the maximum and minimum singular value of A, respectively. For a symmetric positive definite matrix A, this simplifies to $\|A\| = \lambda_{\max}(A)$ and $\|A^{-1}\| = 1/\lambda_{\min}(A)$.

[65] Because Higham is considering general matrices, he states the result [Hig96, form. (7.26)] with $m = n + 1$. However, his proof shows that $m - 1$ is the length of the inner products that actually have to be evaluated in the matrix-vector product. For sparse matrices this length is bounded by the number of nonzero entries per row.

Table 7.4. *Residual-based a posteriori estimates of the relative error.*

n	`tol`	# iter.	Est. based on (7.11)	Est. based on (7.12)
20,000	10^{-11}	14	$1.22 \cdot 10^{-12}$	$1.25 \cdot 10^{-12}$
	10^{-15}	17	$1.22 \cdot 10^{-15}$	$3.47 \cdot 10^{-14}$
200,000	10^{-11}	14	$1.36 \cdot 10^{-12}$	$1.40 \cdot 10^{-12}$
	10^{-15}	17	$1.83 \cdot 10^{-15}$	$4.46 \cdot 10^{-14}$
2,000,000	10^{-11}	14	$1.48 \cdot 10^{-12}$	$1.53 \cdot 10^{-12}$
	10^{-15}	17	$1.53 \cdot 10^{-15}$	$5.42 \cdot 10^{-14}$

For the problem at hand, we have $A_n = |A_n|$,

$$m \leqslant 2\lfloor \log_2 n \rfloor + 4,$$

and by the numerical evidence of Table 7.2 we may assume that $\|A_n^{-1}\| \leqslant 1$. Thus, the following simple MATLAB code estimates the relative error of $\hat{x}_1$:

```
m = 2*floor(log2(n))+4; u = eps/2; gamma = m*u/(1-m*u);
r = b-A*x;
err = (norm(r)+gamma*norm(A*abs(x)+abs(b)))/x(1);
```

Table 7.4 shows a comparison between the results obtained by using (7.11) or (7.12). The influence of round-off errors only becomes visible at very small tolerances. However, the calculation for $n = 20{,}000$ and `tol` $= 10^{-15}$ allows us to claim with substantially increased confidence that Problem 7 is solved by a number in the error interval

$$0.72507\,83462\,68401\,2 \pm 3.47 \cdot 10^{-14},$$

that is, to 12 digits by

$$x_1 \doteq 0.72507\,83462\,68.$$

In the same fashion, the a posteriori estimates for dimensions $n = 200{,}000$ and $n = 2{,}000{,}000$ are good for 12 correct digits.

7.4.2 Validation: Interval Arithmetic

Interval arithmetic is yet another possibility, in the realm of finite-precision arithmetic, to deal with the simple residual-based estimate (7.11); that is, if $\|A_n^{-1}\| = \lambda_{\min}^{-1}(A_n) \leqslant 1$,

$$\|x - \hat{x}\| \leqslant \|A_n^{-1}\| \cdot \|r\| \leqslant \|r\|.$$

One simply calculates an inclusion of $\|r\| = \|b - A_n\hat{x}\|$ using intervals. This way, we will be able to validate the correctness of a certain number of digits; let us see how many before we will prove at the end of this section that indeed $\lambda_{\min}(A_n) \geqslant 1$—at least for all those dimensions n, which could ever be considered for an explicit calculation. Here is

Table 7.5. *Validation results.*

n	Interval solution	# correct digits	Time
20,000	$0.72507\,83462\,68^{4047}_{3977}$	12	0.12s
200,000	$0.72508\,09785\,29^{2024}_{1932}$	12	1.5s
2,000,000	$0.72508\,12561\,325^{275}_{179}$	13	21s

MATLAB/INTLAB code accomplishing the task:

```
x1 = midrad(x(1),norm(b-A*intval(x)))
```

It is important to note that interval arithmetic is used only after we have calculated $\hat{x}$ by a traditional numerical method. Here is how a full run would look:

A MATLAB/INTLAB Session *(cont. of session on p. 148)*

```
>> tol = 1e-15; max_ite = 28;
>> D=spdiags(diag(A),0,n,n);
>> [x,fail,res,ite] = pcg(A,b,tol,max_ite,D);
>> if not(fail), x1 = midrad(x(1),norm(b-A*intval(x))), end

intval x1 = 7.2507834626840__e-001
```

By default, INTLAB outputs correctly rounded digits and gives 14 of those here. Using the command `infsup(x1)` changes the output to the interval solution itself,

$$x_1 \in 0.72507\,83462\,68^{4047}_{3977}.$$

According to the way of counting correct digits that we have adopted in this book, this amounts to 12 correct digits only:

$$x_1 \doteq 0.72507\,83462\,68.$$

The validation results for higher dimensions are displayed in Table 7.5. The cost of validation appears to be negligible: for dimension $n = 2{,}000{,}000$ we need 12 minutes to generate A_n, 17 minutes to solve with the diagonally PCG method for $\texttt{tol} = 10^{-15}$, and just 21 seconds to validate the correctness of 13 digits, all this using IEEE double-precision arithmetic.

Likewise, we can use *Mathematica*'s built-in interval arithmetic to validate the high-precision results that we mentioned at the end of §7.3.1. The pretty printer `IntervalForm` can be found in Appendix C.5.1.

A *Mathematica* Session *(cont. of session on p. 158)*

```
x[[1]] + Interval[{-1, 1}]Norm[b - A.Interval/@x]//IntervalForm
```

$0.72507834626840116746868771925116096886918059447950895787816476920777318999459628357359239278647820 20^{55127}_{44295}$

This validation of the 100 digits from the *Mathematica* session on p. 158 runs in less than eight seconds. Again we can play the game to the extreme and also validate the 10,000-digit run, investing just a further half-minute of run time.

A Rigorous Lower Bound for the Minimal Eigenvalue. We close this section with the promised proof of a rigorous lower bound for the minimal eigenvalue.

Lemma 7.1 *The minimal eigenvalue of the matrix* A_n *satisfies*

$$1 \leqslant \lambda_{\min}(A_n) \leqslant 2$$

for at least the dimensions n in the range $1 \leqslant n \leqslant 2^{100}$.

Proof. We proceed in two steps. In the first step we perform an eigenvalue calculation for a very specific dimension that allows us, in the second step, to estimate the minimal eigenvalue for a broad range of dimensions. The code for the interval calculations using MATLAB/INTLAB can be found in Appendix C.4.2, for *Mathematica*, in Appendix C.5.2.

Step 1. We begin with the claim that

$$\lambda_{\min}(A_{1142}) \in 1.120651470_{854673}^{922646}. \tag{7.13}$$

The dimension 1142 is small enough to calculate numerically all the eigenvalues and eigenvectors of A_{1142} in less than a minute using MATLAB's `eig` or *Mathematica*'s `Eigensystem` command. Now, the point is, that for a symmetric matrix A these numerical values can be validated after the fact by interval methods. Namely, from perturbation theory [Ste01, Thm. 1.3.8/2.1.3] we conclude that, if $(\hat{\lambda}, \hat{x})$ is any approximate, finite-precision eigenvalue–eigenvector pair, we have that there is a true eigenvalue λ of A satisfying the a posteriori error estimate

$$|\lambda - \hat{\lambda}| \leqslant \|\hat{\lambda}\,\hat{x} - A\hat{x}\|/\|\hat{x}\|.$$

It is a simple task to calculate a rigorous estimate of the right-hand side using interval arithmetic. Doing this for all eigenvalue–eigenvector pairs with MATLAB/INTLAB results in the estimate (7.13).

Step 2. Given $1 \leqslant n \leqslant n_*$ we observe that A_n is the $n \times n$ principal submatrix of A_{n_*}; that is, A_{n_*} can be partitioned as

$$A_{n_*} = \left(\begin{array}{c|c} A_n & B \\ \hline B^T & C \end{array}\right)$$

with some $n \times (n_* - n)$-matrix B and $(n_* - n) \times (n_* - n)$-matrix C. We use the variational characterization [HJ85, Thm. 4.2.2] of the minimal eigenvalue of a symmetric matrix,

$$\lambda_{\min}(A_n) = \min_{\|x\|=1} x^T A x, \tag{7.14}$$

to arrive at the upper bound $\lambda_{\min}(A_{n_*}) \leqslant \lambda_{\min}(A_n) \leqslant \lambda_{\min}(A_1) = 2$.

On the other hand, for a likewise partitioned vector $z^T = (x^T|y^T)$ with $\|z\|^2 = \|x\|^2 + \|y\|^2 = 1$ we obtain, using (7.14) once more, the estimate

$$z^T A_{n_*} z = x^T A_n x + 2x^T By + y^T Cy \geqslant \lambda_{\min}(A_n)\|x\|^2 - 2\|B\| \cdot \|x\| \cdot \|y\| + \lambda_{\min}(C)\|y\|^2$$

$$\geqslant \lambda_0\|x\|^2 - 2\alpha\,\|x\| \cdot \|y\| + \lambda_1\|y\|^2 \geqslant \begin{pmatrix} \|x\| \\ \|y\| \end{pmatrix}^T \underbrace{\begin{pmatrix} \lambda_0 & -\alpha \\ -\alpha & \lambda_1 \end{pmatrix}}_{=F_2} \begin{pmatrix} \|x\| \\ \|y\| \end{pmatrix} \geqslant \lambda_{\min}(F_2),$$

where $\lambda_0 \leqslant \lambda_{\min}(A_n)$ and $\lambda_1 \leqslant \lambda_{\min}(C)$ are suitable lower bounds and

$$\|B\| \leqslant \sqrt{\|B\|_1 \cdot \|B\|_\infty} = \alpha$$

is an upper bound of the spectral norm of B (cf. [Hig96, form. (6.19)]) that is simple to evaluate. Minimizing over all vectors z yields by (7.14)

$$\lambda_{\min}(A_{n_*}) \geqslant \lambda_{\min}(F_2).$$

By (7.13), the particular choice $n = 1142$ and $n_* = 2^{100}$ gives

$$\lambda_{\min}(A_{1142}) \geqslant 1.120651470854673 = \lambda_0, \quad \|B\|_1 = \lceil \log_2 1142 \rceil = 11, \quad \|B\|_\infty = 100,$$

that is, $\alpha^2 = 1100$, and by Gershgorin's theorem [HJ85, Thm. 6.1.1] $\lambda_{\min}(C) \geqslant p_{1143} - 100 = 9121 = \lambda_1$. The eigenvalue problem for the 2×2-matrix F_2 is just a quadratic equation, which can be easily solved using interval arithmetic,

$$\lambda_{\min}(F_2) = \frac{2(\lambda_0\lambda_1 - \alpha^2)}{\lambda_0 + \lambda_1 + \sqrt{4\alpha^2 + (\lambda_1 - \lambda_0)^2}} \in 1.00003743527186_2^6.$$

Summarizing, we have $\lambda_{\min}(A_n) \geqslant \lambda_{\min}(A_{n_*}) \geqslant 1$ for $1 \leqslant n \leqslant n_* = 2^{100}$. □

We remark that a result such as the preceding lemma depends very specifically on the actual numbers involved. For instance, the family of matrices $\hat{A}_n = A_n - 1.122 I_n$, where I_n denotes the unit matrix of dimension n, looks essentially like A_n itself. However, $\hat{A}_n$ is positive definite for the dimensions $1 \leqslant n \leqslant 129$ only. In fact, it is not at all clear whether A_n is positive definite for *all* n, or only whether the minimal eigenvalue is uniformly bounded from below. Such questions could be answered, if at all, by giving sense to $\lim_{n\to\infty} A_n$ as a linear operator A_∞ on an appropriate infinite-dimensional sequence space.

7.5 Solving It Exactly

One might think that solving the linear system of equations (7.1) for x is overkill if one is just interested in the first component x_1. There is a classic method, Cramer's rule, that addresses this component directly [HJ85, §§0.8.2/0.8.3]:

$$x_1 = (A^{-1})_{11} = \frac{\det A^{11}}{\det A}. \tag{7.15}$$

Here, A^{11} denotes the minor of A made by deleting the first row and the first column of A,

$$A = \left(\begin{array}{c|ccc} * & * & \cdots & * \\ \hline * & & & \\ \vdots & & A^{11} & \\ * & & & \end{array}\right).$$

However, we have to remember that the best numerical methods for calculating a determinant are based on matrix factorizations, which are therefore as expensive as solving a linear system of equations by a direct method. Moreover, determinants are prone to severe over- or underflow. In fact, we will see later in this section that the two determinants in (7.15) are 97,389-digit integers. For these reasons, and for its numerical instability, Cramer's rule is rightly banned from the canon of numerical analysis [Hig96, §1.10.1].

The assessment of the use of (7.15) changes completely if we want to make an attempt at finding x_1 *exactly*. After all, it is simply a rational number. Why not calculate it?

Upon hearing about the contest, the task of producing the exact solution to Problem 7 was taken on by the following three people from the development team of LinBox[66] as a challenge to illustrate the capabilities of their library: Jean-Guillaume Dumas, William Turner, and Zhendong Wan. Fortunately, the dimension $n = 20{,}000$ chosen by Trefethen turned out to be just right, challenging but feasible:

> This problem is just on the edge of feasibility on current machines. A problem of size a small multiple of this would not be solvable due to time and/or memory resource limitations. [DTW02, p. 2]

There are essentially two approaches to the exact rational solution. One calculates (7.15) using *congruential methods* for the determinants, and the other solves the linear system (7.1) using *p-adic methods*.[67]

7.5.1 The Congruential Approach

Basically, the congruential method for calculating the determinant of an integer matrix A goes in two steps. First, one obtains the determinant modulo certain pairwise different primes $q_1, \ldots, q_r$:

$$\det A \equiv d_j \pmod{q_j}.$$

Next, by Chinese remaindering [vzGG99, §5.4] one combines the results $d_1, \ldots, d_r$ into a single remainder d for the composite modulus $m = q_1 \cdots q_r$,

$$\det A \equiv d \pmod{m}.$$

If m is large enough, namely, $m > 2|\det A|$, one can reconstruct the desired determinant from its remainder $d \in \{0, \ldots, m-1\}$ by

$$\det A = \begin{cases} d & \text{if } d < m/2, \\ d - m & \text{if } d > m/2. \end{cases}$$

[66]A C++ template library for exact, high-performance linear algebra computation with sparse and structured matrices over the integers and over finite fields: `http://www.linalg.org/`

[67]The web page for this book has a bare-bones implementation of these methods in *Mathematica*, which can be used on small n to develop a better understanding of what follows.

The point is that the modular determinants d_j can be calculated *in parallel* using fast-hardware word-sized integer arithmetic, if one restricts the q_j to 31-bit primes. There are 50,697,537 of those, enough for calculating determinants of about 10^9 digits in length. Arbitrary-length integer arithmetic is then needed only for the comparatively minor task of the Chinese remaindering.

To make it work, we need some useful practical bounds for determinants such as those given by the Hadamard inequalities [HJ85, Thm. 7.8.1, Cor. 7.8.2]. Since we know by Lemma 7.1 that the matrix at hand is positive definite, we can use the simplest one, namely, the product of the diagonal entries:

$$0 < \det A \leqslant p_1 \cdots p_{20000} \doteq 8.8210 \cdot 10^{97388};$$
$$0 < \det A^{11} \leqslant p_2 \cdots p_{20000} \doteq 4.4105 \cdot 10^{97388}. \qquad (7.16)$$

To satisfy $m = q_1 \cdot \cdots \cdot q_r > 2 \cdot 8.8210 \cdot 10^{97388}$ we take the smallest $r = 10{,}784$ of the 31-bit primes, the first one being $q_1 = 1{,}073{,}741{,}827$, the last one $q_r = 1{,}073{,}967{,}703$.

The task of calculating the determinant of A modulo a prime q is equivalent to calculating the determinant of A in the *prime field* $\mathbb{F}_q$ of congruence classes modulo q. Now, a calculation of the determinant can be based on an LU-factorization of A as matrices over $\mathbb{F}_q$.[68] However, as we know from §7.2 for the problem at hand, a factorization cannot really take advantage of the sparsity of A: run time scales with the dimension n as $O(n^3)$, and the memory consumption scales as $O(n^2)$.

As with the iterative methods of §7.3, let us look for a Krylov space method that calculates the determinant by addressing A only via matrix-vector products. A particularly economical such method was invented in 1986 by Wiedemann [Wie86].

Wiedemann's Randomized Algorithm for Modular Determinants. Up to sign, the determinant $\det A \in \mathbb{F}_q$ of a matrix $A \in \mathbb{F}_q^{n\times n}$ is the constant term of the characteristic polynomial χ_A,

$$\det A = (-1)^n \chi_A(0).$$

Rather than aiming at the characteristic polynomial, Wiedemann's algorithm aims at the minimal polynomial μ_A of A: the normalized polynomial of smallest degree such that $\mu_A(A) = 0$. It is well known [HJ85, Cor. 3.3.4] that the minimal polynomial μ_A divides the characteristic polynomial χ_A, which has degree n. Thus, we arrive at

$$\deg \mu_A = n \quad \Longrightarrow \quad \mu_A = \chi_A \quad \Longrightarrow \quad \det A = (-1)^n \mu_A(0).$$

On the other hand, if $\deg \mu_A < n$, the minimal polynomial of A by itself does not yield the determinant. But still, the calculation of $\det A$ can be reduced to the minimal polynomial of an appropriately transformed matrix; cf. [Wie86, §V]. However, we do not need to consider this further complication, since for the matrix at hand the condition $\deg \mu_A = n$ turns out to be true.

Now, Wiedemann's algorithm does not calculate μ_A directly, but proceeds by randomly and uniformly selecting two vectors $u, v \in \mathbb{F}_q^n$ and looking for the normalized

[68] Essentially, this approach is used in *Mathematica* if one calls `Det[A,Modulus->q]`.

polynomial $\mu^A_{u,v}$ of smallest degree such that $u^T\mu^A_{u,v}(A)v = 0$. We easily see that $\mu^A_{u,v}$ divides the minimal polynomial μ_A (cf. [vzGG99, p. 320]) and conclude once again that

$$\deg \mu^A_{u,v} = n \implies \mu^A_{u,v} = \mu_A = \chi_A \implies \det A = (-1)^n \mu^A_{u,v}(0).$$

If we write $\mu^A_{u,v}$ as

$$\mu^A_{u,v}(t) = t^k + c_{k-1}t^{k-1} + \cdots + c_1 t + c_0,$$

we observe that $(c_0, \ldots, c_{k-1})$ is equivalently characterized as the shortest sequence in $\mathbb{F}_q$ for which a *linear recurrence* holds with $a_j = u^T A^j v \in \mathbb{F}_q$:

$$c_0 a_j + c_1 a_{j+1} + \cdots + c_{k-1} a_{j+k-1} + a_{j+k} = 0, \qquad j = 0, 1, 2, \ldots.$$

Given at least the first $2k$ values of the sequence $a_0, a_1, a_2, \ldots$ as input, the coefficients $(c_0, \ldots, c_{k-1})$ can efficiently be calculated by an algorithm of Berlekamp and Massey [KS91, p. 31], which was originally invented in 1965 for the decoding of BCH codes and bears interesting relations to Padé approximations and the extended Euclidean algorithm for polynomials [vzGG99, Chap. 7]. In the problem at hand we know that $k \leqslant n$ and, we obtain $\mu^A_{u,v}$ after applying the Berlekamp–Massey algorithm to the $2n$ field values:

$$u^T v, u^T Av, u^T A^2 v, \ldots, u^T A^{2n-1} v,$$

which are calculated by $2n - 1$ recursive applications of the matrix-vector product.

It has been shown [KS91, §2] that the probability of $\mu^A_{u,v} \neq \mu_A$ is below $2n/q$. Thus, with the parameters of the problem at hand, namely, $n = 20{,}000$ and a 31-bit prime $q > 2^{30}$, the probability of a failure is less than 0.0038%, or one failure in 26,843 tries. In the very unlikely event of such a failure, detected by $\deg \mu^A_{u,v} < n$, we simply repeat the calculation with new vectors u, v.

For the matrix at hand, Wiedemann's algorithm has an expected run time of order $O(n^2 \log^\kappa n)$ for some constant $\kappa > 0$ and consumes $O(n)$ memory only.

The Solution. On a 1 GHz PC, Wiedemann's algorithm as implemented in LinBox needs approximately 30 minutes to arrive at one minimal polynomial over a word-size prime field. Because one has to perform $r = 10{,}784$ such calculations for each of the two determinants in (7.15), we can predict a total CPU time of about 10,784 hours, or 450 days. Dumas, Turner, and Wan used a heterogeneous cluster of 182 processors[69] in parallel and obtained the solution in about 4 days of elapsed time:

$$x_1 = \frac{3101640749121412476978542 \,\langle\langle 97{,}339 \text{ digits}\rangle\rangle\, 3312289187417983612357075}{4277662910613638374648603 \,\langle\langle 97{,}339 \text{ digits}\rangle\rangle\, 8142829807013006012935182}.$$

Courtesy of Jean-Guillaume Dumas, the full-length 97,389 digits of the numerator and denominator can be found on the web page for this book; here we have displayed only the first and last 25 digits. We observe that the Hadamard bounds (7.16) are off the actual result by just about a factor of 2.

[69] 96 Pentium®-III at 735 MHz, 6 Pentium-III at 1 GHz, and 20 Sun Ultra-450 at 4 × 250 MHz.

7.5.2 The p-adic Approach

Following the work of Dixon [Dix82] one starts, given a prime p and a positive integer k, with a p-adic approximation $f_k \in \mathbb{Z}_{p^k}$ of x_1, that is, $f_k \equiv x_1 \pmod{p^k}$. If the exponent k is sufficiently large, we can reconstruct the rational number x_1 from the integer f_k. The approximations $f_1, f_2, \ldots, f_k$ are obtained iteratively; each step of the iteration, the *Hensel lifting*, requires the solution of a linear system over the prime field $\mathbb{F}_p$ with the given matrix A,

$$A\Delta x_j \equiv r_j \pmod{p}. \tag{7.17}$$

Now, these systems can be solved comparatively fast, once we have an LU-decomposition of A over the prime field in hand.[70]

The Iteration: Hensel Lifting. The jth iterate $x^{(j)}$ will be a vector of integers from $\mathbb{Z}_{p^j}$ that satisfies the linear system

$$Ax^{(j)} \equiv b \pmod{p^j}. \tag{LIN$_j$}$$

We start the iteration by calculating the LU-decomposition of A modulo the prime p and solve (LIN$_1$) for $x^{(1)}$. If we have arrived at $x^{(j)}$, we know that p^j divides $b - Ax^{(j)}$. Therefore, we may calculate a residual vector r_j over the prime field, such that

$$b - Ax^{(j)} \equiv p^j \cdot r_j \pmod{p^{j+1}}.$$

Reuse of the LU-decomposition allows us to efficiently solve the linear system (7.17). Now, we see that $x^{(j+1)} \equiv x^{(j)} + p^j \cdot \Delta x_j \pmod{p^{j+1}}$ solves the desired linear system (LIN$_{j+1}$) defining the next iterate.

The Rational Number Reconstruction. Given integers $r, t \in \mathbb{Z}$ and $f \in \mathbb{Z}_{p^k}$ such that

$$\frac{r}{t} \equiv f \pmod{p^k}, \qquad |r|, |t| \leqslant \frac{p^{k/2}}{2}, \tag{7.18}$$

there is an efficient algorithm [Dix82, p. 139] that reconstructs the rational number r/t from f. To this end, let s_i/t_i be the ith convergent to the continued fraction for f/p^k and put $r_i = t_i f - s_i p^k$. Now, if j is the first index such that $|r_j| < p^{k/2}$, then $r/t = r_j/t_j$.

As is the case for the Berlekamp–Massey algorithm in §7.5.1, this reconstruction algorithm is related to the extended Euclidean algorithm, now for integers, and goes in essence back to classic work of Thue in 1902; cf. [vzGG99, §5.10].

The Solution. For the problem at hand, Dumas, Turner, and Wan chose the prime p as the largest 50-bit prime, namely $p = 1{,}125{,}899{,}906{,}842{,}597$. They calculated the LU-decomposition of A in the prime field $\mathbb{F}_p$ using 64-bit integer arithmetic, occupying

[70]The use of the LU-decomposition modulo p to iterate for solutions of higher p-adic precision (modulo p^k) is reminiscent of taking a machine-precision matrix factorization as a preconditioner for an iterative method that aims at higher numerical precision; cf. footnote 63.

$20{,}000^2 \cdot 64$ bits = 3 GB of memory. On a 750 MHz Sun UltraSPARC with 8 GB of memory, this took 5.5 days. Now, it follows from the Hadamard bound (7.16) and the requirements (7.18) of the rational number reconstruction that a number k of

$$\lceil 2 \log_p (2 \cdot 8.8210 \cdot 10^{97388}) \rceil = 12941$$

lifting steps will do the job. With less than a minute per lifting step, LinBox needed about 7 days to complete the lifting with a 194,782-digit integer f satisfying $f \equiv x_1 \pmod{p^k}$. Finally, the reconstruction step yields the rational solution x_1 itself in a matter of about a minute.

Running several different algorithms on different machines for the same large-scale problem is always a severe test for software and hardware. Having passed this test, the resulting rational number x_1 was in complete agreement with the result of §7.5.1. And needless to say, the first 10,000 digits of its decimal expansion agree with the validated numerical high-precision solution of §7.4.2.

A Comparison. In comparing the complexity of the congruential approach with that of the p-adic one, we observe that even though the latter needs less total CPU time by a factor of 36, the former takes a factor of 3 less elapsed time. This is because the iterative p-adic approach does not lend itself to a straightforward parallel implementation. Moreover, the p-adic approach puts far more severe memory requirements on the machine.

Chapter 8

In the Moment of Heat

Folkmar Bornemann

In general the theorems concerning the heating of air ... extend to a great variety of problems. It would be useful to revert to them when we wish to foresee and regulate temperature with precision, as in the case of green-houses, drying-houses, sheep-folds, work-shops, or in many civil establishments, such as hospitals, barracks, places of assembly.

In these different applications we must attend to accessory circumstances which modify the results of analysis ..., but these details would draw us away from our chief object, which is the exact demonstration of general principles.

—Joseph Fourier [Fou78, §91, p. 73]

Joseph Fourier himself, who was concerned with keeping wine cellars cool during summer months, solved just this kind of problem.

—Lloyd N. Trefethen [Tre02, p. 3]

Problem 8

> *A square plate $[-1, 1] \times [-1, 1]$ is at temperature $u = 0$. At time $t = 0$ the temperature is increased to $u = 5$ along one of the four sides while being held at $u = 0$ along the other three sides, and heat then flows into the plate according to $u_t = \Delta u$. When does the temperature reach $u = 1$ at the center of the plate?*

We will discuss three approaches to the problem. In §8.1 we obtain 5 correct digits by using a commercial, general-purpose finite element code. In §8.2, standard finite differences and extrapolation yield—justified by a posteriori error estimates—at least 12 significant digits. Finally, in §8.3 analytical techniques based on Fourier series help to simplify the problem to a scalar transcendental equation. This equation can be efficiently solved for 10,000 significant digits whose correctness can be proven by interval analysis. A truncation of the equation leads to the simple approximation $t_* \doteq 4\pi^{-2} \log \eta$ to 11 digits, where η is the unique

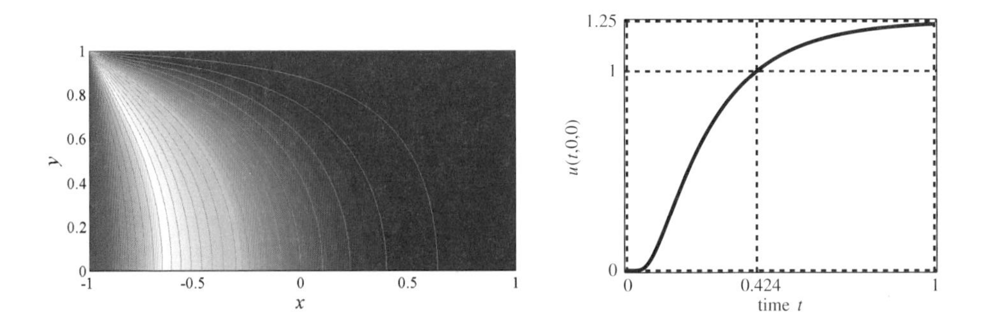

Figure 8.1. *Left: solution* $u(0.424, x, y)$ *with contour lines plotted at* $u = 0.2, \ldots, 4.8$ *in steps of* 0.2. *For symmetry reasons only the upper half-square is shown. Right: solution* $u(t, 0, 0)$ *for* $0 \leqslant t \leqslant 1$.

solution of $3\pi\sqrt{5}\eta^9 - 60\eta^8 + 20 = 0$ subject to the condition $\eta > 1$. Here and for the rest of the chapter, t_* denotes the time we are looking for, defined by $u(t_*, 0, 0) = 1$.

8.1 A First Look: Using Software

Being a plain partial differential equation problem, Problem 8 is a good candidate to probe the capabilities of general-purpose software for the solution of partial differential equations. We take *FEMLAB*®, a commercial software package that provides finite element methods for so-called multiphysics applications.

A time-dependent PDE like the heat equation under consideration is solved in *FEMLAB* using the *method of lines*: a finite element discretization in space reduces the problem to a stiff ordinary differential equation, which in turn is approximated by a state-of-the-art variable-order, variable-step-size solver [DB02]. The left part of Figure 8.1 shows the solution $u(t, x, y)$ at about the time $t \approx t_*$.

To solve for a certain moment in time requires an approximation that is continuous in time. Most ODE solvers offer such a kind of *dense output*. For ease of implementation, and because it is sufficient for the accuracies to be achieved anyway, we use cubic spline interpolation. The values of $u(t, 0, 0)$ so obtained are shown in the right part of Figure 8.1 as a function of time. We observe that $u(t, 0, 0)$ apparently approaches the steady state 5/4, which we will confirm in §8.3. This explains the peculiar choice of the number 5 in the formulation of Problem 8: it is the smallest integer for which the question makes sense.

Assessment of the achieved accuracy is a delicate issue. A single run will not be sufficient because *FEMLAB* does not provide a posteriori error estimates. Instead, we compare the results of several runs using different tolerances of the ODE solver and different-order finite elements at about the same number of degrees of freedom (d.o.f.). Because it is highly improbable that the runs will agree on nonconverged digits, it is safe to take those digits as being correct to which the runs agree after rounding,

$$t_* \doteq 0.42401.$$

Table 8.1. *Details of the* FEMLAB *runs (ATOL = RTOL = TOL).*

Element	#nodes	#triangles	#d.o.f.	TOL	#t-steps	t_*
2nd-order	2913	5632	11457	10^{-6}	310	0.42401 14848
				10^{-7}	423	0.42401 16087
4th-order	753	1408	11457	10^{-6}	318	0.42401 16447
				10^{-7}	434	0.42401 17517

We will later show that these 5 digits are correct. In fact, the FEM solutions of Table 8.1 are all correct to 6 digits. Each of the four runs shown took about 10 minutes on a 2 GHz PC. The ordinary differential equation solver was MATLAB's `ode15s` [SR97], a stiff variable-order, variable-step-size multistep integrator based on the numerical differentiation formulas. The *FEMLAB* scripts of the two runs with TOL $= 10^{-6}$ can be found on the web page for this book.

After comparing with the results of the following sections, we might ask ourselves, why contrary to our expectations, the finite element method result using second-order Lagrange elements is closer to the solution than the one using fourth-order elements. It is because of the jumps of the Dirichlet boundary condition such as at the upper left corner of Figure 8.1: these jumps cause a drop of the global regularity of the solution such that fourth-order elements are not necessarily more accurate.

8.2 Solving It Numerically: Richardson Extrapolation

We will show that an accurate numerical solution of Problem 8 is possible using standard finite differences. In fact, the simplest possible such discretization, that is, first-order forward differences in time and second-order central differences in space, can be efficiently accelerated by Richardson extrapolation (see Appendix A, p. 235). Since Richardson extrapolation supplies us with a posteriori estimates of the discretization error, and round-off errors can be controlled by a *running error analysis*, we will be able to provide very good *scientific reasons*—though not a rigorous mathematical proof—for being certain of the correctness of at least 12 digits.

To fix notation, we denote the square by $\Omega = (-1, 1) \times (-1, 1)$, its boundary by

$$\partial\Omega = \Gamma_0 \cup \Gamma_1, \qquad \Gamma_1 = \{(x, y) \in \partial\Omega : x = -1\}.$$

We consider the initial boundary value problem for the heat equation,

$$u_t = \Delta u, \qquad u(t, \cdot, \cdot)|_{\Gamma_0} = 0, \; u(t, \cdot, \cdot)|_{\Gamma_1} = 5, \; u|_{t=0} = 0,$$

for times $t \in [0, T]$. We will vary the end time T so we can solve for $u(T, 0, 0) = 1$. The solution of this equation will be denoted by $T = t_*$.

By choosing positive integers m, n, we define a time step $\tau = T/m$ and a mesh size $h = 2/n$ in space. The finite difference approximation $u_h(t, x, y)$ of $u(t, x, y)$ at grid points

$$(t, x, y) \in \big((0, T] \times \Omega\big) \cap \big(\tau\mathbb{Z} \times h\mathbb{Z}^2\big)$$

is calculated by the recurrence [Tho95, form. (4.2.5)],

$$\begin{aligned} u_h(t,x,y) = u_h(t-\tau,x,y) - \frac{\tau}{h^2}\big(&4u_h(t-\tau,x,y) \\ &-u_h(t-\tau,x-h,y) - u_h(t-\tau,x+h,y) \\ &-u_h(t-\tau,x,y-h) - u_h(t-\tau,x,y+h)\big), \end{aligned}$$

using the boundary or initial values whenever it is necessary. As is well known, for being stable this recursion requires a bound on the time step [Tho95, form. (4.3.9)],

$$\tau/h^2 \leqslant \frac{1}{4}.$$

Such a bound is called a *CFL condition* for Courant, Friedrichs, and Lewy. We choose $\tau = c_T h^2$, with a constant Courant number $c_T \leqslant 1/4$ depending on T. In Appendix C.3.2, the reader will find the short MATLAB program `heat` that realizes this simple finite difference approximation for more general boundary and initial data.

For smooth solutions u the discretization is known to be of second order [Tho95, §4.3]:

$$\max_{(x,y)\in\Omega\cap h\mathbb{Z}^2} |u_h(T,x,y) - u(T,x,y)| = O(h^2).$$

We need to know more; the well-founded application of Richardson extrapolation for error estimation and convergence acceleration requires the validity of an asymptotic expansion

$$u_h(t,x,y) = u(t,x,y) + \sum_{k=1}^{m} e_k(t,x,y)\, h^{\gamma_k} + O(h^{\gamma_{m+1}}) \tag{8.1}$$

with $\gamma_1 < \gamma_2 < \cdots < \gamma_{m+1}$. It is known (cf. [Ste65, §3.3]) that the discretization under consideration has $\gamma_k = 2k$ and the asymptotic expansion (8.1) holds for all $m \in \mathbb{N}$.

Writing $e_* = e_1(T,0,0)$, $u_* = u(T,0,0)$, and $u_{*,h} = u_h(T,0,0)$ we obtain

$$u_{*,h} - u_* = e_* h^2 + O(h^4), \qquad u_{*,2h} - u_{*,h} = 3e_* h^2 + O(h^4).$$

A comparison yields an a posteriori estimate[71] for the relative discretization error of $u_h(T,0,0)$, namely,

$$\frac{u_{*,h} - u_*}{u_*} = \underbrace{\frac{u_{*,2h} - u_{*,h}}{3u_{*,h}}}_{=\epsilon_h} + O(h^4).$$

Table 8.2 shows the results together with the estimate ϵ_h of the relative discretization error for $T = 0.424$. The last column of Table 8.2 nicely reflects that the discretization is of second order. We learn that on the 127×127 grid with 8192 time steps—hence calculating u_h on 132,128,768 grid points—the approximation $u_h(T,0,0)$ is an approximation of $u(T,0,0)$ just good to about 4 correct digits.

[71] Such estimates are crucial for the success of state-of-the-art extrapolation codes with order and step-size control for systems of ordinary differential equations [DB02, §5.3].

Table 8.2. *A posteriori error estimate ϵ_h for $u_h(T,0,0)$ at $T = 0.424$ with Courant number $c_T = \tau/h^2 = 0.212$.*

# time steps	Space grid	h	$u_h(T,0,0)$	ϵ_h	ϵ_{2h}/ϵ_h
32	7×7	1/4	1.016605320	—	—
128	15×15	1/8	1.004170946	$4.1 \cdot 10^{-3}$	—
512	31×31	1/16	1.001034074	$1.0 \cdot 10^{-3}$	4.0
2048	63×63	1/32	1.000248105	$2.6 \cdot 10^{-4}$	4.0
8192	127×127	1/64	1.000051504	$6.6 \cdot 10^{-5}$	4.0

We have to increase the accuracy of the discretization by *convergence acceleration*. The idea is simple; if we substract the error estimate, we obtain a discretization of increased order,

$$u'_{*,h} = u_{*,h} + \frac{1}{3}(u_{*,h} - u_{*,2h}) = u_* + O(h^4).$$

This new approximation inherits the asymptotic expansion (8.1),

$$u'_{*,h} = u_* + \sum_{k=2}^{m} e'_{*,k} h^{2k} + O(h^{2m+2}),$$

now starting with the $O(h^4)$ term. Completely analogously to what we have done for $u_{*,h}$, we can construct an a posteriori error estimate and repeat the process. This is Richardson extrapolation for a sequence of grids with discretization parameters h, $h/2$, $h/4$, $h/8$, etc. It is fairly obvious how to generalize for arbitrary sequences; a good compromise between stability and efficiency is $h = 1/(2n)$ for $n = n_{\min}, n_{\min}+1, \ldots, n_{\max}$. In Appendix C.3.2 the reader will find the short MATLAB program `richardson` that implements this general extrapolation technique with a posteriori estimation of the relative discretization error. Additionally, it contains a *running error analysis* [Hig96, §3.3] of the amplification of round-off errors.

A MATLAB Session

```
>> u = inline('heat([0,0],[1,1],t,0,0,[5,0,0,0],h)','t','h');
>> t = 0.424; order = 2; tol = 5e-14;
>> nmin = 4; [val,err,ampl] = richardson(tol,order,nmin,u,t);
>> val, err = max(err,ampl*eps)

val = 9.997221678853678e-001
err = 8.187827236734361e-005

>> nmin = 8; [val,err,ampl] = richardson(tol,order,nmin,u,t);
>> val, err = max(err,ampl*eps)

val = 9.999859601012047e-001
err = 4.734306967445346e-014
```

The run time on a 2 GHz PC is less than a second for each of the two runs. The extrapolation of the first run is based on eight grids, $128 \times 15 \times 15, \ldots, 968 \times 43 \times 43$; the estimate of the relative error shows that only the leading three digits are correct. This is in accordance with the second run, based on just four grids, $512 \times 31 \times 31$, $648 \times 35 \times 35$, $800 \times 39 \times 39$, $968 \times 43 \times 43$. Here, the error estimate yields the correctness of the leading 13 digits, a result that can be proven to indeed be correct by the methods of §8.3.2.

We fix the parameters of the second run and solve for the t_* that satisfies $u(t_*, 0, 0) = 1$. This can be done using MATLAB's root-finder `fzero`, which is based on the secant method. The run time on a 2 GHz PC is about 5 seconds.

A MATLAB Session *(cont. of session on previous page)*

```
>> u1 = inline('richardson(tol,order,nmin,u,t)-1','t',...
                'tol','order','nmin','u');
>> options = optimset('TolX',tol);
>> t = fzero(u1,t,options,tol,order,nmin,u)

t = 4.240113870336946e-001

>> [val,err,ampl] = richardson(tol,order,nmin,u,t);
>> val, err = max(err,ampl*eps)

val = 9.999999999999942e-001
err = 4.735414794862723e-014
```

Based on the a posteriori error estimates, to the best of our numerical knowledge, the approximation of t_* is correct to 12 digits,

$$t_* \doteq 0.42401\,13870\,33.$$

For the reader who requires more rigor, the analytical method of the next section will *prove* that this claim is correct.

8.3 Solving It Analytically: Fourier Series

Traditionally, initial boundary value problems for the heat equation on rectangular domains are solved by *Fourier series*. Indeed, Fourier himself extensively studied the trigonometric series Now named after him in his seminal 1822 treatise *Théorie analytique de la chaleur* [Fou78], in which he introduced the heat equation for the first time.

For Problem 8 the calculations simplify considerably if we exploit the symmetries: thanks to the kindness of the proposer who happened to choose a square region and looked at its center point, we can turn it into a one-dimensional problem. To this end we first observe that we could equally well heat any of the four sides of the square plate, always getting the same solution at the center. By linearity, adding those solutions shows that heating of all four sides gives four times the original solution at the center. Thus, imposing the constant boundary condition

$$u(t, \cdot, \cdot)|_{\partial\Omega} = \tfrac{5}{4}$$

leads to the *same* solution at the center as in the original formulation of the problem. Next, by transforming $u(t, x, y) = 5/4 - \hat{u}(t, x, y)$, we obtain the homogeneous heat equation with homogeneous Dirichlet boundary conditions,

$$\hat{u}_t = \Delta \hat{u}, \qquad \hat{u}(t, \cdot, \cdot)|_{\partial\Omega} = 0, \tag{8.2}$$

subject to the initial value $\hat{u}(0, x, y) = \alpha^2$, where $\alpha^2 = 5/4$. Finally, we reduce the dimension of the problem by observing that $\hat{u}(t, x, y) = v(t, x)v(t, y)$ solves the initial boundary value problem (8.2), if v is the solution of

$$v_t(t, x) = v_{xx}(t, x), \qquad v(t, -1) = v(t, 1) = 0, \tag{8.3}$$

subject to the initial value $v(0, x) = \alpha$. The time t_* determined by the equation $u(t_*, 0, 0) = 1$ can therefore be obtained by solving $1/4 = \hat{u}(t_*, 0, 0) = v^2(t_*, 0)$, that is,

$$v(t_*, 0) = \tfrac{1}{2}. \tag{8.4}$$

Now we concentrate on the construction of $v(t, x)$ and follow Fourier [Fou78, §333] in observing that

$$v_k(t, x) = e^{-((k+1/2)\pi)^2 t} \cos((k + 1/2)\pi x)$$

constitutes a particular solution of (8.3) for any $k = 0, 1, 2, \ldots$. Therefore, the trigonometric series

$$v(t, x) = \sum_{k=0}^{\infty} c_k e^{-((k+1/2)\pi)^2 t} \cos((k + 1/2)\pi x)$$

with bounded coefficients c_k solves (8.3) for $t > 0$ as well. This can be seen by differentiating term-by-term, which is permissible because of the exponential decay of the terms of the series. The coefficients c_k have to be chosen in such a way that the trigonometric series matches the initial conditions,

$$\alpha = v(0, x) = \sum_{k=0}^{\infty} c_k \cos((k + 1/2)\pi x), \qquad -1 < x < 1. \tag{8.5}$$

We can infer the coefficients

$$c_k = \alpha \frac{4(-1)^k}{\pi(2k + 1)} \tag{8.6}$$

from one of Fourier's major identities [Fou78, §177],

$$\frac{\pi}{4} = \sum_{k=0}^{\infty} \frac{(-1)^k}{2k + 1} \cos((2k + 1)x), \qquad -\frac{\pi}{2} < x < \frac{\pi}{2}. \tag{8.7}$$

A modern proof[72] of (8.6) would be based on the orthogonality of the trigonometric monomials $\cos((2k+1)x)$ in L^2, a technique that Fourier himself basically employs to prove other identities [Fou78, §§220–224]. The reader will find such a proof of (8.6) in §10.4, when we discuss similar results for Problem 10.

Summarizing, we obtain

$$v(t,0) = \frac{4\alpha}{\pi}\sum_{k=0}^{\infty}\frac{(-1)^k e^{-((k+1/2)\pi)^2 t}}{2k+1} = \frac{2\alpha}{\pi}\,\theta\big(e^{-\pi^2 t}\big),$$

where θ denotes a function that is closely related to the classical theta functions[73],

$$\theta(q) = 2q^{1/4}\sum_{k=0}^{\infty}\frac{(-1)^k}{2k+1}q^{k(k+1)}.$$

Hence, the problem at hand, that is, solving (8.4) for t_*, is equivalent to solving the transcendental equation

$$\theta\big(e^{-\pi^2 t_*}\big) = \frac{\pi}{2\sqrt{5}}. \tag{8.8}$$

Figure 8.2 shows that there is a unique solution $0 < t_* \doteq 0.424$.

8.3.1 Solving the Transcendental Equation

Mathematica and Maple provide commands to deal numerically with infinite series. These commands are based on extrapolation methods. *Mathematica*'s and Maple's root-finders, based on the secant method, can therefore readily be used to solve (8.8) for t_*.

---

[72] Fourier [Fou78] offered at least three proofs for his identity (8.7); in §§171–177 by solving an infinite system of linear equations, in §§179–180 by using identities for trigonometric polynomials. In §189 he used the trigonometric identity $\pi/2 = \arctan(z) + \arctan(1/z)$, which is valid for $\mathrm{Re}\, z > 0$, expanded the two occurrences of arctan into the Taylor series,

$$\frac{\pi}{2} = \sum_{k=0}^{\infty}\frac{(-1)^k}{2k+1}\left(z^{2k+1} + z^{-2k-1}\right), \tag{*}$$

and inserted $z = e^{ix}$, which finally yields the result. One has to be careful with convergence issues to justify the argument; Fourier himself drily remarked [Fou78, p. 154], "the series of equation (*) is always divergent, and that of equation (8.7) is always convergent." Not quite true, but for his time this showed a remarkable awareness of the subtleties of convergence.

[73] Indeed, the relation of θ to Jacobi's θ_1 function [BB87, §2.6] is given by

$$\theta(q) = \int_0^{\pi/2}\theta_1(z,q)\,dz, \qquad \theta_1(z,q) = 2q^{1/4}\sum_{k=0}^{\infty}(-1)^k q^{k(k+1)}\sin((2k+1)z).$$

This is no coincidence, as it is classically known that the Green's function of initial boundary value problems of the one-dimensional heat equation is expressible in terms of theta functions [WW96, §21.4] and [Joh82, p. 221].

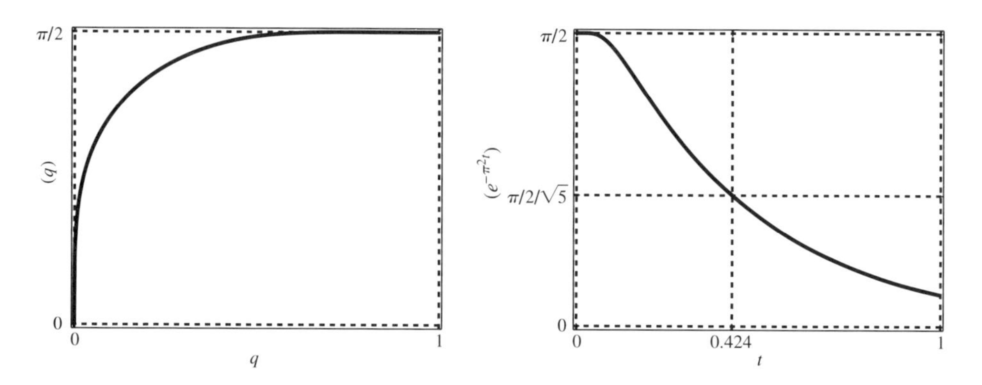

Figure 8.2. *The functions* $\theta(q)$ *and* $\theta(e^{-\pi^2 t})$.

A Maple Session

```
> theta:=q->2*q^(1/4)*sum((-1)^k/(2*k+1)*q^(k*(k+1)),k=0..infinity):
> Digits:=16: fsolve(theta(exp(-Pi^2*t))=Pi/2/sqrt(5),t=0.4);

                        0.4240113870336884
```

A *Mathematica* Session

```
θ[q_Real] := 2q^(1/4) NSum[(-1)^k/(2k + 1) q^(k(k+1)), {k, 0, ∞}]

t* ==
  (t/.FindRoot[θ[e^(-π^2 t)] == π/(2√5), {t, 0.4, 0.5},
       AccuracyGoal → MachinePrecision])
t* == 0.4240113870336884
```

It is quite reassuring that the two programs agree to all digits given. For the precision required by these sessions *Mathematica* uses machine numbers, Maple software arithmetic. Both probably use different algorithms; in any case the algorithms are implemented differently. If we increase the precision, the two programs keep agreeing and confirm the solutions of lower precision. Thus we have good scientific reasons to claim that the following 100 digits of t_* are correct:

$$t_* \doteq 0.42401\,13870\,33688\,36379\,74336\,68593\,25645\,12477\,62090\,66427$$
$$47621\,97112\,49591\,33101\,76957\,56369\,22970\,72442\,29447\,70112.$$

For very high precision, relying on the commands for numerical evaluation of infinite series soon becomes inefficient. Instead we can exploit the fast exponential decay of the

Table 8.3. *Run time versus precision in the solution of the transcendental equation.*

Digits of t_*	Run time	$k_{\max}$	$k^*_{\max}$
10	0.67 ms	1	1
100	8.9 ms	6	6
1,000	0.18 s	22	22
10,000	22 s	73	73

series for $\theta(q)$ at $q = e^{-\pi^2 t_*} \approx 0.015$. We use the truncated sum

$$\theta^{(k_{\max})}(q) = 2q^{1/4} \sum_{k=0}^{k_{\max}} \frac{(-1)^k}{2k+1} q^{k(k+1)},$$

choosing $k_{\max}$ as the smallest possible value for which an increase $k_{\max} \to k_{\max} + 1$ no longer changes the result of the root-finder at the required precision. Table 8.3 shows some timings for *Mathematica* on a 2 GHz PC.

There is a simple estimate of the truncation error that easily allows us to bound the value of $k_{\max}$. In fact, for an alternating series whose terms tend monotonically to zero in absolute value, it is well known that the truncation error is bounded by the first neglected term,

$$\left|\theta^{(k)}(q) - \theta(q)\right| \leqslant \frac{q^{(k+3/2)^2}}{k+3/2}, \qquad 0 < q < 1.$$

Asymptotically, the right-hand side of this estimate equals ϵ for

$$k \sim \sqrt{\frac{|\log \epsilon|}{|\log q|}} - \frac{3}{2}.$$

Now, Figure 8.2 shows that the slope of $\theta(e^{-\pi^2 t})$ at t_* is approximately -1, meaning that the truncation error transforms to roughly the same error in the solution of the transcendental equation. Summarizing, we get the asymptotic upper bound

$$k_{\max} \lesssim k^*_{\max} = \left\lceil \frac{\sqrt{\#\,\text{digits} \cdot \log(10)}}{\pi\sqrt{t_*}} - \frac{3}{2} \right\rceil.$$

This simple formula gives excellent results, as can be seen from a glance at Table 8.3. For 11 significant digits we obtain $k_{\max} = 1$, which leads to the simple approximation mentioned in the introduction of this chapter.

8.3.2 Using Interval Arithmetic

Alternating series such as the one defining $\theta(q)$ converge to a value that is always *enclosed* between two successive partial sums; for $0 < q < 1$ we get

$$\theta^{(1)}(q) < \theta^{(3)}(q) < \theta^{(5)}(q) < \cdots < \theta(q) < \cdots < \theta^{(4)}(q) < \theta^{(2)}(q) < \theta^{(0)}(q).$$

Based on this observation, we can elegantly enclose the solution t_* by interval root-finding methods without having to explicitly estimate the actual *size* of the truncation error. From the truncated series we construct a family of interval maps

$$\theta^{(k)}[Q] = \text{convex hull of } \left(\theta^{(k-1)}(Q) \cup \theta^{(k)}(Q)\right), \qquad Q \subset [0,1],$$

which enclose θ increasingly tighter, constituting a *filtration* of θ,

$$\theta^{(k)}[Q] \supset \theta^{(k+1)}[Q] \supset \theta(Q), \qquad \lim_{k\to\infty} \theta^{(k)}[Q] = \theta(Q).$$

This family can be calculated using standard tools of an interval arithmetic package; implementations can be found in Appendix C.4.2, for MATLAB/INTLAB, in C.5.2 for *Mathematica*.

Playing the same trick for the derivative of θ in *Mathematica*, or using the automatic differentiation capability in MATLAB/INTLAB, allows us to calculate an enclosure of t_* by the interval Newton iteration of §4.5.[74] By construction, any choice of k leads to such an enclosure of t_*; the tightness increases with larger k, however.

A MATLAB/INTLAB Session

```
>> f = inline('theta(exp(-pi^2*t),k)-pi/2/sqrt(5)','t','k'); kt = [];
>> for k=1:4, kt = [kt; k IntervalNewton(f,infsup(0.4,0.5),k)]; end
>> kt
intval kt =
   1   4.240_____________e-001
   2   4.24011387033_____e-001
   3   4.24011387033688_e-001
   4   4.24011387033688_e-001
```

A *Mathematica* Session

```
Table[
    {k, t* == IntervalForm@IntervalNewton[(θ[Exp[-π^2#], k] - π/(2√5))&, {0.4, 0.5}]},
    {k, 4}]//TableForm
```

1 $\quad t_* == 0.4240^{427176438009}_{113824891349}$

2 $\quad t_* == 0.42401138703^{36890}_{26777}$

3 $\quad t_* == 0.42401138703368^{90}_{79}$

4 $\quad t_* == 0.42401138703368^{90}_{79}$

Thus, using machine numbers only, we have validated the correctness of 15 digits,

$$t_* \doteq 0.42401\,13870\,33688.$$

If we spend 5 minutes on a 2 GHz PC, *Mathematica* allows us to prove—based on the interval map $\theta^{(74)}[Q]$—the correctness of the 10,000 digits reported on in Table 8.3. However, as

[74]Implementations can be found in Appendix C.4.3 for MATLAB/INTLAB, in Appendix C.5.3 for *Mathematica*.

already discussed in §4.5, there is no need to use the interval method from the beginning. Instead we could start using the method of §8.3.1 and obtain an approximation of t_* that is very likely correct to the requested number of digits. By ϵ-inflation we put a small interval around that approximation having diameter of a few units of the last digit. Now, we apply the interval Newton method and obtain a validated enclosure. For 10,000 digits, this technique of approximation first, validation last shows the correctness within a minute computing time. Thus, the price to pay for a validated solution is just a factor of 3.

8.4 Harder Problems

To review and compare the different approaches to Problem 8, we think about possible changes to the problem that make it harder or impossible to solve by one of the approaches. Three aspects of the problem were used to varying extents: the spatial geometry was two-dimensional, the shape was rectangular, and the boundary values were particularly simple. There are three generalizations immediately at hand:

- general polygon in two dimensions
- more general boundary values
- n-dimensional box

Table 8.4 tells us whether the methods of this chapter can be extended.

Let us comment on one aspect of this table. *Finite differences and extrapolation* is a general method that is good for medium relative accuracies. However, if the domain is not a box, one has to exercise some care in discretizing the boundary. The discretization error might then have an asymptotic expansion in several incompatible powers of h^γ, making a generalization not straightforward.

Table 8.4. *Extendibility of the various methods for Problem 8.*

Method	Precision	Polygon	Boundary values	nD-box
§8.1: finite elements	Low	✓	✓	✓
§8.2: extrapolation	Medium	(✓)	✓	✓
§8.3: separation	High	—	—	✓

Chapter 9

Gradus ad Parnassum

Dirk Laurie

It seems to be expected of every pilgrim up the slopes of the mathematical Parnassus, that he will at some point or other of his journey sit down and invent a definite integral or two towards the increase of the common stock.

—James Joseph Sylvester [Syl60]

Because it is there.

—George Leigh Mallory, when asked why he wanted to climb Mount Everest.

Problem 9

The integral $I(\alpha) = \int_0^2 \big(2 + \sin(10\alpha)\big)x^\alpha \sin\big(\alpha/(2-x)\big)\,dx$ *depends on the parameter* α. *What is the value* $\alpha \in [0, 5]$ *at which* $I(\alpha)$ *achieves its maximum?*

9.1 A First Look

We can write $I(\alpha)$ as a product: $I(\alpha) = p(\alpha)q(\alpha)$ with

$$p(\alpha) = 2 + \sin(10\alpha), \qquad q(\alpha) = \int_0^2 f_1(x,\alpha)\,dx, \qquad f_1(x,\alpha) = x^\alpha \sin\left(\frac{\alpha}{2-x}\right).$$

To get a rough graph of I without the need to think very hard, we use the midpoint rule to approximate q, because it is very simple and avoids the endpoints. So

$$q(\alpha) \approx h \sum_{k=1}^{2n} f_1\left(\left(k - \frac{1}{2}\right)h, \alpha\right), \quad h = \frac{1}{n}.$$

The graph in Figure 9.1, with α going from 0 to 5 in steps of 0.02 (251 points) and $h = 0.0001$, takes only a few seconds to calculate on my workstation.

The left third of the graph of $I(\alpha)$ looks smooth, and in fact to the accuracy of my video screen is the same as a previous graph (not shown here) calculated with $h = 0.0005$. The rest of it looks rougher, because the larger α becomes, the faster does the integrand oscillate

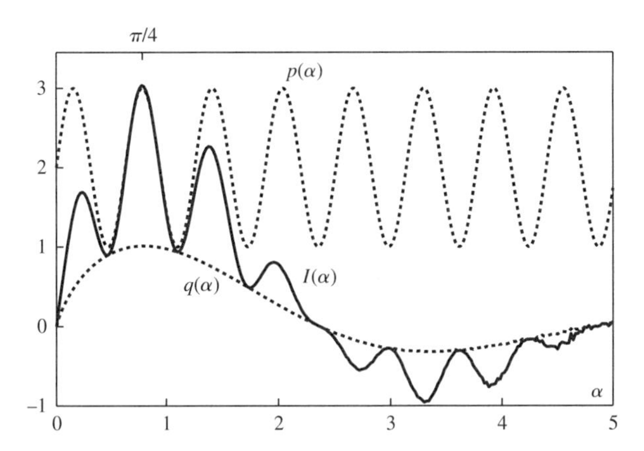

Figure 9.1. *Graph of $p(\alpha)$ (the sine wave) and the midpoint rule approximations to $I(\alpha)$ (the solid line) and $q(\alpha)$.*

near the right endpoint, and the less accurate does the midpoint rule become. Fortunately, it is not necessary to achieve great accuracy for $\alpha > 1$, as the required maximum is evidently not in that region.

It so happens that p has a relative maximum, and q a global maximum, very close to each other. However, the maximum of q is fairly flat, and therefore the position of the maximum is mainly determined by p. The second maximum of p occurs at $\alpha = \pi/4$, which gives our first approximation:

$$\alpha_{\text{opt}} \approx \pi/4 \doteq 0.785.$$

9.2 Accurate Evaluation of the Integral

Figure 9.3 shows what $f_1(x, \pi/4)$ looks like. We have a clearly defined subproblem: evaluate this oscillating integral accurately for α near $\pi/4$. Several general approaches to a very similar subproblem are discussed in detail in Chapter 1, where the integral itself is the object of the exercise. In order to avoid repetition and to concentrate on the special features of Problem 9, at this stage we just pick the one that is easiest to use, namely, contour integration.

We need to express the integrand in terms of an analytic function, which is easy enough: define

$$f_0(z, \alpha) = z^\alpha e^{i\alpha/(2-z)} = e^{\alpha(\log z + i/(2-z))}, \tag{9.1}$$

then $f_1(x, \alpha) = \text{Im} f_0(x, \alpha) = \text{Re}(-if_0(x, \alpha))$. To make the integrand decay exponentially for $z \to 2$, $i\alpha/(2 - z)$ should have a negative real part along the contour, which implies that z should have a positive imaginary part. A parabolic contour similar to the one used in Problem 1 works here too:

$$z(t) = t + it(2-t), \qquad z'(t) = 1 + 2i(1-t), \qquad t \in [0, 2]. \tag{9.2}$$

Figure 9.3 is a graph of $f_2(t, \pi/4)$, where

$$f_2(t, \alpha) = \text{Re}\left(-if_0(z(t, \alpha))z'(t)\right).$$

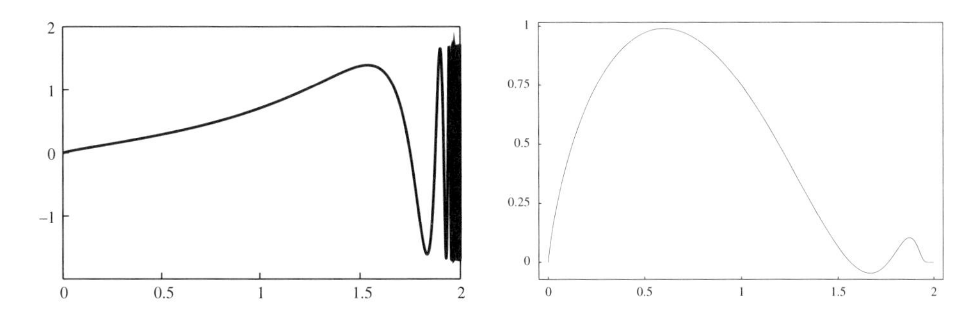

Figure 9.2. *Graph of $f_1(x, \pi/4)$ (left) and of $f_2(t, \pi/4)$ (right).*

The function $f_2(t, \pi/4)$ is obviously easier to integrate with respect to t than $f_1(t, \pi/4)$. The oscillations have not disappeared, but their amplitude has been damped exponentially. But there still is a tricky little difficulty. The factor x^α has also not disappeared, and since the interesting values of α are less than 1, this factor has an infinite derivative at 0. Quadrature routines based on polynomial behavior (like Romberg's) will fail or at best be intolerably slow.

There are several ways out of this difficulty: one could use a Gauss–Jacobi formula or a modified Romberg algorithm that takes into account the precise nature of the singularity at 0 (see Appendix A, Table A.6); one could apply a further transformation of the variable t to reach an integral for which the trapezoidal rule is optimal (one such is described in §9.4); or one could use a good general-purpose automatic quadrature program.

When in search of a quick answer, the latter option is undeniably tempting. The industry standard, 20 years after its publication, is still QUADPACK [PdDKÜK83]. It is not necessary to leave the convenience of Octave, since its `quad` function is simply a link to QUADPACK's routine `QAG`, which combines a robust strategy of adaptive interval-halving with a basic formula of high degree and a conservative, safety-first approach to error estimation.

An Octave Session *(computing $q(\pi/4)$)*

```
>> function y=f2(t)
>>    global alfa
>>    z=t+i*t.*(2-t); dz=1+2i*(1-t);
>>    y=real(-i*exp(alfa*(log(z)+i./(2-z))).*dz);
>> end
>> global alfa; alfa=pi/4;
>> quad_options('abs',1e-10); quad_options('rel',0);
>> [ans,ierr,neval,error]=quad('f2',0,2)

   ans = 1.01123909053353
   ierr = 0
   neval = 651
   error =  3.85247389544929e-13
```

If we request a tighter tolerance, quad uses more function evaluations, but the answer does not change except for fluctuations at round-off level. This is typical behavior: QUADPACK really is almost paranoid in its approach to error estimation.

We are now able to evaluate $I(\alpha)$ to nearly machine accuracy. Has the back of Problem 9 been broken? Let's proceed to the optimization problem.

9.3 The Maximization Problem

There is a large literature on finding the extremum of a numerically defined function, dealing with matters such as making sure that an interval bracketing the extremum is maintained and that the algorithm is useful even when one starts far from the extremum. We don't need any of that here, since the function we are trying to maximize is very well behaved, and we know to graphical accuracy where the maximum is.

A simple algorithm [Pow64] for maximizing a function without using derivatives is the method of quadratic interpolation. In the case of a function $f(x)$ of one variable, the method is particularly simple. We start with three initial values x_1, x_2, x_3, ordered such that $f(x_1) < f(x_2) < f(x_3)$. At stage n, let x_{n+1} be the position of the maximum of the parabola through the three points $(x_j, f(x_j))$, $j = n-2, n-1, n$. One can think of this method as an analogue of the secant method (see p. 187) for solving a single nonlinear equation.

Let $h_n = x_{n+1} - x_n$. The following formula gives a convenient way of implementing the computation:

$$d_n = \frac{f(x_{n+1}) - f(x_n)}{h_n}, \qquad \theta_n = \frac{d_n}{2(d_{n-1} - d_n)},$$

$$h_n = \theta_{n-1} h_{n-2} + \left(\theta_{n-1} - \frac{1}{2}\right) h_{n-1}.$$

We start at $n = 3$; when $h_n = 0$ or if $f(x_{n+1}) \leqslant f(x_n)$ we stop, taking x_n as the result; otherwise we increment n and continue. This no-brakes implementation would not be suitable in a general-purpose routine, but on an easy optimization problem like this one, it works very well.

Actually, taking x_{n-1} as the result is often better. The reason is that d_n approaches an indeterminate $0/0$ form, and if h_{n-1} is too small, the calculation of h_n suffers badly from errors caused by cancellation. For a thorough discussion on the trade-off between errors caused by approximation and by cancellation, see the discussion on numerical differentiation in §9.7.

Applying this algorithm to the function I with initial values $\alpha_1 = \pi/4 - 0.01$, $\alpha_2 = \pi/4 - 0.005$, $\alpha_3 = \pi/4$, we obtain in IEEE double precision the results given in Table 9.1.

There is something highly disconcerting about Table 9.1. We have printed out the values $I(\alpha_n)$ to 17 significant digits, one more than the precision of the arithmetic deserves, so that it can be seen that the last two values are indeed different. In fact, they differ in the very last bit. Two distinct machine numbers near 3 cannot differ by anything smaller on this computer. Yet the two values of α agree to only 9 digits. It would actually be a little optimistic to claim that

$$\alpha_{\text{opt}} \doteq 0.78593\,3674,$$

Table 9.1. *Optimization by quadratic interpolation.*

n	α_n	$I(\alpha_n)$
1	**0.77**53981633974483	**3.0**278091521555970
2	**0.780**3981633974483	**3.032**0964090785236
3	**0.785**3981633974483	**3.0337**172716005982
4	**0.78593**75909509202	**3.03373258**56662893
5	**0.785933**7163733469	**3.033732586485**3986
6	**0.7859336743**674070	**3.0337325864854936**
7	**0.785933674**1864730	**3.03373258648549**45

since the calculation of I is so complicated; surely we cannot expect values of I to be accurate to the last bit.

We have calculated the value $I(\alpha_{\text{opt}})$ of the extremum to 15 digits, but unlike Problems 4 and 5, that is not good enough here: we are asked for its position α_{opt}, which is always harder. Let us analyze the reason for this.

In the neighborhood of a relative maximum α_0, we have

$$I(\alpha_0 + h) = I(\alpha_0) + \frac{1}{2}h^2 I''(\alpha_0) + O(h^3),$$

since $I'(\alpha_0) = 0$. Suppose that we are able to evaluate I to machine accuracy in IEEE double precision, about 16 digits. Then if α has 8 correct digits, we have h of the order of 10^{-8} and h^2 of the order of 10^{-16}, which is at machine round-off level; that is, any observed difference between $I(\alpha)$ and $I(\alpha_0)$ is spurious. We can therefore not expect to get more than 8 correct digits by a procedure based on evaluating $I(\alpha)$ for various values of α and stopping when we find no further improvement. Note that α_6 is indeed more accurate than α_7.

Figure 9.3 shows a graph of $I(\alpha)$ as a function of n for $\alpha = \alpha_0 + nh$, $\alpha_0 = 0.785933674$, $h = 10^{-10}$, $n = -500, -499, \dots, 499, 500$, using the contour integral formula for I evaluated by `quad`. The points are plotted as small dots that are not connected by lines. The graph is very far from the mathematical idealization of a one-dimensional curve in a two-dimensional plane; in fact, it looks as if it were drawn using a soft blunt pencil on coarse paper, not the way that highly accurate drawings are made. This appearance is caused by pseudorandom behavior of the round-off error. From the graph we can identify an interval of length perhaps $2 \cdot 10^{-9}$ containing the maximum, but that is the best we can do by just comparing two function values. (One can do better by judiciously using more function values; see §9.7.)

Unlike the case of Problem 2, where there is no alternative to working in higher precision, there is a standard way out of this difficulty. It is the familiar calculus trick of finding a zero of $I'(\alpha)$, together with extra information to make sure that the correct maximum is being found—in this case, the information that α_0 is near $\pi/4$.

To find $I'(\alpha)$, we need to differentiate under the integral sign. This would have been problematic for the original definition of I, since the standard theorem (see, e.g., [Apo74, p. 167]) on differentiation under the integral sign requires continuity with respect to x on

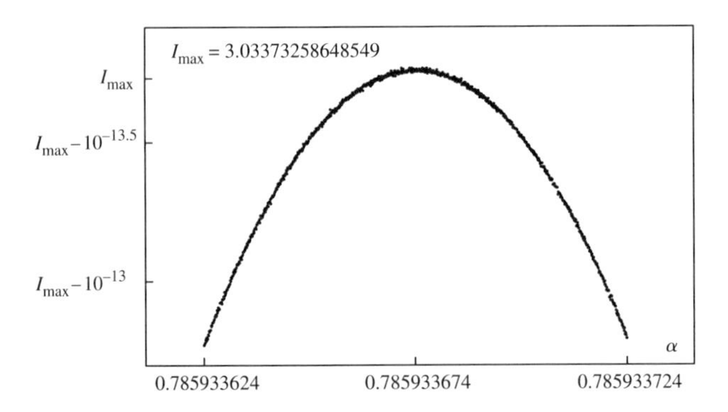

Figure 9.3. *Close-up graph of numerically evaluated values of* $I(\alpha)$.

[0, 2], which we do not have. There is no such problem along the contour (9.2), since

$$f_2^{(n)}(t,\alpha) = \frac{\partial^n f_2(t,\alpha)}{\partial \alpha^n} = \mathrm{Re}\left(-i f_0(z(t),\alpha)\left(\log z(t) + \frac{i}{(2-z(t))}\right)^n z'(t)\right), \quad (9.3)$$

which stays bounded over the closed interval.

Figure 9.4 shows a graph of $f_2'(t, \pi/4)$. The logarithmic term does not substantially add to the difficulty of the integral, and `quad` still copes easily with the problem of finding $q'(\alpha)$.

The zero-finding problem, by itself, holds no terrors. The behavior of I is parabola-like in the neighborhood of our excellent first approximation for the zero, and therefore I' is nearly linear. All methods for solving a nonlinear equation $I'(\alpha) = 0$ can be written in the form

$$\alpha_{n+1} = \alpha_n - \frac{I'(\alpha_n)}{m_n}, \quad (9.4)$$

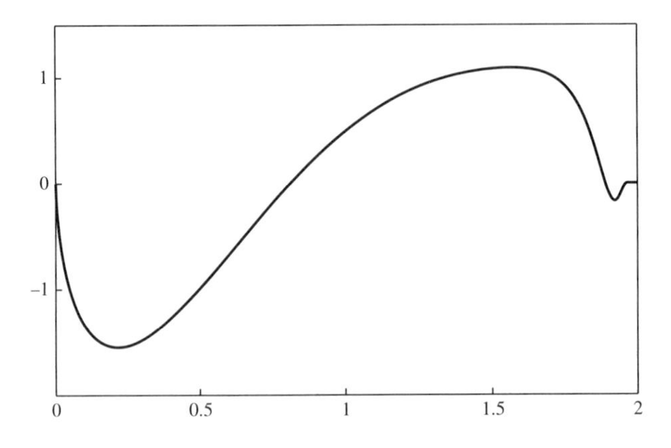

Figure 9.4. *Graph of* $f_2'(t, \pi/4)$.

Table 9.2. *Secant method applied to solve $I'(\alpha_{\text{opt}}) = 0$.*

n	α_n	$I'(\alpha_n)$
1	**0.78**03981633974483	**0.**5910382447996654
2	**0.785**3981633974483	**0.0**571975138044098
3	**0.785933**8804239271	−**0.0000**220102325740
4	**0.7859336743**534198	−**0.0000000003**255850
5	**0.7859336743503714**	−**0.00000000000000**12

where m_n is some approximation to the slope of the graph of I' in the neighborhood of the root. The simplest possible method is to take a constant value of m_n. Since $I(\alpha)$ is so dominated by the term $p(\alpha) = 2 + \sin(10\alpha)$, the slope of I' near the maximum is approximately $p''(\pi/4) = -100$; since p and q are both concave, the slope is actually a little steeper. The simple fixed-point iteration

$$\alpha_{n+1} = \alpha_n + 0.01 I'(\alpha_n)$$

is good enough to give a little better than one digit per iteration. The lower bound 100 for $|I''|$ near the maximum also tells us that when $|I'(\alpha_n)| = \epsilon$, the error in α_n is less than 0.01ϵ.

A slightly more sophisticated method for solving the nonlinear equation is the secant method, which uses two consecutive values to get a good approximation

$$m_n = \frac{I'(\alpha_n) - I'(\alpha_{n-1})}{\alpha_n - \alpha_{n-1}} \tag{9.5}$$

to the slope of the graph of I' at almost no cost. The result of the secant iteration, starting with initial values $\alpha_1 = \pi/4 - 0.005$, $\alpha_2 = \pi/4$ is given in Table 9.2. Allowing for round-off error in the numerical integration, we are confident that

$$\alpha_{\text{opt}} \doteq 0.78593\,36743\,5037$$

to 14 digits of accuracy.

Problem 9 is solved, but of course we would like to have another method for confirmation. Since clearly the behavior of the nonlinear solver does not depend on how the function values are obtained, for the next two sections we devote our attention to other methods for calculating $q(\alpha)$ and $q'(\alpha)$. Our benchmark is now the evaluation of

$$q'(\tfrac{\pi}{4}) \doteq 0.01906583793480.$$

The Octave routine `quad` applied to $f_2'(\cdot, \pi/4)$, with the same settings as before, gives the value 0.0190658379348022 using 693 function evaluations.

9.4 Double-Exponential Formulas

It is well known that the trapezoidal and midpoint rules for numerical integration, humble though their origins might be, are extremely effective when the integrand is smooth and periodic. In particular, this is the case when all derivatives of the integrand vanish at both endpoints of the interval. One would therefore expect that the trapezoidal rule should be particularly effective for integrals over $(-\infty, \infty)$; and in fact, it can be shown (see, e.g., [LB92, p. 48] or the discussion on p. 70 of this book) that when $g(z)$ is analytic in the strip $|\mathrm{Im} z| < d$, and a_0 is some fixed real number,

$$\left| \int_{-\infty}^{\infty} g(t)\, dt - h \sum_{k=-\infty}^{\infty} g(kh + a_0) \right| \leqslant \frac{\|g\|}{e^{2\pi d/h} - 1}, \tag{9.6}$$

where the exact meaning of the norm $\|g\|$ is not needed for this discussion.

From (9.6), we see that the following conditions are necessary for the trapezoidal rule to be efficient:

1. The doubly infinite sum must converge fast.
2. h must be small in relation to d.

The double-exponential method was developed by Mori in collaboration with several other Japanese mathematicians over many years. The basic idea is to transform one's original integral $\int_a^b f(x)\, dx$ to $\int_{-\infty}^{\infty} g(t)\, dt$ by the transformation $x = w(t)$, so that $g(t) = w'(t) f(w(t))$, in such a way that $|g(t)|$ behaves for large $|t|$ like $\exp(-c \exp(|t|))$ for some constant c, so that no matter what accuracy one is aiming at, the sum behaves in practice like a finite sum. Not only does the procedure yield a very effective integration formula, but it is also robust against endpoint singularities. For a survey of the method, see [MS01]. We will only enlarge on the practical application of the method here; in §3.6.1 the reader can find a detailed discussion on the issues that influence the convergence behavior of trapezoidal sums.

For example, when a and b are both finite, the standard transformation is

$$w(t) = (a + b)/2 + (b - a)/2 \tanh(\sinh ct)$$

for some positive constant c. The exact value of c is not crucial, but $c = \pi/2$ is recommended. This transformation is suitable for a wide range of functions, including functions that become infinite at the endpoints, but it is overkill for functions that already decay exponentially at one or both endpoints, such as the function in (9.3). The appropriate transformation in the case where f decays exponentially at b is

$$w(t) = (a + b)/2 + \big((b - a)/2\big) \tanh\left(t - e^{-ct}\right)$$

for some positive constant c.[75] Again, the value of c is not crucial, and in this case $c = 1$ is the usual choice.

For all the double-exponential formulas, it is important to be careful near the endpoints to avoid overflow, underflow, and unnecessary loss of precision. In particular, if the integrand becomes infinite at an endpoint, say at a, then $w(t) - a$ should be analytically simplified to avoid cancellation and the routine that evaluates the integrand should be written to take that

[75] This formula does not appear explicitly in [MS01] but can readily be derived by applying to [MS01, form. (1.15)] the same reasoning that transforms [MS01, form. (1.16)] into [MS01, form. (1.17)].

information into account. In the case of the function defined in (9.3), the integration interval is $[0, 2)$ and there is exponential decay at the right endpoint, so the transformation function is

$$w(t) = \frac{2}{1 + e^{2(e^{-t}-t)}}. \tag{9.7}$$

An Octave Session *(computing $q(\alpha)$ and its first two derivatives at $\alpha = \pi/4$)*

The range $-5.3 \leqslant t \leqslant 4.2$ has been chosen so that $g(t)$ does not underflow, and the step size $h = 1/16$ was found by repeatedly halving the step size, starting at $h = 1/2$, until two successive answers agree to more than eight digits. That is sufficient, because each halving of the step size doubles the number of correct digits.

```
>> function y=df2(t,n)   % n-th derivative of f2
>>    global alfa
>>    z=t+i*t.*(2-t); dz=1+2i*(1-t); f=log(z)+i./(2-z);
>>    y=-i*exp(alfa*f).*dz;
>>    for k=1:n, y=f.*y; end
>>    y=real(y);
>> end
>> global alfa; alfa=pi/4;
>> h=1/16; t=-5.3:h:4.2;
>> x=2./(1+exp(2*(exp(-t)-t)));
>> w=h*sech(t-exp(-t)).^2.*(1+exp(-t));
>> for n=0:2, wf=w.*df2(x,n); q=sum(wf), sum(abs(wf)>eps*abs(q)), end

   q = 1.01123909053353
   ans = 78
   q = 0.0190658379348029
   ans = 82
   q = -1.89545525464014
   ans = 82
```

The number of function evaluations is certainly much less than that of `quad`. It is often the case that the double-exponential technique, when it is applicable, outperforms just about any other automatic integration algorithm. In fairness to `quad`, it must be said that it can handle integrands that are not analytic on the open interval, which double-exponential methods cannot do.

9.5 Transformation to a Fourier Integral

The usual double-exponential method fails when the integrand is highly oscillatory. In particular, it cannot cope with the function $f_1(\cdot, \pi/4)$. The solution in the previous section, of using a different contour in the complex plane, is not always available (for example, software for evaluating the integrand at complex arguments might be lacking).

A technique that is often useful is to transform the integration interval to $[0, \infty)$ in an attempt to get regularly spaced zeros. In this case, we make the substitution

$$x(t) = \frac{2t}{1+t}, \quad x'(t) = \frac{2}{(1+t)^2},$$

giving

$$q(\alpha) = \int_0^\infty x^\alpha(t)x'(t)\sin\bigl(\alpha(1+t)/2\bigr)\,dt \tag{9.8}$$

and

$$q'(\alpha) = 1/2\int_0^\infty (1+t)x^\alpha(t)x'(t)\cos\bigl(\alpha(1+t)/2\bigr)\,dt + \int_0^\infty \log x(t)x^\alpha(t)x'(t)\sin\bigl(\alpha(1+t)/2\bigr)\,dt. \tag{9.9}$$

The integral $q(\alpha)$ is absolutely convergent, so there is no problem with differentiation under the integral sign. Figure 9.5 shows a graph of the integrand appearing in (9.8) and (9.9) with $\alpha = \pi/4$.

In both cases, the integrand is 0 when $t = 0$, although one cannot easily see that at the scale of this graph. Even so, the integral is not all that easy by standard methods. The same singularities are still present at $t = 0$ as in the case of the contour integral, and the decay of the integrand is not exponential. Truncation of the infinite interval is not appropriate.

However, the integrand belongs to a quite common family of integrands for which very effective methods are known. Integrals of the form

$$F(s;\omega,\theta) = \int_0^\infty s(x)\sin(\omega x + \theta\pi)\,dx \tag{9.10}$$

arise in Fourier analysis (the cases $\theta = 0$ and $\theta = \frac{1}{2}$ being the familiar sine and cosine transforms) and have in consequence been studied in much greater depth than general oscillatory integrals, and many numerical methods are known. We will discuss only one here, a recent one due to Ooura and Mori [OM99], which is an ingenious variation of the double-exponential technique. As in the case of other double-exponential methods, the singularity at $t = 0$ is handled effortlessly.

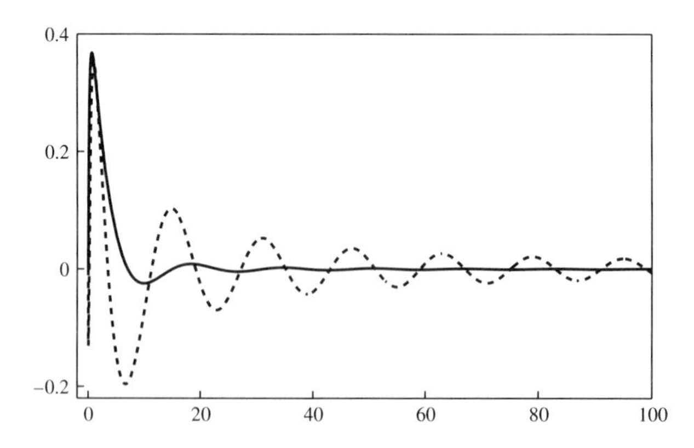

Figure 9.5. *Graph of integrands after the transformation by $x(t) = 2t/(1+t)$. The solid line is the integrand in (9.8), and the dotted line, the integrand in (9.9).*

The basic idea is the following: when transforming the integrand in (9.10) by $x = w(t)$ to obtain a transformed integrand $g(t) = w'(t)f(w(t))$, where $f(x) = s(x)\sin(\omega x + \theta\pi)$, do not aim at double-exponential decay for $g(t)$ for all t sufficiently large, but do so *only for the quadrature points* t_k. Thus, we require that $|g(t_k)|$ behave for large $|t_k|$ like $\exp(-c\exp(|t_k|))$. That is clearly equally sufficient for fast convergence of the sum. In the case of the Fourier integral (9.10), the crucial fact is that the sine term has evenly spaced zeros. If the quadrature points t_k tend to those zeros double-exponentially for large k, then $g(t_k)$ will have the required behavior.

Suppose that we can find a function ϕ such that $\phi(t) \to 0$ as $t \to -\infty$, and $\phi(t) \to t$ as $t \to \infty$, the limiting behavior in both cases being double-exponential. Then for constants $c_1 > 0$ and c_2, $w(t) = c_1\phi(t - c_2)$ also maps the real axis to $(0, \infty)$ and $w(t) \to 0$ as $t \to -\infty$, but now $w(t) \to c_1(t - c_2)$ as $t \to \infty$, the limiting behavior in both cases being double-exponential. Thus, taking $a_0 = 0$ in (9.6) so that $t_k = kh$, the function f needs to be evaluated at points that tend double-exponentially to $c_1(kh - c_2)$. Since $f(t) = 0$ when $\omega t + \theta\pi = k\pi$, we choose c_1 and c_2 such that $\omega c_1(kh - c_2) + (\theta - k)\pi = 0$ identically in k, giving

$$w(t) = \frac{\pi}{\omega h}\phi\,(t - \theta h)\,. \tag{9.11}$$

A family of functions ϕ with the desired property is

$$\phi(t) = \frac{t}{1 - \exp(-(2t + \alpha(1 - e^{-t}) + \beta(e^t - 1)))},$$

where α and β are positive numbers that may depend on h. After considerable theoretical investigation and practical experimentation, Ooura and Mori [OM99] recommended

$$\beta = \frac{1}{4}, \quad \alpha = \beta\left(1 + \frac{\log(1 + \pi/h)}{4h}\right)^{-1/2}.$$

The motivation for this somewhat esoteric formula for α is that when s has a singularity at some complex number z_0, the image $w^{-1}(z_0)$ of z_0 is bounded away from the real axis, so that the quantity d/h in (9.6) varies in inverse proportion to h. (The more obvious function $\phi(t) = t/(1 - \exp(-k\sinh t))$, for some $k > 0$, does not have this desirable property; see [OM99].)

The implementation of the method must be done with great care to avoid unnecessary loss of accuracy. There are two pitfalls for the unwary:

1. The function $e_1(x) = (e^x - 1)/x$ must be evaluated to full precision for small x. This is a well-known problem, and in fact standard IEEE arithmetic provides the function $2^x - 1$ (with the aid of which $e_1(x)$ can easily be evaluated) in hardware; but few programming languages have a library routine for e_1. There are two more or less satisfactory workarounds: either one can put $e_1(x) = 2e^{x/2}\sinh(x/2)/x$ and rely on the presumably good accuracy of the built-in hyperbolic sine, or one can use the

algorithm (due to W. Kahan)

$$y = e^x, \qquad e_1(x) = \frac{y-1}{\log y}.$$

Of course, in the trivial case when x is so close to 0 that $y = 1$ in machine arithmetic, then also $e_1(x) = 1$. For an explanation of why the second method works, see [Hig96, §1.14.1].

2. The value of $\sin(\omega w(kh) + \theta\pi)$ when k is large must be calculated to high relative precision. This cannot be achieved by passing the large number $\omega w(kh) + \theta\pi \approx k\pi$ (as $k \to \infty$) to the built-in sine routine of one's favorite language, since the reduction of the argument modulo 2π leaves a number with high relative error.

Assuming that we have a routine for e_1, we can compute its derivative

$$e_1'(x) = e_1(x) - e_2(x), \qquad \text{where} \quad e_2(x) = (e_1(x) - 1)/x.$$

Then we have, using logarithmic differentiation,

$$\phi(t) = \frac{1}{v(t)e_1(-tv(t))}, \qquad \text{where} \quad v(t) = 2 + \alpha e_1(-t) + \beta e_1(t);$$

$$\phi'(t) = \phi(t)\left(-\frac{v'(t)}{v(t)} + (tv'(t) + v(t))\left(1 - \frac{e_2(-tv(t))}{e_1(-tv(t))}\right)\right).$$

The expressions for $\phi(t)$ and $\phi'(t)$ are well behaved even when t is very close to 0, provided that e_2 can be accurately evaluated for small arguments. This is a more difficult task than in the case of $e_1(t)$, but fortunately it is not really necessary in this particular instance. By (9.11), ϕ is evaluated at $(k-\theta)h$ for integer values of k. By (9.8) and (9.9), $\theta = \alpha/2\pi \approx 1/8$ for the integrals involving the sine and $\theta \approx 5/8$ for the integral involving the cosine, neither of which is critically close to an integer; therefore, the argument $t_k - \theta h = (k - \theta)h$ at which ϕ is to be evaluated cannot for any integer k become critically close to 0.

Turning to the second problem, we need to evaluate

$$\begin{aligned} \sin(\omega w(kh) + \theta\pi) &= \sin\left(\frac{\pi(\phi(w_k) - w_k)}{h} + \pi k\right) \\ &= (-1)^k \sin\left(\frac{\pi(\phi(w_k) - w_k)}{h}\right), \end{aligned} \tag{9.12}$$

where $w_k = (k-\theta)h$. It suffices to compute $\phi(t) - t$ to high relative accuracy when $t \gg 0$, which is easy, since

$$\phi(t) - t = \frac{te^{-tv(t)}}{1 - e^{-tv(t)}}.$$

Note that the right-hand side of (9.12) should only be used for fairly large values of t, say at the stage where $|\phi(t) - t| < 0.1$; for the others we stay with $\sin(\omega w(kh) + \theta\pi)$.

We now have all the ingredients necessary for the Ooura–Mori method, not without a good deal of labor. But it has been instructive to see the difference between the elegant mathematical description of a great idea and the tortuous and finicky details that lie behind its meticulous numerical implementation.

An Octave Session *(calculating* $q = q(\pi/4)$ *and* $dq = q'(\pi/4)$*)*

All the precautions above have already been packaged into a subroutine

```
q_ossinf(M,omega,theta)
```

whose code may be found on the web page for this book. The subroutine calculates the points t_k and weights w_k such that

$$F(s;\omega,\theta) \approx \sum w_k s(t_k),$$

where the integration step size is $h = \pi/M$. The choice $M = 16$ is the result of experiment, starting as before at $M = 2$ and doubling M until the results agree to more than 8 digits.

```
>> function y=intfun (t, alpha)
>> x=2*t./(1+t); dx=2./(1+t).^2; y=dx.*x.^ alpha;
>> end
>> alfa=pi/4; [ts,ws]=q_ossinf(16,alfa/2,1/8);
>> [tc,wc]=q_ossinf(16,alfa/2,5/8);
>> wf=ws.*intfun(ts,alfa); q=sum(wf), sum(abs(wf)>eps*q)

   q = 1.01123909053353
   ans = 49

>> x=2*ts./(1+ts);
>> wf1=wc.*intfun(tc,alfa).*(1+tc)/2; wf2=ws.*log(x).*intfun(ts,alfa);
>> dq=sum(wf1)+sum(wf2), sum(abs(wf1)>eps*dq), sum(abs(wf2)>eps*dq)

   dq = 0.0190658379347639
   ans = 51
   ans = 51
```

Thus, only 102 points, involving no complex arithmetic, are required at this level of precision to evaluate $q(\pi/4)$ and $q'(\pi/4)$. This compares well with the 82 complex points required by the usual double-exponential formula.

9.6 An Analytical Surprise

When one commands a computer algebra package like Maple or *Mathematica* to evaluate the integral $q(\alpha)$, one gets just what one might expect from a problem that, after all, is a Trefethen challenge problem: nothing happens, the integral is returned unchanged. But if we help just a little, asking instead for the equivalent $\int_0^2 (2-x)^\alpha \sin(\alpha/x)\,dx$, we get a surprise. Both packages can do the integral analytically—no numerical integration is necessary!

A Maple Session

```
> int( (2-x)^alpha * sin(alpha/x), x=0..2);
```

$$\frac{1}{4}\sqrt{\pi}\alpha\,\Gamma(\alpha+1)$$
$$\left(\frac{2^{-1+\alpha}\sqrt{\pi}\alpha\,\text{hypergeom}\left([-1/2\,\alpha+1/2,1-1/2\,\alpha],[2,3/2,3/2],-1/16\,\alpha^2\right)}{\Gamma(\alpha)}\right.$$
$$+2\sqrt{2}\sum_{k=0}^{\infty}\frac{\pi^{-3/2}}{2^{-3/2-\alpha+2k}\Gamma(2+2k)\,\Gamma(1+2k)}\left((-1)^{2k}\left(-\psi\left(\frac{1}{2}+k\right)+\pi\,\tan(\pi\,k)\right.\right.$$
$$+\psi\left(\frac{1}{2}-\frac{1}{2}\alpha+k\right)+\pi\,\tan\left(\frac{1}{2}\pi\,\alpha-\pi\,k\right)+\psi\left(-\frac{1}{2}\alpha+k\right)$$
$$\left.-\pi\,\cot\left(\frac{1}{2}\pi\,\alpha-\pi\,k\right)-2\,\psi(1+k)-\psi\left(\frac{3}{2}+k\right)-4\,\ln(2)+2\,\ln(\alpha)\right)$$
$$2^{-4k}\alpha^{2k}\sec(\pi\,k)\cos\left(\frac{1}{2}\pi\,\alpha-\pi\,k\right)\sin\left(\frac{1}{2}\pi\,\alpha-\pi\,k\right)\left(2^{2k}\right)^2$$
$$\left.\left.\left(-\frac{1}{2}-\frac{1}{2}\alpha+k\right)\Gamma(-1-\alpha+2k)\right)\right)$$

(The result is not a screen shot, but Maple's own LaTeX output, edited in various ways to make it readable.) Here $\psi(z)=\Gamma'(z)/\Gamma(z)$ is the digamma function. The formula is rather daunting, although we can immediately see quite a few simplifications that can be made: for example, $(-1)^{2k}=1$ and $\pi\tan(\pi k)=0$. Still, Maple can, of course, evaluate it. With `Digits:=50` we obtain:

$$q(\pi/4)\doteq 1.0112\,39090\,53353\,25262\,70537\,50657\,49498\,85803\,05492\,49392$$

which, it is encouraging to note, agrees with the 15 digits we earlier got from `quad`.

A *Mathematica* Session

```
Simplify[∫_0^2 (2 - x)^α Sin[α/x] dx, α > 0] // TraditionalForm
```

$$\sqrt{\pi}\;\Gamma(\alpha+1)\,G_{2,4}^{3,0}\left(\frac{\alpha^2}{16}\,\middle|\,\begin{matrix}\frac{\alpha+2}{2},\frac{\alpha+3}{2}\\ \frac{1}{2},\frac{1}{2},1,0\end{matrix}\right)$$

To understand how a computer program managed to arrive at this formula involving *Meijer's G function*, see [MT03, eqs. (07.34.21.0084.01) and (07.34.03.0055.01)]. *Mathematica*, too, can evaluate its own formula, and if we ask for enough digits, the numerical result agrees with that of Maple.

According to [Luk75, form. 5.3.1(1)], the $G_{2,4}^{3,0}$ instance of Meijer's G function is defined by

$$G_{2,4}^{3,0}\left(z \,\middle|\, \begin{matrix} (\alpha+2)/2,\ (\alpha+3)/2 \\ 1/2,\ 1/2,\ 1,\ 0 \end{matrix}\right)$$
$$= \frac{1}{2\pi i}\int_C \frac{\Gamma^2(1/2-s)\Gamma(1-s)}{\Gamma(1+s)\Gamma(1+\alpha/2-s)\Gamma(3/2+\alpha/2-s)}\, z^s\, ds,$$

where the contour C is a loop starting and ending at $+\infty$ and encircling all poles of $\Gamma^2(1/2-s)\Gamma(1-s)$ once in the negative direction.

The contour integral can be evaluated in terms of residue series. The integrand has double poles at $s = 1/2+n, n = 0, 1, 2, \ldots$, and simple poles at $s = 1+n, n = 0, 1, 2, \ldots$. The calculation of the residues at these poles is straightforward, though tedious in the case of the double poles. (Alternatively, one may replace one factor $\Gamma(1/2-s)$ in the integrand by $\Gamma(1/2+\delta-s)$, whereupon the poles at $s = 1/2+n$ become simple, at the expense of an additional set of simple poles at $s = 1/2+\delta+n$, $n = 0, 1, 2, \ldots$. Then by [Luk75, form. 5.3.1(5)] the three residue series equal a sum of three generalized hypergeometric functions ${}_2F_3$, in which one should take limits as $\delta \to 0$.) Omitting further details we present the final result for $q(\alpha)$ in the following hypergeometric-function form:

$$q(\alpha) = 2^{\alpha+1}\pi\alpha z \sum_{n=0}^{\infty} \frac{(1/2-\alpha/2)_n(1-\alpha/2)_n}{(2)_n((3/2)_n)^2\, n!}(-z)^n$$
$$+2^{\alpha+2}\pi\cot(\pi\alpha) z^{1/2} \sum_{n=0}^{\infty} \frac{(-\alpha/2)_n(1/2-\alpha/2)_n}{(1/2)_n(3/2)_n(n!)^2}(-z)^n$$
$$-2^{\alpha+1} z^{1/2} \sum_{n=0}^{\infty} \frac{(-\alpha/2)_n(1/2-\alpha/2)_n}{(1/2)_n(3/2)_n(n!)^2}(-z)^n\,(\log z + h_n),$$

where $z = \alpha^2/16$,

$$h_n = \psi(-\alpha/2+n) + \psi(1/2-\alpha/2+n) - \psi(1/2+n) - \psi(3/2+n) - 2\psi(1+n),$$

and $(a)_n$ is the Pochhammer symbol, defined for $a \in \mathbb{C}$ by

$$(a)_0 = 1, \quad (a)_n = \prod_{m=0}^{n-1}(a+m) \quad \text{for} \quad n = 1, 2, 3, \ldots.$$

The first and second series are equal to the generalized hypergeometric functions

$${}_2F_3(1/2-\alpha/2, 1-\alpha/2; 2, 3/2, 3/2; -z), \qquad {}_2F_3(-\alpha/2, 1/2-\alpha/2; 1/2, 3/2, 1; -z),$$

respectively, while the third series is expressible as the derivative of a generalized hypergeometric function, viz.,

$$\frac{\Gamma(1/2)\Gamma(3/2)}{\Gamma(-\alpha/2)\Gamma(1/2-\alpha/2)}$$
$$\times \frac{d}{dt}\left(\sum_{n=0}^{\infty}\frac{\Gamma(-\alpha/2+n+t)\Gamma(1/2-\alpha/2+n+t)}{\Gamma(1/2+n+t)\Gamma(3/2+n+t)\Gamma^2(1+n+t)}(-1)^n z^{n+t}\right)\Bigg|_{t=0}.$$

It has been verified that the above expansion for $q(\alpha)$ agrees with the Maple result on p. 194.

From the point of view of programming rather than mathematical elegance, the formula for $q(\alpha)$ can be optimized to give

$$q(\alpha) = 2^{\alpha}\alpha\left(2\pi z\sum_{n=0}^{\infty}(-z)^n s_n + \sum_{n=0}^{\infty}(-z)^n t_n\right); \tag{9.13}$$
$$s_n = \prod_{m=1}^{n}\frac{(\alpha-2m)(\alpha-2m+1)}{m(m+1)(2m+1)^2};$$
$$t_n = \left(1-\log\frac{\alpha}{2}-\psi(\alpha+1)-2\gamma+u_n\right)\prod_{m=1}^{n}\frac{(\alpha-2m+1)(\alpha-2m+2)}{m^2(2m-1)(2m+1)};$$
$$u_n = \sum_{m=1}^{n}\left(\frac{1}{\alpha-2m+1}+\frac{1}{\alpha-2m+2}+\frac{1}{m}+\frac{1}{2m-1}+\frac{1}{2m+1}\right),$$

where $z = \alpha^2/16$, and $\gamma \doteq 0.5772156649$ is Euler's constant. Only one evaluation of the digamma function is required, $\psi(\alpha+1)$, and, somewhat surprisingly, the gamma function does not appear in the formula at all. These formulas have been implemented in the PARI/GP file `meijerg.gp`, given on the web page for this book. Both series converge quite fast for α near $\pi/4$. The terms behave asymptotically like $(-z)^n/(n!)^2$, and since $z \approx 0.04$, one already gets 16 digits with six terms.

9.7 Accurate Numerical Differentiation

The representation of q in terms of the G function and the availability of the series (9.13) give a promising approach to the accurate solution of Problem 9, but there is a difficulty: we know that the optimization method, by itself, yields only half the precision to which function values can be computed. We can work in higher precision, true. Or we can differentiate the result of the *Mathematica* session on p. 194, requiring term-by-term differentiation of (9.13).[76] The resulting expression contains the trigamma function $\psi_1 = \psi'$, which *Mathematica* and Maple have but PARI/GP does not.

Another approach is to find a zero of I' without analytical differentiation by using numerical derivatives for q'. The basic idea is very simple: for some step size h,

$$q'(\alpha) = \frac{q(\alpha+h)-q(\alpha)}{h} + O(h). \tag{9.14}$$

[76] This approach has been used to solve Problem 9 to 10,000 digits of accuracy; see Appendix B.

Unfortunately, we cannot shrink h indefinitely because the cancellation in the numerator throws away significant digits. If $h = 10^{-t}$, the hoped-for number of correct digits (in exact arithmetic, that is) is approximately t and the number of cancelled digits is also approximately t. If we work in d-digit arithmetic, there are $d - t$ uncontaminated digits remaining. So the best we can do is to achieve an accuracy of about $\min(t, d - t)$ digits. The optimal choice is $t = d/2$, which is what we could achieve before—no progress yet.

But one can do better by centered differences: replacing $O(h)$ by $c_1 h + O(h^2)$ in (9.14), and averaging the cases with step sizes h and $-h$, we get

$$q'(\alpha) = \frac{q(\alpha + h) - q(\alpha - h)}{2h} + O(h^2).$$

The number of digits lost to cancellation is much the same as before, but the hoped-for number of correct digits is now $2t$. We can get $\min(2t, d - t)$ digits, so the optimal choice is $t = d/3$, giving $2d/3$ correct digits.

One level of Richardson extrapolation (see Appendix A, p. 235), for which we need to replace $O(h)$ in (9.14) by $c_1 h + c_2 h^2 + c_3 h^3 + O(h^4)$, gives us a further improvement:

$$q'(\alpha) = \frac{4}{3}\frac{q(\alpha + h) - q(\alpha - h)}{2h} - \frac{1}{3}\frac{q(\alpha + 2h) - q(\alpha - 2h)}{4h} + O(h^4).$$

Now the hoped-for number of correct digits is $4t$, we can get $\min(4t, d - t)$ digits, and the optimal choice is $t = d/5$, giving $4d/5$ correct digits.

We can continue along these lines, with more sophisticated numerical differentiation formulas, but there is a law of diminishing returns. We have already 80% of the working precision. Another level of extrapolation will give about $6d/7$ correct digits, about 86% of the working precision. Figure 9.6 shows a graph of the actual error, using the series formula (9.13) for $q(\alpha)$, plotted against h.

The lowest error is approximately where we predicted it should be. One can clearly see the irregular effect of round-off when h is too small, and the smooth effect of truncation

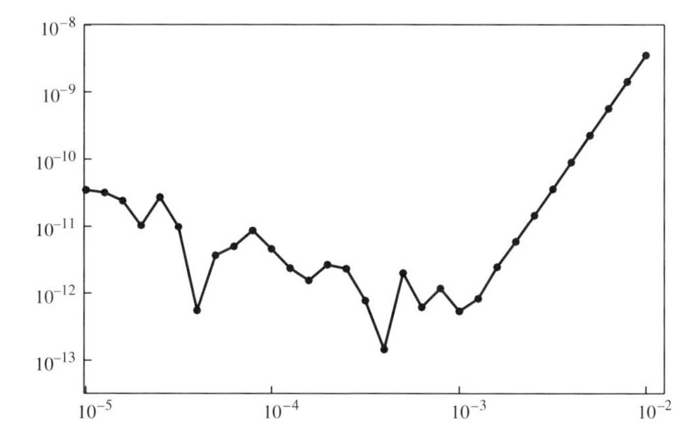

Figure 9.6. *Error in numerical differentiation, using central differences and one step of extrapolation, as a function of step size.*

Table 9.3. *Secant method with numerical differentiation.*

n	α_n	$I'(\alpha_n)$
1	**0.78**03981633974483	**0.**5910382447969147
2	**0.785**3981633974483	**0.0**571975138027900
3	**0.785933**8804239131	−**0.0000**220102314507
4	**0.7859336743**534163	−**0.000000000**3278613
5	**0.7859336743503**467	**0.00000000000**19298
6	**0.7859336743503**647	−**0.00000000000**25024
7	**0.7859336743503**545	**0.000000000000**5199
8	**0.7859336743503**563	−**0.00000000000**12696

when h is too large. Note that it is better to take h a little too small than a little too large. On this log–log scale, the approximately straight line when the error is dominated by $O(h^4)$ is approximately four times as steep as the jagged line when the error is dominated by $O(h^{-1})$.

Using the secant method and the numerical derivative for q', we obtain the results given in Table 9.3.

Thirteen digits are correct, and the 14th is close. This is a little better than we have a right to expect; the reason is that near the optimal value of α, a small change in α affects q' (which we cannot compute very accurately) only by about 0.06 of the amount by which it affects p' (which we can compute very accurately). Thus, the error in I' induced by the inaccurate value of q' is only about 0.06 of the error in q'.

Note that the numerically differentiated values do not converge to zero. In general, in a numerical calculation, round-off errors (if only you notice their presence!) are your friends; they warn you when a number is not very accurate.

Chapter 10

Hitting the Ends

Folkmar Bornemann

Monte Carlo is an extremely bad method; it should be used only when all alternative methods are worse.
—Alan Sokal [Sok97, p. 132]

Separation of variables is of very limited utility but when it works it is very informative.
—Jeffrey Rauch [Rau91, p. 211]

Problem 10

A particle at the center of a 10×1 *rectangle undergoes Brownian motion (i.e., two-dimensional random walk with infinitesimal step lengths) until it hits the boundary. What is the probability that it hits at one of the ends rather than at one of the sides?*

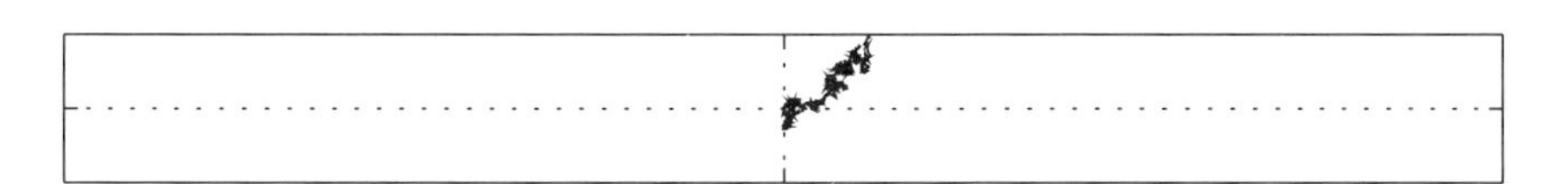

Figure 10.1. *A sample path hitting at the upper side.*

The *ends* are the short sides, of length 1; the *sides* are the long ones, having length 10. A glance at Figure 10.1 will convince the reader that this yields a probability p that is much smaller than the other way around with probability $1 - p$. Indeed, as we will see later, $p \approx 4 \cdot 10^{-7}$. Thus, 10 digits of $1 - p$ correspond to just 4 digits of p; calculating the smaller probability to 10 significant digits is much more demanding.

The Menu for a Seven-Course Meal. Problem 10 is an extremely rich topic and offers many interesting mathematical dishes. The *hors d'oeuvre*, served in §10.4, explores the power of a simple stochastic algorithm and gets the order of magnitude of p. The *deterministic soup*, served in §10.2, reformulates the problem as one about a partial differential equation. The *numerical analysis course*, served in §10.3, applies standard finite differences and convergence acceleration to obtain—assured by tools from scientific computing—at least 10 significant digits. The *real analysis course*, served in §10.4, leaves the algorithmic approaches behind and switches to analytical techniques. We get a representation of p as an exponentially fast converging alternating series, allowing us to prove the correctness for any number of significant digits. As a by-product we get the simple approximation

$$p \doteq \tfrac{8}{\pi} e^{-5\pi} \text{ to 13 significant digits.}$$

The *complex analysis course*, served in §10.5, uses conformal mappings to represent p as the solution of a scalar transcendental equation. The *formula platter*, served in §10.6, offers assorted mature samples from elliptic function theory; especially worth mentioning is a closed-form solution based on Jacobi's elliptic modular function λ,

$$p = \tfrac{2}{\pi} \arcsin \sqrt{\lambda(10i)},$$

that can be evaluated to 10,000 significant digits in a matter of seconds. The *dessert*, served in §10.7, explores the number $\lambda(10i)$ with the help of some wonderful results by Ramanujan, culminating in a closed-form solution in terms of elementary functions,

$$p = \tfrac{2}{\pi} \arcsin \left((3 - 2\sqrt{2})^2 (2 + \sqrt{5})^2 (\sqrt{10} - 3)^2 (5^{1/4} - \sqrt{2}\,)^4\right).$$

Bon appetit!

10.1 A First Look: Why Not Monte Carlo?

At a first look, a stochastic problem might be well served by a randomized algorithm of Monte Carlo type. Such an algorithm is quickly designed and implemented if we only know that Brownian motion is *isotropic*: started at the center of a circle, the probability distribution of hitting a point of that circle is *uniform*. This property, which we will prove in Lemma 10.2, immediately suggests the following algorithm; in Appendix C.1 the reader will find a short C implementation of it:

> *Single run:* Start at the center of the rectangle and take the largest circle that fits into the rectangle. Go to a random point on that circle. Repeat until the point is within distance h of the boundary. Count as a hit of the ends or the sides, accordingly.
>
> *Statistics:* Repeat the single run N times and take the relative frequency of hits of the ends as an approximation of the probability p.

Figure 10.2 shows some typical runs. The algorithm, called *walk on spheres*, has been known since about 1956 and—for reasons that will become clear in the next section—its n-dimensional generalization has been used as a pointwise solver for Laplace's equation [DR90].

Since the algorithm yields an approximation of the form $p \approx k/N$ with $k \in \mathbb{N}_0$, we observe that for smaller p we have to choose N larger to get significant results. To obtain a feel for the problem, we will look at the 10×1 rectangle, the 1×1 rectangle, and the $\sqrt{3} \times 1$ rectangle. For symmetry reasons, the former obviously has $p = 1/2$; the probability

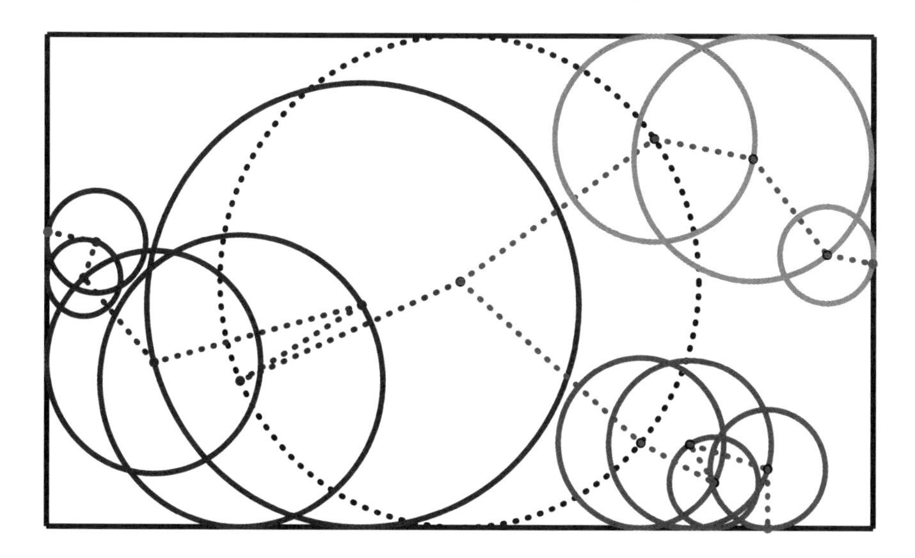

Figure 10.2. *Three runs of the random walk on spheres: $\sqrt{3} \times 1$ rectangle.*

p of the latter will be explicitly determined in §10.4.1 and can be used for a "reality check" of a given method.

The application of a Monte Carlo method should always be accompanied by a careful statistical assessment of the results; cf. [Sok97]. However, since Table 10.1 shows that we utterly fail to get 10 digits anyway, we are less demanding and simply infer the coarse approximations

$$p|_{\sqrt{3} \times 1 \text{ rectangle}} \approx 0.166, \qquad p|_{10 \times 1 \text{ rectangle}} \approx 4 \cdot 10^{-7}. \tag{10.1}$$

Table 10.1 nicely shows that h enters the run-time complexity just logarithmically; the error is affected by some power of h; cf. [DR90, p. 131]. The bulk of the error is of a statistical nature, which restricts the accuracy of Monte Carlo methods in general.

Table 10.1. *Walk on spheres: results of some experiments.*

Rectangle	N	h	p	Run time
1×1	10^7	10^{-4}	$4.997816 \cdot 10^{-1}$	26 s
	10^8	10^{-4}	$4.999777 \cdot 10^{-1}$	4 min 25 s
$\sqrt{3} \times 1$	10^7	10^{-4}	$1.667260 \cdot 10^{-1}$	29 s
	10^8	10^{-4}	$1.666514 \cdot 10^{-1}$	4 min 54 s
10×1	10^7	10^{-2}	$4 \cdot 10^{-7}$	15 s
	10^7	10^{-4}	$2 \cdot 10^{-7}$	29 s
	10^8	10^{-2}	$3.7 \cdot 10^{-7}$	2 min 38 s
	10^8	10^{-4}	$4.1 \cdot 10^{-7}$	4 min 57 s

Loosely speaking, a Monte Carlo algorithm constructs by means of a *simple* stochastic process, which we have called a "single run" above, a random variable X, whose expectation value is just the answer to the problem at hand, $E(X) = p$. Drawing N independent samples X_k we take the sample mean as an approximation of p,

$$S_N = \frac{1}{N}\sum_{k=1}^{N} X_k.$$

Now, the *mean quadratic error* of the Monte Carlo algorithm is just the variance of S_N and is readily related to the variance of X,

$$E\left(|S_N - p|^2\right) = \sigma^2(S_N) = \tfrac{1}{N}\sigma^2(X).$$

The natural concept of error is therefore the concept of *absolute error* as opposed to the concept of correct digits, that is, relative error. Mostly, as with the walk on spheres, the simple stochastic process is characterized by $\sigma(X) \approx 1$. Thus, $N = 10^8$ gives an absolute error of about 10^{-4}, an estimate that is reflected in the 3 digits asserted for the $\sqrt{3} \times 1$ rectangle in (10.1). For the 10×1 rectangle, 10 digits of $p \approx 4 \cdot 10^{-7}$ would require N as large as $N \approx 10^{33}$. This is completely out of the question if we note that the totality of arithmetical operations ever carried out by man and machine is estimated to be just 10^{24}; cf. [CP01, p. 4].

We are lucky: there are much more accurate methods than Monte Carlo available if we reformulate the problem as a *deterministic* one.

10.2 Making It Deterministic

What is the mathematical definition of Brownian motion? Either we address it directly using the language of stochastic calculus and talk about continuous-time stochastic processes and Wiener measures, or we approach it as the limit of a two-dimensional simple random walk, as already indicated in Trefethen's formulation of Problem 10. We will follow the latter, less technical approach [Zau89, §1.3]: we define the probability of hitting the ends for the random walk and take the limit of vanishing step length, as is schematically depicted in Figure 10.3.

Let n be a positive integer and consider the lattice $L_h = h\mathbb{Z} \times h\mathbb{Z}$ with step length $h = 1/2n$. For a simple random walk the transition probability from a lattice point to one of its nearest neighbors is $1/4$. We decompose the boundary of the rectangle $R = (-5,5) \times (-1/2, 1/2)$ into two parts: the "ends" $\Gamma_1 = \{(x,y) \in \partial R : x = \pm 5\}$ and the "sides" $\Gamma_0 = \partial R \backslash \Gamma_1$. Rather than speaking of hitting one part of the boundary before another, it is more convenient to view the boundary as *absorbing*. The particle simply stops or ceases to exist at the boundary.

Sometimes in mathematics a problem becomes easier if we try to solve for more. Here, we try to determine the probability $u_h(x,y)$ that a particle starting at an arbitrary lattice point $(x,y) \in R \cap L_h$—not just the center $(0,0)$ of the rectangle—reaches Γ_1.

Since the probability that the particle reaches Γ_1 from the point (x,y) can be expressed in terms of the probability that it moves to any of its nearest neighbors and reaches Γ_1 from

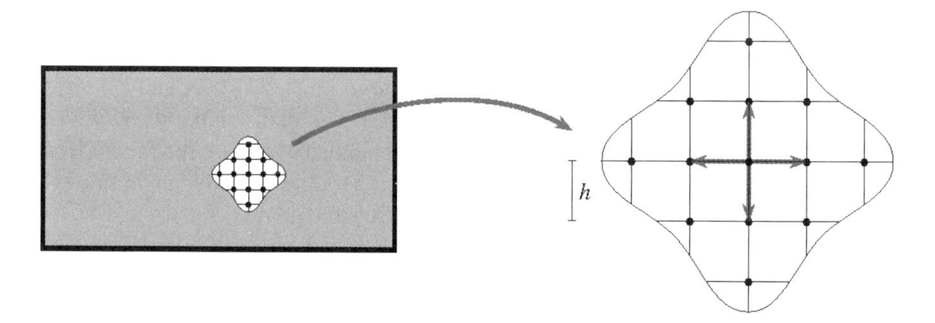

Figure 10.3. *Approximating Brownian motion with a simple random walk.*

there, we obtain the *partial difference equation*

$$u_h(x, y) = \frac{1}{4}\big(u_h(x+h, y) + u_h(x-h, y) + u_h(x, y+h) + u_h(x, y-h)\big)$$

with the absorbing boundary conditions $u_h|_{\Gamma_1} = 1$, $u_h|_{\Gamma_0} = 0$. Problem 10 is solved by the limit $p = \lim_{h\to 0} u_h(0, 0)$. The difference equation can be recast as a linear equation in $\mathbb{R}^{(2n-1)(20n-1)}$; in §10.3 on numerical methods we will essentially solve it for several small values of n and extrapolate to the limit $n \to \infty$.

However, the limit is also accessible to analytical methods. Actually, the difference equation is quite familiar, as we can see by writing it in the equivalent form

$$\frac{u_h(x+h, y) - 2u_h(x, y) + u_h(x-h, y)}{h^2} + \frac{u_h(x, y+h) - 2u_h(x, y) + u_h(x, y-h)}{h^2} = 0.$$

We recognize the well-known five-point discretization of *Laplace's equation*

$$\Delta u(x, y) = \frac{\partial^2}{\partial x^2} u(x, y) + \frac{\partial^2}{\partial y^2} u(x, y) = 0 \qquad \text{on } R$$

with the Dirichlet boundary conditions $u|_{\Gamma_1} = 1$, $u|_{\Gamma_0} = 0$. As we will see in §10.3, the discretization converges pointwise, $u_h(0, 0) \to u(0, 0)$; hence we have $p = u(0, 0)$.

Such boundary value problems have been extensively studied in the literature on potential theory in the plane; cf. [Hen86, §15.5] or [Neh52, §I.10]. Let $\Omega \subset \mathbb{R}^2$ be a bounded domain and let the boundary be the disjoint union

$$\partial\Omega = \Gamma_0 \cup \Gamma_1 \cup \{t_1, \ldots, t_n\},$$

where Γ_0 and Γ_1 denote nonempty relatively open subsets of $\partial\Omega$ having only finitely many components. The solution u of the Dirichlet boundary value problem

$$\Delta u = 0 \quad \text{on } \Omega, \qquad u|_{\Gamma_1} = 1, \quad u|_{\Gamma_0} = 0, \tag{10.2}$$

is called the *harmonic measure* of Γ_1 with respect to Ω. Some care has to be exercised concerning the right notion of a *solution*. In addition to (10.2), that is, that u is harmonic and takes the boundary values, we require that u be *continuous* on $\Omega \cup \Gamma_0 \cup \Gamma_1$ and *bounded* on Ω. The finitely many points $\{t_1, \ldots, t_n\}$ are thus possible points of discontinuity. *Uniqueness* of a solution follows from the *maximum principle* for bounded harmonic functions [Hen86, Lem. 15.4c]; *existence* will be shown for rectangles in §10.4. Generally, a solution exists [Hen86, Prop. 15.4b] if no component of $\mathbb{R}^2 \setminus \Omega$ reduces to a point.

We are now able to restate Problem 10 compactly in a deterministic fashion:

What value does the harmonic measure of the ends of a 10×1 *rectangle take at the center?*

10.3 Solving It Numerically

We will demonstrate that an efficient and accurate numerical solution of Problem 10 is possible using standard finite differences and acceleration by Richardson extrapolation. If we exercise some care and use tools of scientific computing such as *a posteriori error estimation*, we can provide very good *scientific reasons*—though not a rigorous mathematical proof—for being certain of the correctness of at least 10 digits.

We study a general $2a \times 2b$ rectangle, the "ends" now understood as the two parallel sides of length $2b$, and consider the following Dirichlet problem for Laplace's equation:

$$\Delta u = 0 \quad \text{on } R, \qquad u|_{\Gamma_1} = 1, \quad u|_{\Gamma_0} = 0. \tag{10.3}$$

The rectangle and its boundary are denoted by

$$R = (-a, a) \times (-b, b), \quad \Gamma_1 = \{-a, a\} \times (-b, b), \quad \Gamma_0 = (-a, a) \times \{-b, b\}.$$

We discretize the problem using a finite-difference grid R_h on R with $2n_x - 1$ grid points in the x-direction and $2n_y - 1$ grid points in the y-direction,

$$R_h = R \cap \left(h_x\mathbb{Z} \times h_y\mathbb{Z}\right), \qquad h_x = a/n_x, \quad h_y = b/n_y.$$

To obtain a nearly uniform grid, $h_x \approx h_y$, we choose $n_x = r \cdot n$ and $n_y = s \cdot n$ with $n \in \mathbb{N}$ such that the aspect ratio $\rho = a/b$ is approximated as $\rho \approx r/s$ with small integers r, s. We define $h = 1/n$ as the discretization parameter. The discrete boundary $\Gamma_h = \partial R \cap \left(h_x\mathbb{Z} \times h_y\mathbb{Z}\right)$ decomposes into the two parts

$$\Gamma_{1,h} = \{(x, y) \in \Gamma_h : x = \pm a\}, \qquad \Gamma_{0,h} = \Gamma_h \setminus \Gamma_{1,h}.$$

The five-point discretization of (10.3) is the linear equation, for $(x, y) \in R_h$,

$$\begin{aligned} h_x^{-2} u_h(x + h_x, y) + h_x^{-2} u_h(x - h_x, y) & \\ + h_y^{-2} u_h(x, y + h_y) + h_y^{-2} u_h(x, y - h_y) & \\ - 2\left(h_x^{-2} + h_y^{-2}\right) u_h(x, y) &= 0 \end{aligned}$$

with $u_h : R_h \cup \Gamma_h \to \mathbb{R}$ a grid function that satisfies the boundary conditions

$$u_h|_{\Gamma_{1,h}} = 1, \qquad u_h|_{\Gamma_{0,h}} = 0.$$

After elimination of the boundary points, we get a large linear system of dimension $N = (2n_x - 1)(2n_y - 1)$ for the values of u_h at the interior grid points R_h, compactly written as

$$A_h x_h = b_h. \tag{10.4}$$

Here, $A_h \in \mathbb{R}^{N \times N}$ is a symmetric positive definite, sparse matrix having five nonzero diagonals. There are various methods at hand to solve (10.4); cf. [Dem97, p. 277, Table 6.1]. The optimal run-time complexity of $O(N)$ is achieved by the iterative multigrid method. Fairly easy to code in MATLAB are two direct methods: sparse Cholesky decomposition of complexity $O(N^{3/2})$ and the FFT-based fast Poisson solver of complexity $O(N \log N)$; cf. [Dem97, §6.7]. In Appendix C.3.2 the reader will find the short MATLAB program `poisson` that realizes the two solvers. For a grid with $N = 2559 \times 255 = 652{,}545$ unknowns, on a 2 GHz PC, the sparse Cholesky decomposition needs about 60 seconds, the fast Poisson solver, just 2 seconds.

How accurate is the approximation $p_h = u_h(0,0) \approx p$? Before we discuss the question of the discretization error, we have to realize that because of round-off errors the linear solver will not really calculate u_h, but some *perturbed* grid function $\hat{u}_h$ instead. Since the linear system is rather badly conditioned, this numerical error can be of considerable size.

10.3.1 Assessment of the Linear Solver

A forward-stable linear solver [Hig96, §7.6], such as the Cholesky decomposition or the fast Poisson solver, is guaranteed to stay within the a priori error estimate

$$\|u_h - \hat{u}_h\|_\infty \leqslant c\,\kappa_\infty(A_h)\,\epsilon_M\,\|u_h\|_\infty, \tag{10.5}$$

where c is a constant that depends on the solver, $\kappa_\infty(A_h)$ is the condition number of the matrix A_h, and ϵ_M denotes the machine precision or unit round-off.[77] It is known that in two dimensions the condition number scales like $\kappa_\infty(A_h) \propto N$; cf. [Hac92, Thm. 4.4.1]. Now, for the problem with aspect ratio $\rho = a/b = 10$, we have $\|u_h\|_\infty = 1$ and $p \approx 4 \cdot 10^{-7}$. Thus, on an $N = 2559 \times 255$ grid, using IEEE double precision with $\epsilon_M \approx 10^{-16}$ and assuming reasonable constants, at the center of the rectangle we would just get some estimate like

$$|\hat{p}_h - p_h| \lesssim 10^{-4} |p_h|.$$

If this estimate were sharp, only about 4 digits would be correct: a dramatic setback in view of the task to deliver 10 correct digits.

A MATLAB Session

```
>> h = 1/128; % N = 2559 x 255
>> p = poisson([0,0],[10,1],0,[1,1,0,0],'Cholesky',h)

p = 3.838296382528924e-007
```

[77] In the literature the unit round-off is often denoted by u. To avoid confusion with the solution of Laplace's equation (10.3) we have changed the notation for the purposes of this section.

Table 10.2. *Estimated relative error of the linear solver, $\rho = a/b = 10$.*

Grid	159×15	319×31	639×63	1279×127	2559×255
Cholesky	$8.3 \cdot 10^{-14}$	$3.9 \cdot 10^{-13}$	$1.5 \cdot 10^{-12}$	$6.7 \cdot 10^{-12}$	$2.6 \cdot 10^{-11}$
Fast Poisson	$5.4 \cdot 10^{-9}$	$1.3 \cdot 10^{-8}$	$1.3 \cdot 10^{-7}$	$2.9 \cdot 10^{-7}$	$2.0 \cdot 10^{-6}$

```
>> p = poisson([0,0],[10,1],0,[1,1,0,0],'FFT',h)

p = 3.838296354378100e-007
```

Both methods solve the same linear equation, but they agree to only about eight digits on a $N = 2559 \times 255$ grid. Are both linear solvers affected, or just one?

To answer this question we have to pursue a finer analysis, using the a posteriori error estimator that we introduced in §7.4.1. The results are shown in Table 10.2. For the Cholesky solver, we observe an estimated relative error of $\hat{p}_h$ that scales approximately like $0.4N\epsilon_M$, nicely reflecting the theoretical prediction (10.5) of the normwise relative error. The estimated relative error of the fast Poisson solver is five orders of magnitude larger. Thus, most probably, the first number of the above MATLAB session is correct to at least ten digits, the second one—as indicated by the difference—to only about eight.

The lesson to be learned is that we should use the sparse Cholesky solver here and stay with small grids.

10.3.2 Discretization Error and Extrapolation

The five-point discretization is of second order. We need to know more; the well-founded application of Richardson extrapolation (see Appendix A, p. 235) for convergence acceleration and error estimation requires the validity of an asymptotic expansion

$$u_h(x, y) = u(x, y) + \sum_{k=1}^{m} e_k(x, y)\, h^{\gamma_k} + O(h^{\gamma_{m+1}}) \tag{10.6}$$

with $\gamma_1 < \gamma_2 < \cdots < \gamma_{m+1}$. It is well known (cf. [MS83, Thm. 4.2.1]) that the five-point discretization of two-dimensional Poisson problems on smooth domains with smooth data has $\gamma_k = 2k$, if the boundary is appropriately dealt with. For polygonal domains and boundary conditions with jumps, like problem (10.3), the asymptotic expansion (10.6) does not hold uniformly in (x, y). Folklore states that the asymptotic expansion for interior points can be severely polluted by boundary effects.

There is a less well known, extremely careful analysis of the Dirichlet problem for Laplace's equation on a rectangle by Hofmann; [Hof67, Thm. 2] allows one to determine the exponents γ_k from the boundary data. In our case of harmonic measures, [Hof67, p. 309, Rem. 1] shows that

$$\gamma_k = 2k, \qquad k \in \mathbb{N},$$

and the asymptotic expansion (10.6) holds for all $m \in \mathbb{N}$,

$$u_h(x, y) = u(x, y) + \sum_{k=1}^{m} e_k(x, y)\, h^{2k} + O(h^{2m+2}). \tag{10.7}$$

The estimate holds uniformly in h, but not so in (x, y). Near to the boundary, the constants involved may be large.

Writing $e_* = e_1(0, 0)$ we obtain

$$p_h - p = e_* h^2 + O(h^4), \qquad p_{2h} - p_h = 3e_* h^2 + O(h^4).$$

A comparison yields an *a posteriori* estimate[78] for the relative discretization error of p_h, namely,

$$\frac{p_h - p}{p} = \underbrace{\frac{p_{2h} - p_h}{3p_h}}_{=\epsilon_h} + O(h^4).$$

Table 10.3 shows the results of the Cholesky solver together with the estimate of the relative discretization error for the same grids as in Table 10.2. Since for those grids we know from the latter table that the Cholesky solver is correct to at least the given ten digits, we therefore no longer distinguish between $\hat{p}_h$ and p_h. The last column of Table 10.3 nicely reflects that the discretization is of second order. We learn that on the 2559×255 grid p_h is an approximation of p just good to about 3 correct digits. Further refinement would not pay off, for reasons of computer memory and run time and because the error of the linear solver would start touching the 10th digit.

We have to increase the accuracy of the discretization by *convergence acceleration*. The idea is simple; if we substract the error estimate, we obtain a discretization of increased order,

$$p'_h = p_h + \tfrac{1}{3}(p_h - p_{2h}) = p + O(h^4).$$

Table 10.3. *A posteriori error estimate ϵ_h for p_h (Cholesky solver).*

Grid	h	p_h	ϵ_h	ϵ_{2h}/ϵ_h
159×15	$1/8$	$4.022278462 \cdot 10^{-7}$	—	—
319×31	$1/16$	$3.883130701 \cdot 10^{-7}$	$1.2 \cdot 10^{-2}$	—
639×63	$1/32$	$3.848934609 \cdot 10^{-7}$	$3.0 \cdot 10^{-3}$	4.0
1279×127	$1/64$	$3.840422201 \cdot 10^{-7}$	$7.4 \cdot 10^{-4}$	4.0
2559×255	$1/128$	$3.838296383 \cdot 10^{-7}$	$1.8 \cdot 10^{-4}$	4.0

[78] Such estimates are crucial for the success of state-of-the-art extrapolation codes with order and step-size control for systems of ordinary differential equations [DB02, §5.3].

This new approximation inherits the asymptotic expansion (10.7),

$$p'_h = p + \sum_{k=2}^{m} e'_{*,k} h^{2k} + O(h^{2m+2}),$$

now starting with the $O(h^4)$ term. Completely analogously to what we have done for p_h, we can construct an a posteriori error estimate and repeat the process. This is Richardson extrapolation for a sequence of grids with discretization parameters h, $h/2$, $h/4$, $h/8$, etc. It is fairly obvious how to generalize for arbitrary sequences; a good compromise between stability and efficiency is $h = 1/(2n)$ for $n = n_{\min}, n_{\min}+1, \ldots, n_{\max}$. In Appendix C.3.2 the reader will find the short MATLAB program `richardson` that implements this general extrapolation technique with a posteriori estimation of the relative discretization error. Additionally, it contains a *running error analysis* [Hig96, §3.3] of the amplification of the relative error that was produced by the linear solver. By Table 10.2, the error of the linear solver is below $3.9 \cdot 10^{-13}$.

A MATLAB Session

```
>> f = inline('poisson([0,0],[10,1],0,[1,1,0,0],sol,h)','sol','h');
>> order = 2; nmin = 4; tol = 1e-11;
>> [p,err,ampl] = richardson(tol,order,nmin,f,'Cholesky');
>> p, err = max(err,ampl*3.9e-13)

p =   3.837587979250745e-007
err = 3.325183129259237e-011
```

The run time on a 2 GHz PC is less than half a second. The extrapolation is based on the grids 159×15, 199×19, 239×23, 279×27, 319×31. The estimated relative error is $3.3 \cdot 10^{-11}$. Thus, to the best of our numerical knowledge, the approximation of p is correct to at least 10 digits,

$$p|_{\rho=10} \doteq 3.8375\,87979 \cdot 10^{-7},$$

and Problem 10 is solved.

The analytical method of the next section will *prove* that the numerical solution given by the MATLAB session above is in fact good for 12 digits. We overestimated the error by just two orders of magnitude—not bad at all.

10.4 Solving It Analytically I: Separation of Variables

A traditional analytical technique for solving boundary value problems such as (10.3) is *separation of variables*. Essentially, this technique is applicable if the underlying geometry is rectangular, the differential equation is homogeneous, and the boundary conditions on two opposite sides are homogeneous.

Separation of variables tries the tensor-product Ansatz

$$u(x, y) = \sum_{k=0}^{\infty} v_k(x) \cdot w_k(y)$$

with the idea that every term of the sum solves the differential equation and that the sum is responsible for matching the boundary data. Plugging $v_k(x)\, w_k(y)$ into Laplace's equation gives

$$v_k''(x)\, w_k(y) + v_k(x)\, w_k''(y) = 0, \text{ i.e., } \quad v_k''(x)/v_k(x) = -w_k''(y)/w_k(y) = \lambda_k.$$

Since the ratio λ_k depends only on x and simultaneously only on y, it depends on neither and is a *constant*. The *homogeneous* boundary conditions at Γ_0 are easily matched if we require $w_k(-b) = w_k(b) = 0$. Hence, any $w_k \not\equiv 0$ solves the second-order *eigenvalue problem*

$$w_k''(y) + \lambda_k w_k(y) = 0, \qquad w_k(-b) = w_k(b) = 0. \tag{10.8}$$

Because the boundary values at $x = \pm a$ are *even* functions of y, it suffices to look for even solutions $w_k(y)$. It is readily checked that all those solutions are given by

$$w_k(y) = \cos\big((k + 1/2)\pi y/b\big), \qquad \lambda_k = \big((k+1/2)\pi/b\big)^2, \qquad k \in \mathbb{N}_0.$$

To match the boundary conditions at Γ_1, we calculate the coefficients $(c_k)_{k\geqslant 0}$ of the series representation

$$\sum_{k=0}^{\infty} c_k\, w_k(y) = 1, \qquad y \in (-b, b).$$

Orthogonality and completeness of the eigenfunctions w_k yield

$$c_k \int_{-b}^{b} w_k^2(y)\, dy = \int_{-b}^{b} w_k(y)\, dy, \text{ i.e., } \qquad c_k\, b = \frac{4(-1)^k}{\pi(2k+1)}\, b.$$

This way we obtain for v_k the boundary value problem

$$v_k''(x) - \lambda_k v_k(x) = 0, \qquad v_k(-a) = v_k(a) = c_k = \frac{4\,(-1)^k}{\pi\,(2k+1)}.$$

A short calculation, or resorting to *Mathematica*'s `DSolve` or Maple's `dsolve` command, shows that the solution is given by

$$v_k(x) = \frac{4\,(-1)^k}{\pi(2k+1)} \frac{\cosh\big((k+1/2)\pi x/b\big)}{\cosh\big((k+1/2)\pi a/b\big)}.$$

Summarizing, the solution of (10.3) reads

$$u(x, y) = \frac{4}{\pi} \sum_{k=0}^{\infty} \frac{(-1)^k}{2k+1} \frac{\cosh\big((k+1/2)\pi x/b\big)}{\cosh\big((k+1/2)\pi a/b\big)} \cos((k+1/2)\pi y/b),$$

which is a Fourier series with respect to y. In principle, there are two approaches to showing that this really is the bounded solution of (10.3). One approach [Rau91, §5.7] uses the completeness of the eigenfunctions of (10.8) in $L^2(0, b)$ and proves the success of the method a priori. The other approach [Zau89, §4.4] exploits the exponentially fast convergence of the resulting series to prove the boundedness, differentiate term by term, and show that it solves the equation a posteriori.

At the center of the rectangle, the solution $p = u(0,0)$ simplifies to

$$p = \frac{4}{\pi}\sum_{k=0}^{\infty}\frac{(-1)^k}{2k+1}\operatorname{sech}\big((2k+1)\pi\rho/2\big), \tag{10.9}$$

where we denote the aspect ratio of the rectangle by $\rho = a/b$. Now, this is an alternating series, whose terms are in absolute value monotonically decreasing to zero; consequently the series converges to a value that is always enclosed between two successive partial sums. Thus, merely a few terms of the exponentially fast converging series will give an accurate result together with a proven error bound. For $\rho = 10$, already the first two terms give the estimate

$$\begin{aligned} 3.8375\,87979\,25122\,8\cdot 10^{-7} &\doteq \tfrac{4}{\pi}\operatorname{sech}(5\pi) - \tfrac{4}{3\pi}\operatorname{sech}(15\pi) \\ &< p|_{\rho=10} < \tfrac{4}{\pi}\operatorname{sech}(5\pi) \doteq 3.8375\,87979\,25125\,8\cdot 10^{-7}. \end{aligned}$$

The numbers given were computed using MATLAB's IEEE double-precision machine arithmetic. Because the library routine for `sech` computes a result within machine precision, many would say that the argument qualifies as a *proof*[79] that the first 14 correct digits of p are

$$p|_{\rho=10} \doteq 3.8375\,87979\,2512\cdot 10^{-7}.$$

The first term of the series, $\frac{4}{\pi}\operatorname{sech}(\rho\pi/2)$, allows for the further simplified approximation

$$p = \frac{4}{\pi}\operatorname{sech}(\rho\pi/2) + O(e^{-3\pi\rho/2}) = \frac{8}{\pi}e^{-\pi\rho/2} + O(e^{-3\pi\rho/2}).$$

For $\rho \geqslant 1$, Figure 10.5 shows that $p \approx \frac{8}{\pi}e^{-\pi\rho/2}$ is good to plotting accuracy, while

$$p|_{\rho=10} \doteq \frac{8}{\pi}e^{-5\pi} \text{ to 13 significant digits.}$$

10.4.1 Intermezzo à la Cauchy

Exchanging the roles of a and b gives by symmetry the probability that the particle reaches the sides rather than the ends. Clearly, the two probabilities add up to 1:

$$p|_{\rho=a/b} + p|_{\rho=b/a} = 1.$$

Using the series representation (10.9) we obtain the remarkable identity

$$\sum_{k=0}^{\infty}\frac{(-1)^k}{2k+1}\Big(\operatorname{sech}\big((2k+1)\pi\rho/2\big) + \operatorname{sech}\big((2k+1)\pi\rho^{-1}/2\big)\Big) = \frac{\pi}{4}, \tag{10.10}$$

[79] For those readers who need more evidence: a computer-assisted proof taking round-off into account can be constructed along the lines presented in §8.3.2.

which is valid for all $\rho > 0$. The specific case $\rho = 1$ simplifies to

$$\sum_{k=0}^{\infty} \frac{(-1)^k}{2k+1} \operatorname{sech}\big((2k+1)\pi/2\big) = \frac{\pi}{8},$$

while the limit $\rho \to \infty$ recovers the Leibniz series

$$\sum_{k=0}^{\infty} \frac{(-1)^k}{2k+1} = \frac{\pi}{4}.$$

It is interesting to note that the identity (10.10), and the additional explicit value $p|_{\rho=\sqrt{3}}$, were obtained as early as 1827 by Cauchy [Cau27] as an application of his residue calculus—in the same memoir in which he for the first time used contour integrals along circles in a systematic way [Smi97, §5.10]. The identity was later rediscovered by Ramanujan and can be found as an entry in his notebooks [Ber89, Entry 15, p. 262].

Lemma 10.1 (Cauchy 1827 [Cau27, form. (86/87)]).

(i) *For* $\eta \in \mathbb{C} \setminus \mathbb{R}$

$$\sum_{k=0}^{\infty} \frac{(-1)^k}{2k+1} \left(\sec\left((2k+1)\pi\eta/2\right) + \sec\left((2k+1)\pi\eta^{-1}/2\right)\right) = \frac{\pi}{4}.$$

(ii) *For* $\eta \in \mathbb{C} \setminus \mathbb{R}$

$$\sum_{n=1}^{\infty} \frac{(-1)^{n-1}}{n} \left(\csc(n\pi\eta) + \csc(n\pi\eta^{-1})\right) = \frac{\pi}{12}(\eta + \eta^{-1}).$$

Proof. Note that both series are absolutely and therefore unconditionally convergent.

(i) The function $\sec(\pi z)\sec(\pi\eta z)$ tends to zero uniformly in $\arg z$ when $|z| = k$, $z \in \mathbb{C}$, and $k \to \infty$ through integral values; consequently, by Cauchy's residue theorem [Hen74, Thm. 4.7a], the sum of the residues of

$$\frac{\pi \sec(\pi z)\sec(\pi\eta z)}{z}$$

at all of its poles is zero. The determination of the residues is straightforward using the conventional formulas [Hen74, form. (4.7-9/10)], or by using *Mathematica*'s `Residue` or Maple's `residue` command. The residue at the simple pole $z = k + 1/2$, with $k \in \mathbb{Z}$, is

$$\frac{(-1)^{k-1}}{k+1/2} \sec((k+1/2)\pi\eta);$$

the residue at the simple pole $z = (k+1/2)/\eta$, with $k \in \mathbb{Z}$, is

$$\frac{(-1)^{k-1}}{k+1/2} \sec((k+1/2)\pi/\eta);$$

and the residue at the simple pole $z = 0$ is π. Hence

$$\sum_{k\in\mathbb{Z}} \frac{(-1)^k}{k+1/2}(\sec((k+1/2)\pi\eta) + \sec((k+1/2)\pi/\eta)) = \pi,$$

which is the result stated.

(ii) The function $\csc(\pi z)\csc(\pi\eta z)$ tends to zero uniformly in $\arg z$ when $|z| = n+1/2$, $z \in \mathbb{C}$, and $n \to \infty$ through integral values; consequently, by Cauchy's residue theorem [Hen74, Thm. 4.7a], the sum of the residues of

$$\frac{\pi\csc(\pi z)\csc(\pi\eta z)}{z}$$

at all of its poles is zero. The residue at the simple pole $z = \pm n$, with $n \in \mathbb{N}$, is

$$\frac{(-1)^n}{n}\csc(n\pi\eta);$$

the residue at the simple pole $z = \pm n/\eta$, with $n \in \mathbb{N}$, is

$$\frac{(-1)^n}{n}\csc(n\pi/\eta);$$

and the residue at the pole $z = 0$ of order 3 is

$$\frac{\pi}{6}(\eta+\eta^{-1}).$$

Hence

$$2\sum_{n=1}^{\infty}\frac{(-1)^{n-1}}{n}\left(\csc(n\pi\eta) + \csc\left(\frac{n\pi}{\eta}\right)\right) = \frac{\pi}{6}(\eta+\eta^{-1}),$$

which is the result stated. □

If we put $\eta = i\rho$ in (i) we recover our remarkable identity (10.10). Cauchy's relation (ii) evaluates to a new specific value of p if we put $\eta = e^{i\pi/3} = (1+i\sqrt{3})/2$. Because $\eta + \eta^{-1} = 1$ and

$$\csc\left(n\pi e^{i\pi/3}\right) + \csc\left(n\pi e^{-i\pi/3}\right) = \begin{cases} 0, & n = 2k, \\ 2(-1)^k \operatorname{sech}((2k+1)\pi\sqrt{3}/2), & n = 2k+1, \end{cases}$$

we obtain from (ii), as Cauchy [Cau27, form. (110)] did in 1827,

$$\sum_{k=0}^{\infty}\frac{(-1)^k}{2k+1}\operatorname{sech}((2k+1)\pi\sqrt{3}/2) = \frac{\pi}{24}, \quad \text{i.e.,} \quad p|_{\rho=\sqrt{3}} = \frac{1}{6}. \tag{10.11}$$

This is a remarkable, unexpectedly explicit result, which we can formulate as:

The probability that a particle undergoing Brownian motion reaches the ends of a $\sqrt{3}\times 1$ rectangle from the center is $1/6$.

10.5 Solving It Analytically II: Conformal Mapping

The *conformal transplant* of a harmonic function, that is, the pull-back under a conformal one-to-one mapping, is again harmonic. For simply connected regions $\Omega \subset \mathbb{R}^2 = \mathbb{C}$ this is readily proved by observing that any harmonic function $u : \Omega \to \mathbb{R}$ is the real part of an analytic function on Ω [Hen86, Thm. 15.1a]. Therefore, conformal transplants of harmonic measures are again harmonic measures. Even more, a conformal transplant of a Brownian motion is again a Brownian motion.

The idea is now to map the $2a \times 2b$ rectangle conformally and one-to-one to a domain for which the harmonic measure can easily be evaluated at the image of the center. Which domain qualifies as easy enough?

Quite remarkably, for the specific case of a $\sqrt{3} \times 1$ rectangle, starting from a regular hexagon and using only reflections and symmetries, Hersch [Her83] was able to prove the result $p|_{\rho=\sqrt{3}} = 1/6$ of (10.11).[80]

The general case is best dealt with using the unit disk.

Lemma 10.2 *Let Γ_1 be a finite union of open arcs of the unit circle having length $2\pi p$. The harmonic measure of Γ_1 with respect to the unit disk D evaluated in the center of D is p.*

Proof. We give a proof that conveys the geometrical flavor of the problem. Let Γ_1 be a single arc of length $2\pi/n$ for some $n \in \mathbb{N}$. We can cover the circle up to n points by n suitable rotated copies of Γ_1. Adding their harmonic measures just gives the harmonic measure of the circle itself, which is identical to 1. Since the value of the harmonic measure at the center is invariant under rotations, we get $u(0) = 1/n$. Now, patching $m \in \mathbb{N}$ suitable rotated copies of the arc Γ_1 together shows that the harmonic measure of a single arc of length $2\pi m/n$ gives $u(0) = m/n$. Using monotonicity and enclosure by rationals allows us to conclude that a single arc of arbitrary length $2\pi p$ gives $u(0) = p$. Adding the harmonic measures for single arcs finishes the proof for finite unions. □

We now construct a suitable conformal mapping of the unit disk D to the $2a \times 2b$ rectangle R. We place the rectangle in such a way that the real and the imaginary axes are the axes of symmetry. The vertices of the rectangle are then given by

$$(A, B, C, D) = (a + ib, -a + ib, -a - ib, a - ib).$$

By the Riemann mapping theorem [Hen74, Cor. 5.10c] and the reflection principle of Riemann and Schwarz [Hen74, Thm. 5.11b], there is a conformal one–one map $f : D \to R$ that respects the symmetries (cf. Figure 10.4):

$$f(-z) = -f(z), \qquad \overline{f(z)} = f(\overline{z}). \tag{10.12}$$

In particular we have $f(0) = 0$: the center of the rectangle is the image of the center of the disk. By the Osgood–Carathéodory theorem [Hen86, Thm. 16.3a] the conformal mapping

[80] Have a look at his sequence of figures $7, 7', 7'', 11'', 13, 41'', 41$ and admire a proof without words and calculations.

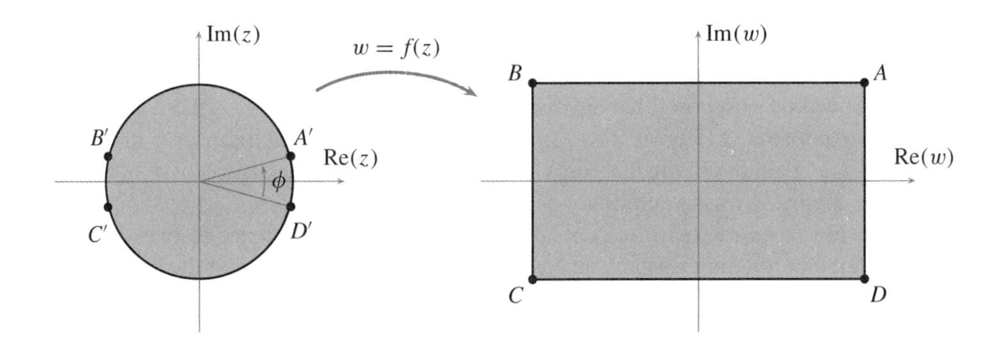

Figure 10.4. *Symmetry-preserving conformal mapping $f : D \to R$.*

f extends as a homeomorphism of the closure of D onto the closure of R. The preimages of (A, B, C, D) are therefore

$$(A', B', C', D') = \left(e^{i\phi/2}, -e^{-i\phi/2}, -e^{i\phi/2}, e^{-i\phi/2}\right)$$

for some angle $0 < \phi < \pi$. By taking Γ_1 as the union of the arc from B' to C' and the arc from D' to A', Lemma 10.2 and the invariance of the value p of the harmonic measure at the center show that

$$\phi = p\pi.$$

The conformal transformation f is explicitly given by the Schwarz–Christoffel formula [Hen86, form. (16.10-1)],

$$\begin{aligned} f(z) &= 2c \int_0^z \frac{d\zeta}{\sqrt{(1-e^{-ip\pi/2}\zeta)(1+e^{ip\pi/2}\zeta)(1+e^{-ip\pi/2}\zeta)(1-e^{ip\pi/2}\zeta)}} \\ &= 2c \int_0^z \frac{d\zeta}{\sqrt{(1-e^{-ip\pi}\zeta^2)(1-e^{ip\pi}\zeta^2)}}. \end{aligned}$$

Here, the integral is taken along any path connecting 0 and z within the disk D. The symmetries (10.12) imply that the unknown parameter c is *real*. Evaluation of $f(A') = A$ along the radial path $\zeta = e^{ip\pi/2}t$ with $t \in [0, 1]$ yields

$$\begin{aligned} a + ib &= 2c \int_0^{e^{ip\pi/2}} \frac{d\zeta}{\sqrt{(1-e^{-ip\pi}\zeta^2)(1-e^{ip\pi}\zeta^2)}} \\ &= 2c\, e^{ip\pi/2} \int_0^1 \frac{dt}{\sqrt{(1-t^2)(1-e^{2ip\pi}t^2)}} = 2c\, e^{ip\pi/2} K(e^{ip\pi}), \end{aligned} \tag{10.13}$$

where $K(k)$ denotes the *complete elliptic integral of the first kind of modulus k*,

$$K(k) = \int_0^1 \frac{dt}{\sqrt{(1-t^2)(1-k^2t^2)}}. \tag{10.14}$$

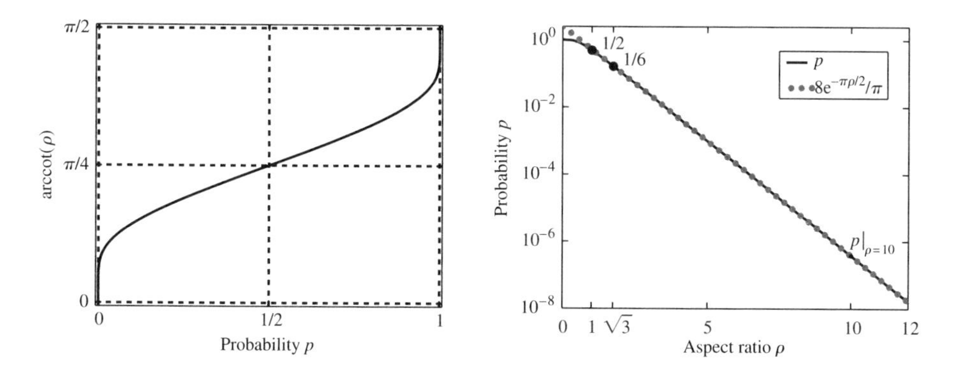

Figure 10.5. *The relation of aspect ratio ρ and probability p.*

Taking the complex arguments of both sides in (10.13) yields a transcendental equation relating the aspect ratio $\rho = a/b$ of the rectangle and the answer to the problem at hand, the probability p,

$$\operatorname{arccot} \rho = \frac{p\pi}{2} + \arg K(e^{ip\pi}). \qquad (10.15)$$

Figure 10.5 shows that given a ratio $\rho > 0$ there is a unique solution $0 < p < 1$.

10.5.1 Solving the Transcendental Equation

Mathematica and Maple provide the complete elliptic integral K for complex arguments. Their root-finders, based on the secant method, can therefore readily be used to solve (10.15) for p at $\rho = 10$.

A Maple Session

```
> p:= rho->fsolve(
>   arccot(rho)=P*Pi/2+argument(EllipticK(exp(I*P*Pi))),P=1e-3):
> Digits:= 16: p(10);

                    0.3837587979251226e-6
> Digits:= 17: p(10);

                    0.38375879792512261e-6
```

A *Mathematica* Session

```
p[ρ_, prec_] := Block[{$MinPrecision = prec},
   p/.FindRoot[ArcCot[ρ] == pπ/2 + Arg[EllipticK[e^(2 i p π)]],
      {p, 10^-8, 10^-3}, AccuracyGoal → prec, WorkingPrecision → prec]]

{Precision == #, p* == p[10, #]}&/@{MachinePrecision, 17}//TableForm
Precision == MachinePrecision   p* == 3.837587979228946 × 10^-7
Precision == 17                 p* == 3.8375879792512261 × 10^-7
```

Note that *Mathematica* evaluates K as a function of the *parameter* $m = k^2$.

At a precision of 17 digits, both programs agree to all digits given. A comparison with the method of §10.4 reveals that all the digits are in fact correct. At 16 digits of precision, however, Maple gives 16 correct digits, but *Mathematica*, only about 10.

What has happened? At 16-digit precision *Mathematica* uses IEEE double-precision machine numbers and stops monitoring numerical errors. It is now the user's responsibility to not use numerically unstable formulas and algorithms. The loss of 6 digits points to a severe numerical instability in using the transcendental equation (10.15) naively.

In fact, evaluating $\arg K(e^{ip\pi})$ for small p by just using the expression *as given* is unstable. A look at Figure 10.5 shows that $\arg K(e^{ip\pi})$ is zero for $p = 0$ and depends very sensitively on small p. However, if the intermediate quantity $e^{ip\pi} \approx 1$ is calculated, we irretrievably lose some of the valuable information on the final digits of p: a kind of "hidden" cancellation effect.

The instability can be avoided if we have a closer look at how $K(k)$ is actually calculated. Essentially, all efficient implementations use the relation [BB87, Alg. 1.2(a)]

$$K(k) = \frac{\pi}{2M(1, \sqrt{1-k^2})} \tag{10.16}$$

of the complete elliptic integral K with the exceedingly fast *arithmetic-geometric mean* $M(a, b)$ of Gauss,

$$M(a, b) = \lim_{n\to\infty} a_n = \lim_{n\to\infty} b_n, \quad a_0 = a, \; b_0 = b, \; a_{n+1} = \frac{a_n + b_n}{2}, \; b_{n+1} = \sqrt{a_n b_n}.$$

The iteration converges *quadratically*. The relation (10.16) extends from $0 \leqslant k < 1$ by analytic continuation to all $k^2 \in \mathbb{C}\backslash[1, \infty)$, if the sign of square roots is generally understood to give a positive real part; cf. [Cox84]. For $k = e^{i\phi}$ with $\phi > 0$ small, the cancellation effect becomes plain in $1 - k^2$ and is circumvented by observing

$$1 - e^{2i\phi} = -2i\,\sin(\phi)\,e^{i\phi}, \qquad \sqrt{1 - e^{2i\phi}} = (1 - i)\sqrt{\sin(\phi)}\,e^{i\phi/2}.$$

Summarizing, we obtain the *stabilized* equation

$$\operatorname{arccot}\rho = \frac{p\pi}{2} - \arg M\left(1, (1-i)\sqrt{\sin(p\pi)}\,e^{ip\pi/2}\right).$$

We are now able to calculate 14 correct digits using IEEE double precision.

A *Mathematica* Session *(the session of p. 215 revisited)*

```
p* ==
 (p/.
    FindRoot[
     ArcCot[10] ==
      pπ/2 - Arg[ArithmeticGeometricMean[1, (1 - i) √(Sin[p π]) e^(i p π/2)]],
     {p, 10^-8, 10^-5}, AccuracyGoal → MachinePrecision])
p* == 3.837587979251236 × 10^-7
```

After the easy exercise of implementing the arithmetic-geometric mean in MATLAB, MATLAB's `fzero` command just gives the same result.

10.6 The Joy of Elliptic Functions

Elliptic function theory will supply us with a plethora of wonderful formulas and, in the end, even with a closed-form solution of Problem 10.

We first separate real and imaginary parts in (10.13). We use the conventional notation $k' = \sqrt{1-k^2}$ and $K'(k) = K(k')$.

Lemma 10.3 *For* $0 < \phi < \pi$,

$$e^{i\phi/2} K(e^{i\phi}) = \frac{1}{2}(K'(\sin(\phi/2)) + i\, K(\sin(\phi/2)))$$

holds.

Proof. We give two proofs, one using some of the many useful transformations of elliptic integrals known in the literature and one self-contained, elementary proof that we owe to John Boersma.

(i) Using Landen's transformation [BB87, Thm. 1.2(b)] and Jacobi's imaginary quadratic transformation [BB87, Ex. 3.2.7(d)] we conclude that

$$\begin{aligned} K\left(\frac{k'-ik}{k'+ik}\right) &= K\left(\frac{1-ik/k'}{1+ik/k'}\right) = \frac{1+ik/k'}{2} K'(ik/k') \\ &= \frac{1+ik/k'}{2} k'(K'(k) - iK(k)) = \frac{k'+ik}{2}(K'(k) - iK(k)). \end{aligned}$$

Taking the complex conjugate and putting $k = \sin(\phi/2)$, and therefore $k' = \cos(\phi/2)$, yields the result stated.

(ii) We go back to (10.13), now taking a path that connects 0 and $e^{i\phi/2}$ by first following the real axis from 0 to 1 and then the circular arc $\zeta = e^{i\theta}$ for $0 \leqslant \theta \leqslant \phi/2$,

$$\begin{aligned} e^{i\phi/2} K(e^{i\phi}) &= \int_0^{e^{i\phi/2}} \frac{d\zeta}{\sqrt{(1-e^{-i\phi}\,\zeta^2)(1-e^{i\phi}\,\zeta^2)}} \\ &= \int_0^1 \frac{d\zeta}{\sqrt{1-2\zeta^2\cos\phi+\zeta^4}} + \int_1^{e^{i\phi/2}} \frac{d\zeta}{\sqrt{1-2\zeta^2\cos\phi+\zeta^4}}. \end{aligned}$$

By the substitution $t = 2\zeta/(1+\zeta^2)$, and taking the definition (10.14) of K into account, the first integral reduces to

$$\frac{1}{2}\int_0^1 \frac{dt}{\sqrt{(1-t^2)(1-\cos^2(\phi/2)\,t^2)}} = \frac{1}{2}K(\cos(\phi/2)) = \frac{1}{2}K'(\sin(\phi/2)).$$

Analogously, by the substitution $t = i(1-\zeta^2)/(2\zeta\sin(\phi/2))$, the second integral reduces to

$$\frac{i}{2}\int_0^1 \frac{dt}{\sqrt{(1-t^2)(1-\sin^2(\phi/2)\,t^2)}} = \frac{i}{2}K(\sin(\phi/2)).$$

This proves the result stated. □

Now, equation (10.13) transforms to

$$a = c\,K'(\sin(p\pi/2)), \qquad b = c\,K(\sin(p\pi/2)),$$

or, after elimination of c,

$$\rho = \frac{K'(\sin(p\pi/2))}{K(\sin(p\pi/2))}. \tag{10.17}$$

We could have used this equation instead of (10.15) to solve numerically for p given $\rho = 10$. Note, however, that the same cautionary remarks about numerical instability as in §10.5.1 apply here. Along the same lines, using the arithmetic-geometric mean, a stable variant reads

$$\rho\,M(1, \sin(p\pi/2)) = M(1, \cos(p\pi/2)).$$

Introducing the *elliptic nome* q and the Weierstrassian ratio τ of periods,

$$q = e^{i\pi\tau}, \qquad \tau = i\frac{K'(k)}{K(k)},$$

we deduce from (10.17) the following correspondence between the parameters ρ and p of the problem at hand and the conventional quantities in elliptic function theory:

$$q = e^{-\pi\rho}, \qquad \tau = i\,\rho, \qquad k = \sin(p\pi/2). \tag{10.18}$$

Thus, solving for the elliptic modulus k given the nome q is essentially equivalent to solving for the probability p given the aspect ratio ρ. The sech-series (10.9) of §10.4 accomplishes such a solution; we have thus proven the identity

$$\frac{1}{2}\arcsin k = \sum_{n=0}^{\infty} \frac{(-1)^n}{2n+1}\operatorname{sech}((2n+1)\pi\rho/2) = 2\sum_{n=0}^{\infty} \frac{(-1)^n\,q^{n+1/2}}{(2n+1)(1+q^{2n+1})},$$

which was originally obtained by Jacobi in his seminal 1829 treatise on elliptic integrals and functions [Jac29, p. 108, form. (46)]. By expanding $1/(1+q^{2k+1})$ into a geometric series and changing the order of summation, Jacobi obtained ("quae in hanc facile transformatur") [Jac29, p. 108, form. (47)]

$$\arcsin k = 4\sum_{n=0}^{\infty}(-1)^n \arctan q^{n+1/2},$$

"which," as he proudly exclaimed, "one is obliged to rank among the most elegant formulas" [quae inter formulas elegantissimas censeri debet]. For $\rho = 10$, using the same argument as at the end of §10.4, the first two terms of the series yield the estimate

$$\begin{aligned} 3.8375\,87979\,25122\,9\cdot 10^{-7} &\doteq 8/\pi\,\left(\arctan e^{-5\pi} - \arctan e^{-15\pi}\right) \\ &< p|_{\rho=10} < 8/\pi\,\arctan e^{-5\pi} \doteq 3.8375\,87979\,25131\,6\cdot 10^{-7}. \end{aligned}$$

Hence, $8\arctan(e^{-5\pi})/\pi$ is an approximation to $p|_{\rho=10}$ that is good for 13 correct digits. Or, even simpler, because it agrees to the first 13 digits, $8e^{-5\pi}/\pi$.

In terms of τ the determination of k is accomplished by the classical *elliptic modular function* $k^2 = \lambda(\tau)$. Using the correspondence (10.18) we can therefore write

$$p = \frac{2}{\pi} \arcsin \sqrt{\lambda(i\rho)}. \tag{10.19}$$

Mathematica provides the command `ModularLambda` for the evaluation of $\lambda(\tau)$ that employs exponentially fast converging series like (10.9) in terms of theta functions.

A *Mathematica* Session

```
N[2/π ArcSin[√ModularLambda[10i]], 100]
3.8375879792512261034071331862048391007930055940725095690300227991734
   66068527432765008428456472699 10 × 10^-7
```

All 100 digits given are correct; we could equally well get 10,000 significant digits in about two seconds on a 2 GHz PC.

Whether (10.19) counts as a closed-form solution or not is a matter of taste. It depends on whether we view $\lambda(\tau)$ as a mere notation or as a useful function from the canon with efficient algorithms to evaluate at hand. However, there is another solution that is closed form by anyone's definition and so eliminates this debate entirely.

10.7 Finale à la Ramanujan: Singular Moduli

Abel discovered that certain values of k, called *singular moduli*, led to elliptic integrals that could be algebraically transformed into complex multiples of themselves. This phenomenon came to be called *complex multiplication*. The singular moduli with purely imaginary period ratio τ are given by the equation

$$\frac{K'(k_r)}{K(k_r)} = \sqrt{r}, \qquad r \in \mathbb{Q}_{>0}.$$

In view of (10.17) and the correspondence (10.18), the singular moduli are connected to the problem at hand by

$$p = \tfrac{2}{\pi} \arcsin k_{\rho^2}, \qquad \rho^2 \in \mathbb{Q}.$$

Abel stated as early as 1828, and Kronecker proved in 1857, that the singular moduli are algebraic numbers that can be expressed by radicals over the ground field of rational numbers:[81] *the* classical form of a closed-form solution. His study led Kronecker to his conjecture about abelian extensions of imaginary quadratic fields, Kronecker's famous "Jugendtraum," the subject of Hilbert's 12th problem, which was solved by one of the pearls of 20th century pure

[81] Arithmetically, the singular moduli have a most appropriate name. From a 1937 theorem of Schneider it follows that for algebraic $\tau = i\rho$ with positive imaginary part, other than a quadratic irrational, k is *transcendental* [Bak90, Thm. 6.3].

mathematics, class field theory. The wonderful book [Cox89] of Cox offers an accessible modern introduction to complex multiplication and class field theory, driven by history and a concrete problem of number theory. By the efforts of Weber, Ramanujan, Watson, Berndt, and others, radical expressions of singular moduli k_n for various $n \in \mathbb{N}$ have actually been calculated. Using the explicit values so far known for the probability p from §10.4.1,

$$p|_{\rho=1} = \tfrac{1}{2}, \qquad p|_{\rho=\sqrt{3}} = \tfrac{1}{6},$$

we are readily able to calculate the radical expressions of k_1 and k_3 ourselves:

$$k_1 = \sin(\pi/4) = 1/\sqrt{2}, \qquad k_3 = \sin(\pi/12) = (\sqrt{3}-1)/\sqrt{8};$$

further examples of k_n for small integer n are shown in Table 10.4. To solve Problem 10, we need to know k_{100}. Weber has it in a large list of *class invariants* at the end of his 1891 book [Web91, p. 502]; for n even he employs the class invariant $f_1(\sqrt{-n}) = \sqrt[6]{2}\,\sqrt[12]{k_n'^2/k_n}$ [Web91, p. 149] and gets

$$\sqrt[8]{2}\, f_1(\sqrt{-100}) = x, \qquad x^2 - x - 1 = \sqrt{5}\,(x+1).$$

Certainly, this qualifies as an expression in radicals for k_{100}. It is, however, not particularly attractive and does not display the fact that, for n a positive integer, k_n enjoys further arithmetic properties and is—up to a power of $\sqrt{2}$—a unit[82] in some algebraic number field; cf. Table 10.5. Ramanujan was prolific in creating formulas representing singular moduli as a product of simple units. Here is the famous example from his February 27, 1913, letter to Hardy [BR95, p. 60]:

$$\begin{aligned} &k_{210} \\ &= (\sqrt{2}-1)^2(2-\sqrt{3})(\sqrt{7}-\sqrt{6})^2(8-3\sqrt{7})(\sqrt{10}-3)^2(4-\sqrt{15})^2(\sqrt{15}-\sqrt{14})(6-\sqrt{35}). \end{aligned}$$

Table 10.4. *Some singular moduli k_n [Zuc79, Table 4].*

n	k_n
1	$1/\sqrt{2}$
2	$\sqrt{2}-1$
3	$(\sqrt{3}-1)/(2\sqrt{2})$
4	$3-2\sqrt{2}$
5	$\left(\sqrt{\sqrt{5}-1}-\sqrt{3-\sqrt{5}}\right)/2$
6	$(2-\sqrt{3})(\sqrt{3}-\sqrt{2})$
7	$(3-\sqrt{7})/(4\sqrt{2})$
8	$(\sqrt{2}+1)^2\left(1-\sqrt{2\sqrt{2}-2}\right)^2$
9	$(\sqrt{2}-3^{1/4})(\sqrt{3}-1)/2$
10	$(\sqrt{10}-3)(\sqrt{2}-1)^2$

[82]An algebraic number x is a unit if x and $1/x$ are algebraic integers.

Table 10.5. *Arithmetic properties of k_n [Ber98, p. 184].*

$n \in \mathbb{N}$	$n \equiv 0 \bmod 2$	$n \equiv 1 \bmod 4$	$n \equiv 3 \bmod 8$	$n \equiv 7 \bmod 8$
$2^{m/2}k_n$ is unit	$m = 0$	$m = 1$	$m = 2$	$m = 4$

Hardy later called it "one of the most striking of Ramanujan's results" [Har40, p. 228]. Let us try to accomplish something similar for k_{100}.

Lemma 10.4

$$k_{100} = (3 - 2\sqrt{2})^2\,(2 + \sqrt{5})^2\,(\sqrt{10} - 3)^2\,(5^{1/4} - \sqrt{2}\,)^4.$$

Proof. We follow the method for calculating k_{4n} that Berndt presents in his edition of Ramanujan's notebooks (cf. [Ber98, p. 284, Ex. 9.4]) in proving Ramanujan's expressions for k_4, k_{12}, k_{28}, and k_{60}. It is based on a formula of Ramanujan [Ber98, p. 283, form. (9.4)] relating k_{4n} to the class invariant $G_n = (2k_nk'_n)^{-1/12}$,

$$\begin{aligned} k_{4n} &= \left(\sqrt{G_n^{12} + 1} - \sqrt{G_n^{12}}\right)^2 \left(\sqrt{G_n^{12}} - \sqrt{G_n^{12} - 1}\right)^2 \\ &= \left(G_n^3\sqrt{G_n^6 + G_n^{-6}} - G_n^6\right)^2 \left(G_n^3 - \sqrt{G_n^6 - G_n^{-6}}\right)^2 (G_n^3)^2. \end{aligned}$$

Now, G_{25} was already known by Weber [Web91, p. 500] to be the golden ratio, $G_{25} = (\sqrt{5} + 1)/2$; cf. also [Ber98, p. 190]. Observing $G_{25}^3 = 2 + \sqrt{5}$, $G_{25}^6 = 9 + 4\sqrt{5}$, and $G_{25}^{-6} = 9 - 4\sqrt{5}$ yields

$$\begin{aligned} k_{100} &= \left((2 + \sqrt{5})\,3\,\sqrt{2} - (9 + 4\sqrt{5})\right)^2 \left((2 + \sqrt{5}) - 2\,\sqrt{2}\cdot 5^{1/4}\right)^2 (2 + \sqrt{5})^2 \\ &= (3 - 2\sqrt{2})^2(\sqrt{10} - 3)^2(5^{1/4} - \sqrt{2}\,)^4(2 + \sqrt{5})^2, \end{aligned}$$

which is the result stated. □

Arithmetically, the expression for k_{100} is certainly wonderful. *Numerically*, it is slightly unstable because of cancellation of leading digits; we lose one digit calculating $3 - 2\sqrt{2}$ or $\sqrt{10} - 3$, two digits calculating $5^{1/4} - \sqrt{2}$. Thus, we can predict an overall loss of two digits using this formula; IEEE double precision will give about 14 correct digits. However, the representation of k_{100} via units facilitates a *stabilized* version that involves no subtractions,

$$k_{100} = \frac{1}{(3 + 2\sqrt{2})^2(2 + \sqrt{5})^2(3 + \sqrt{10})^2(\sqrt{2} + 5^{1/4})^4}.$$

Summarizing, the effort in calculating k_{100} is rewarded by the following unexpected *closed-form solution* of Problem 10:

$$\begin{aligned} p|_{\rho=10} &= \frac{2}{\pi}\arcsin\left((3 - 2\sqrt{2})^2(2 + \sqrt{5})^2(\sqrt{10} - 3)^2(5^{1/4} - \sqrt{2}\,)^4\right) \\ &= \frac{2}{\pi}\arcsin\left(\frac{1}{(3 + 2\sqrt{2})^2(2 + \sqrt{5})^2(3 + \sqrt{10})^2(\sqrt{2} + 5^{1/4})^4}\right). \end{aligned}$$

A MATLAB Session

```
>> r1=3-2*sqrt(2); r2=sqrt(5)+2;      % the Ramanujan-style formula
>> r3=sqrt(10)-3;r4=5^(1/4)-sqrt(2);
>> p=2/pi*asin(r1^2*r2^2*r3^2*r4^4)

p = 3.837587979251201e-007

>> r1=3+2*sqrt(2); r2=sqrt(5)+2;      % the stabilized formula
>> r3=sqrt(10)+3;r4=5^(1/4)+sqrt(2);
>> p=2/pi*asin(1/r1^2/r2^2/r3^2/r4^4)

p = 3.837587979251226e-007
```

As predicted, using IEEE double precision, the Ramanujan-style formula gives 14 correct digits. The stabilization leads to 16 correct digits.

10.8 Harder Problems

To review and compare the different approaches to Problem 10, we ask how one might change the problem to make it harder or impossible to solve by one of the approaches. Three aspects of the problem were used to varying extents: the geometry was two-dimensional, the shape was rectangular, and the particle was started at the center, the unique point common to the two axes of symmetry. Hence, besides varying the aspect ratio there are three generalizations immediately at hand:

- the particle starts at some point off-center,
- the domain is a general two-dimensional polygon
- the domain is an n-dimensional box.

Table 10.6 tells us whether or not the methods of this chapter can be extended. Let us comment on some aspects of this table.

The *Monte Carlo method* is a general method that is good for low absolute accuracies only. It is easily coded and the method of choice for high dimensions $n > 4$, for which the "curse of dimension" deprives us of better general methods.

Table 10.6. *Extendibility of the various methods for Problem 10.*

Method	Precision	Off-center	Polygon	nD-box
§10.4: Monte Carlo	Low	✓	✓	✓
§10.3: finite differences	Medium	✓	(✓)	✓
§10.4: separation	High	✓	—	✓
§10.5: conformal mapping	Medium	✓	✓	—
§10.6: elliptic integrals	High	✓	—	—
§10.7: singular moduli	High	—	—	—

Finite differences and extrapolation is a general method that is good for medium relative accuracies. For reasons of run-time efficiency, it is restricted to small dimensions $n = 1, \dots, 4$. If the domain is not a box, one has to exercise some care in discretizing the boundary. The discretization error might then have an asymptotic expansion in several incompatible powers of h^γ, making a generalization not straightforward.

Separation of variables. Because of the exponentially fast converging series, this is suited to high relative accuracies. Geometrically, it is restricted to boxes.

Conformal mapping. The method is inherently two-dimensional. For general polygons, the Schwarz–Christoffel map has to be evaluated numerically, leading to the rich topic of numerical conformal mapping; cf. the recent book of Driscoll and Trefethen [DT02].[83]

Elliptic function theory. For points starting off-center, we have to use a result that is more general than Lemma 10.2 (cf. [Hen86, §15.5, Expl. 2]) and obtain a transcendental equation involving *incomplete* elliptic integrals of the first kind. Alternatively, we could map the rectangle to the upper half-plane using Jacobi's elliptic sine function. The harmonic measure can then be calculated by means of [Hen86, §15.5, Expl. 1].

Singular moduli. Here, all the specifics of Problem 10 played a role; the results do not extend to more general problems.

[83]Driscoll has written a *Schwarz–Christoffel Toolbox* for MATLAB, available for download at `http://www.math.udel.edu/~driscoll/software/SC`, suited to solve Problem 10 within a second:

```
» pol = polygon([10+i -10+i -10-i 10-i]);
» f = center(crdiskmap(pol,scmapopt('Tolerance',1e-11)),0);
» prevert = get(diskmap(f),'prevertex'); p = angle(prevert(1))/pi

p = 3.837587979278246e-007
```

Varying the tolerance from 10^{-8} to 10^{-14} establishes that most likely 10 digits are correct. In fact, we know from §10.4 the correctness of the first 11 digits.

Appendix A

Convergence Acceleration

Dirk Laurie

Analytical methods seem to become more and more in favor in numerical analysis and applied mathematics and thus one can think (and we do hope) that extrapolation procedures will become more widely used in the future.
—Claude Brezinski and Michela Redivo Zaglia [BZ91, p. v]

The idea of applying suitable transformations for the acceleration of the convergence of a series or for the summation of a divergent series is almost as old as analysis itself.
—Ernst Joachim Weniger [Wen89, p. 196]

A.1 The Numerical Use of Sequences and Series

Almost every practical numerical method can be viewed as the approximation of the limit of a sequence

$$s_1, s_2, s_3, \ldots, \tag{A.1}$$

which sometimes arises via the partial sums

$$s_k = \sum_{i=1}^{k} a_i \tag{A.2}$$

of a series, by computing a finite number of its elements. In this discussion we only consider the case in which the elements of the sequence are real numbers (but most of what we have to say goes for complex-valued sequences too) and leave aside the question of vector-valued, matrix-valued, and function-valued, sequences.

The sequence (A.1) and the series (A.2) are theoretically equivalent if we define $s_0 = 0$, but in practice there is some[84] accuracy to be gained when working with series, assuming of course that the a_i can be found to full machine precision, not by the formula $a_i = s_i - s_{i-1}$.

The question whether a series converges is largely irrelevant when the reason for using a series is to approximate its sum numerically.

[84] Not much—see §A.5.3.

Convergence Is Not Sufficient

In the case of convergent sequences for which $\rho = \lim(s - s_{k+1})/(s - s_k)$ exists, the convergence is said to be *linear* if $-1 \leqslant \rho < 1$, *sublinear* or *logarithmic* if $\rho = 1$, and superlinear if $\rho = 0$. There are important series for which ρ does not exist, for example,

$$\frac{1}{\zeta(s)} = \sum_{k=1}^{\infty} \mu(k)\, k^{-s},$$

where ζ is the Riemann zeta function and $\mu(k)$ is the Möbius function (if k is the product of n distinct primes, then $\mu(k) = -1$ for n odd and $\mu(k) = 1$ for n even; otherwise $\mu(k) = 0$) but we shall not have any more to say about them here.

For practical purposes, one requires *fast* convergence, for example, linear convergence with $\rho \ll 1$, or preferably superlinear convergence. A series such as

$$s_k = \sum_{n=1}^{k} \frac{1}{n^2}$$

is useless unless convergence acceleration techniques are applied, because $O(1/\epsilon)$ terms are needed to obtain an accuracy of ϵ.

But even fast convergence is not enough. A series such as

$$s_k = \sum_{n=0}^{k} \frac{x^n}{n!}$$

is useless in floating-point arithmetic for large negative values of x, because the largest term is orders of magnitude larger than the sum. The round-off error in the largest term swamps the tiny sum.

Convergence Is Not Necessary

There is an important class of divergent series, known as *asymptotic* series, which are highly useful numerically. These series typically have partial sums $s_k(x) = \sum_{n=0}^{k} a_n x^n$, where x is a real or complex variable, with the following properties:

- The power series $\sum_{n=0}^{\infty} a_n x^n$ has convergence radius 0.
- For x sufficiently small, the sequence $a_n x^n$ decreases in magnitude until a smallest term is reached and thereafter increases.
- There exists a function $f(x)$ whose formal power series coincides with the given series.
- If the series alternates, $s_k(x) - f(x)$ is smaller in magnitude than $a_k x^k$.

When all these properties hold, an alternating asymptotic series is often more convenient than a convergent series for the purpose of obtaining a good approximate value of $f(x)$ with rigorous error bound: one stops after the smallest term, and the absolute value of that term is the bound. We call the size of that bound the *terminal accuracy* of the asymptotic series.

Even when an asymptotic series is monotonic, the approximation obtained by truncating it after the smallest term may sometimes be good enough for practical purposes, although in this case there is no error bound. Be aware that it is the rule rather than the exception for an asymptotic series to be valid only in certain sectors of the complex plane.

A well-known example of an asymptotic series is the logarithmic form of Stirling's formula for the factorial (see [Olv74, §8.4]),

$$\log(n!) \sim \left(n+\frac{1}{2}\right)\log n - n + \frac{1}{2}\log(2\pi) + \sum_{j=1}^{\infty} \frac{B_{2j}}{2j(2j-1)n^{2j-1}}$$
$$= \left(n+\frac{1}{2}\right)\log n - n + \frac{1}{2}\log(2\pi) + \frac{1}{12n} - \frac{1}{360n^3} + \frac{1}{1260n^5} - \frac{1}{1680n^7} + \cdots,$$

where B_j is the jth Bernoulli number. Since the numbers B_j grow faster than any power of j, the series is divergent for all n. Yet it is very useful for large n and remains valid as the first step in the calculation of .(z) when n is replaced by $z+1$, where z is a complex number not on the negative real axis. Our solution of Problem 5 to 10,000 digits relies on this formula (see §5.7).

Convergence acceleration algorithms are useful for obtaining improved estimates of the limit s of a sequence s_k when

- *the members of the sequence can be computed to high precision;*
- *the behavior of s_k as a function of k is regular enough.*

A.2 Avoiding Extrapolation

There is an important class of series where extrapolation methods can be avoided: those in which the terms (with or without an alternating sign) are explicitly known as analytic functions evaluated at the integers. In such a case one can express the sum as a contour integral. The formulas are (see Theorem 3.6)

$$\sum_{k=1}^{\infty}(-1)^k f(k) = \frac{1}{2i}\int_C f(z)\csc(\pi z)\,dz, \qquad \sum_{k=1}^{\infty} f(k) = \frac{1}{2i}\int_C f(z)\cot(\pi z)\,dz,$$

where

1. the integration contour C runs from ∞ to ∞, starting in the upper half-plane and crossing the real axis between 0 and 1;
2. f decays suitably as $z \to \infty$ and is analytic in the component of the complex plane that contains the integers $1, 2, 3, \ldots$.

There is considerable scope for skill and ingenuity in the selection of the contour and its parametrization. A simple example is given in §1.7, and a more sophisticated one, with a discussion of the issues involved in choosing the contour, in §3.6.2.

The contour integration method is usually more expensive computationally than a suitable extrapolation method, but has the advantage of in principle being able to obtain any desired precision.

A.3 An Example of Convergence Acceleration

Algorithm A.1. Archimedes' Algorithm for Approximating π.

Purpose: To compute for $k = 1, 2, \ldots, m$, the numbers

$$a_k = n_k \sin(\pi/n_k), \quad b_k = n_k \tan(\pi/n_k), \qquad \text{where } n_k = 3 \cdot 2^k,$$

via the intermediate quantities

$$s_k = \csc(\pi/n_k), \qquad t_k = \cot(\pi/n_k).$$

Procedure:

$$t_1 = \sqrt{3}, \quad s_1 = 2, \quad n_1 = 6, \quad a_1 = n_1/s_1, \quad b_1 = n_1/t_1.$$

$$\text{For } k = 2, 3, \ldots, m: \quad \begin{cases} t_k = s_{k-1} + t_{k-1}, \quad s_k = \sqrt{t_k^2 + 1}, \\ n_k = 2n_{k-1}, \quad a_k = n_k/s_k, \quad b_k = n_k/t_k. \end{cases}$$

Archimedes did the whole calculation up to $m = 5$ in interval arithmetic (anticipating its invention by over two thousand years), culminating in the famous inequality

$$3\tfrac{10}{71} < \pi < 3\tfrac{1}{7}.$$

In Table A.1 we give the actual lower bounds $\underline{a}_k$ for a_k and upper bounds $\overline{b}_k$ for b_k that Archimedes got, together with their decimal representation, rounded up or down as appropriate to 6 significant digits, and the machine numbers $\hat{a}_k$, $\hat{b}_k$ that one gets by calculating them in IEEE double precision, rounded to 15 significant digits.

Now one might feel that the 15-digit approximations are not much better than the rational bounds: obtaining $\hat{b}_5 - \hat{a}_5 = 0.00168$ rather than $\overline{b}_k - \underline{a}_k = 0.00193$ seems a poor return for double-precision computation. But the extra precision in the values $\hat{a}_k$ and $\hat{b}_k$ can be put to good use. Since we know that $\sin(\pi x)/x$ has a Taylor series of the form

$$c_0 + c_1 x^2 + c_2 x^4 + \cdots$$

it follows that a_k has a series of the form

$$d_0 + d_1 4^{-k} + d_2 4^{-2k} + \cdots. \tag{A.3}$$

Now the sequence $a'_k = a_k + (a_{k+1} - a_k)/(4^1 - 1)$ has a series starting at 4^{-2k}; the sequence $a''_k = a'_k + (a'_{k+1} - a'_k)/(4^2 - 1)$ has a series starting at 4^{-3k}; etc.

These values are given in Table A.2. It is immediately obvious that the numbers in each column agree to more digits than in the previous column. The final number a''''_1 happens to be correct to all 15 digits, but of course we are not supposed to know that.

Still, by looking at the first row, we are able to assert, on the basis of our knowledge and experience of this extrapolation algorithm, that the limit of the original sequence to 12

Table A.1. *Algorithm A.1 as computed by Archimedes and by a modern computer.*

k	$\underline{a}_k$	$\underline{a}_k$	$\hat{a}_k$	$\overline{b}_k$	$\overline{b}_k$	$\hat{b}_k$
1	3	**3.**00000	**3.**000000000000000	$\frac{918}{265}$	**3.**46416	**3.**46410161513775
2	$\frac{9360}{3013\frac{3}{4}}$	**3.1**0576	**3.1**05828541230249	$\frac{1836}{571}$	**3.**21542	**3.**21539030917347
3	$\frac{5760}{1838\frac{9}{11}}$	**3.1**3244	**3.1**32628613281238	$\frac{3672}{1162\frac{1}{8}}$	**3.1**5973	**3.1**5965994209750
4	$\frac{3168}{1009\frac{1}{6}}$	**3.1**3922	**3.1**39350203046867	$\frac{7344}{2334\frac{1}{4}}$	**3.14**620	**3.14**608621513143
5	$\frac{6336}{2017\frac{1}{4}}$	**3.14**090	**3.141**031950890510	$\frac{14688}{4673\frac{1}{2}}$	**3.14**283	**3.14**271459964537

digits is 3.14159265359, and feel fairly confident that $\hat{a}_1''''$ is probably correct to 14 digits (knowing that our machine works to just under 16 digits,[85] we are not absolutely sure of the 15th digit).

Similarly, the Taylor series for $\tan(\pi x)/x$ also contains only even powers of x, and therefore b_k can be expressed as a series of the form (A.3). Extrapolation can be applied to the sequence $\hat{b}_k$ to obtain the values in Table A.3. The improvement in convergence is less spectacular, and only an optimist would claim more than nine digits on this evidence. With hindsight we can explain the difference in behavior: the coefficients in the Taylor series for $\tan(\pi x)/x$ do not decay exponentially, since tan is not an entire function.

The typical challenge solver[86] would have been happy, but the dialectic mathematician is not satisfied. How can you trust a number about which you do not even know whether it is greater or less than the desired quantity?

Table A.2. *Accelerated sequences derived from lower bounds in the double-precision version of Algorithm A.1.*

k	$\hat{a}_k'$	$\hat{a}_k''$	$\hat{a}_k'''$	$\hat{a}_k''''$
1	**3.141**10472164033	**3.141592**45389765	**3.1415926535**7789	**3.14159265358979**
2	**3.1415**6197063157	**3.14159265**045789	**3.1415926535897**5	
3	**3.14159**073296874	**3.141592653**54081		
4	**3.141592**53350506			

Let us be quite clear on this point: the art of confidently asserting the correctness up to a certain accuracy of a computed number, without being able to prove the assertion, *but still being right*, is not mathematics. But it is science. Richard Feynman said:

[85] $\log_{10} 2^{53} \approx 15.95$.

[86] I can, of course, only speak for one such.

Table A.3. *Accelerated sequences derived from upper bounds in the double-precision version of Algorithm A.1.*

k	$\hat{b}'_k$	$\hat{b}''_k$	$\hat{b}'''_k$	$\hat{b}''''_k$
1	**3.1**32486540518 71	**3.141**65626057574	**3.14159**254298228	**3.14159265**363782
2	**3.141**08315307218	**3.14159**353856967	**3.14159265**320557	
3	**3.1415**6163947608	**3.141592**66703939		
4	**3.14159**072781668			

> Mathematics is not a science from our point of view, in the sense that it is not a *natural* science. The test of its validity is not experiment. [FLS63, p. 3-1]

Scientific computing *is* a science from Feynman's point of view. The test of its validity *is* experiment. Having formulated the theory that the sought-for limit to 12 digits is 3.14159265359, we can test it by computing a_6 and calculating another diagonal of extrapolated values, maybe using higher precision, and by repeating the calculation on a different type of computer.

> *More than any other branch of numerical analysis, convergence acceleration is an experimental science. The researcher applies the algorithm and looks at the results to assess their worth.*

A.4 A Selection of Extrapolation Methods

Strictly speaking, the only difference between interpolation and extrapolation is whether the point at which we wish to approximate a function lies inside or outside the convex hull of the points at which the function is known. In theory and in practice, the difference is profound when (as here) extrapolation is used to estimate a limit to infinity. When doing interpolation, the limiting process is one in which typically a step size h is allowed to approach 0, and the behavior of the function to be interpolated becomes more and more polynomial-like as the step size decreases. When doing extrapolation, the limiting process is one in which more and more terms may be taken into account, but there is no question of the behavior of the partial sum s_n becoming polynomial-like as a function of n. Few functions on $[0, \infty)$, except polynomials themselves, asymptotically behave like polynomials.

In a survey like this one, it is impossible to be exhaustive. I have presented my personal favorite convergence acceleration methods in a unified way. Some of them I like because so often they work so well; others are useful steps on the way to understanding the more sophisticated methods. The emphasis is on computational aspects, not on the theory.

Readers who want a more complete presentation should consult the excellent monograph by Brezinski and Redivo-Zaglia [BZ91], which goes into greater detail than we can, surveys all available theoretical results on acceleration of sequences, gives many more methods, and contains computer software for all of them. Other monographs, each with its own strong points, are those of Wimp [Wim81] and Sidi [Sid03]. The historical development of the field is very ably summarized by Brezinski [Bre00]. A report by Weniger [Wen89] has similar aims to this one, but is much more comprehensive.

Most extrapolation algorithms arise naturally from consideration of a sequence, and little attention has been paid to formulating those as series-to-series transformations.

We shall adopt the following notation for all the methods:

- $s_{k,n}$ is an extrapolated value depending on the elements $s_k, s_{k+1}, \ldots, s_{k+n}$ of the sequence. Thus, $s_{k,0} = s_k$.
- X is a generic extrapolation operator that maps a sequence s_k into a sequence $X(s_k)$.

When we need to display[87] the elements $s_{k,n}$ two-dimensionally, we form the following matrix:

$$S = \begin{pmatrix} s_{1,0} & s_{1,1} & s_{1,2} & \cdots & s_{1,n-2} & s_{1,n-1} \\ s_{2,0} & s_{2,1} & s_{2,2} & \cdots & s_{1,n-2} & \\ \vdots & \vdots & & & & \\ s_{n-1,0} & s_{n-1,1} & & & & \\ s_{n,0} & & & & & \end{pmatrix}.$$

Occasionally, there might be a row number 0, if we decide to treat $s_0 = 0$ as a member in good standing of the given sequence. This is not always a sensible thing to do: for example, in the Archimedes case, we could have started the sequence one term sooner (Archimedes didn't) with the semiperimeter of a triangle, which would require $s_0 = 3\sqrt{3}/2$, not $s_0 = 0$.

The advantage of generating the whole triangular table S is to get more insight into the accuracy and reliability of the extrapolation than just one number, or even one sequence of numbers, would give. This matter is taken up in §A.5.4.

All reasonable extrapolation methods are quasi-linear; that is, they satisfy the relation $X(\lambda s_k + \delta) = \lambda X(s_k) + \delta$, where λ and δ are constants. Most (but not all) of them are based on a model of the form

$$s_k = s + \sum_{j=1}^{m} c_j \phi_{k,j} + \eta_k, \tag{A.4}$$

where the auxiliary columns are ordered so that each $\phi_{k,j+1}$ tends to zero as k increases faster than its predecessor $\phi_{k,j}$, and it is hoped that η_k tends to zero more rapidly than $\phi_{k,m}$. Usually (but not always) the elements of the matrix $\{ = [\phi_{k,j}]$ would be given by a formula of the form

$$\phi_{k,j} = \phi_j(k) \tag{A.5}$$

for certain simple functions ϕ_j.

One can classify extrapolation methods as linear or nonlinear; in the latter case, it is useful to make a further distinction between semilinear and strongly nonlinear methods.

[87] Our notation $s_{k,j}$ corresponds to T_j^k in [Bre00]. There is no universal agreement on how to display the extrapolated values. Some authors put them in a lower triangular matrix so that our rows run along the diagonals. Others emphasize the symmetry of the formula by using a triangular array within which each column is offset downwards by half a step from the previous one.

Linear Extrapolation Methods

These methods are linearly invariant; that is, they satisfy the relation

$$X(\lambda s_k + \mu t_k) = \lambda X(s_k) + \mu X(t_k), \quad \text{where } \lambda \text{ and } \mu \text{ are constants.} \tag{A.6}$$

They typically assume a model of the form (A.4), where the matrix { is known beforehand, independent of the sequence elements, although prior information about the sequence may influence the choice of model.

If in the model equations for $s_k, s_{k+1}, \ldots, s_{k+n}$, we put $m = n$, ignore η_k, and replace the unknown s by $s_{k,n}$, we obtain $n + 1$ equations

$$s_i = s_{k,n} + \sum_{j=1}^{n} c_j \phi_{i,j}, \quad i = k, k+1, \ldots, k+n. \tag{A.7}$$

The various methods differ in the model chosen and the way in which the parameters c_j are eliminated to find $s_{k,n}$.

> *Linear extrapolation methods can only be expected to work when substantial prior knowledge about the behavior of the sequence is exploited.*

In other words: the model must describe the sequence fairly well or the extrapolation method won't deliver.

Semilinear Extrapolation Methods

These typically assume a model of the form (A.4), where the matrix elements have the form $\phi k, j = \alpha k \beta k, j$, where αk is allowed to depend in a quantitative way on the elements of the sequence. The resulting linear systems are then solved by a process similar to (A.7). In particular, extrapolated values are available for all pairs (k, l) that satisfy $k + l \leqslant n$.

Semilinear methods can be extremely effective, even when no a priori information about the asymptotic behavior of the sequence is available.

Strongly Nonlinear Extrapolation Methods

These methods either assume a model of the form (A.4), (A.5) in which the functions ϕ_j depend on unknown parameters that lead to nonlinear equations or in some cases have no explicit model at all, being derived by heuristic reasoning from other methods. Extrapolated values $s_{k,l}$ are only available for some pairs (k, l).

A.4.1 Linear Extrapolation Methods

Having selected the auxiliary functions ϕ_j in (A.5), for each index pair (k, l) one could simply solve the equations (A.7) by brute force—they are, after all, linear. That would require $O(n^5)$ operations to calculate all the entries $s_{k,l}, \ k + l \leqslant n$. But we can do much better than this.

Since it is desirable to have all the $s_{k,j}$ available, linear extrapolation methods should be cast into the form

$$s_{k,j} = s_{k+1,j-1} + f_{k,j}(s_{k+1,j-1} - s_{k,j-1}), \tag{A.8}$$

where the multipliers $f_{k,j}$ do not depend on the original sequence s_k. The geometric picture is

$$\begin{pmatrix} s_{k,j-1} & s_{k,j} \\ s_{k+1,j-1} & \end{pmatrix} \cdot \begin{pmatrix} -f_{k,j} & -1 \\ 1+f_{k,j} & \end{pmatrix} = 0.$$

In most cases, the multipliers are easier to interpret when the model is written as

$$s_{k,j} = s_{k+1,j-1} + \frac{r_{k,j}}{1-r_{k,j}}(s_{k+1,j-1} - s_{k,j-1}). \tag{A.9}$$

It is obvious that if all the multipliers $f_{k,j}$ are chosen with complete freedom, then any extrapolation table S that satisfies $s_{k+1,j-1} \neq s_{k,j-1}$ can be obtained. This would defeat the self-validating property of the extrapolation table. Therefore, it is usual to have a model involving no more than n free parameters, where $n+1$ is the number of available terms of the sequence.

Often it is possible to give a simple formula for the constants $f_{k,j}$, so that the whole extrapolation process takes $O(n^2)$ operations, which is optimal, since there are $O(n^2)$ entries in the triangular table. In the general case, such low complexity is not possible, but there is an ingenious way of organizing Gaussian elimination, known as the E-algorithm, which reduces the operation count from $O(n^5)$ to $O(n^3)$. The general outline can be described in a few sentences. Think of the first step of (A.8) as $s_{:,1} = X_j s_{:,0}$; that is, column 1 of S is obtained by applying an extrapolation operator X_j to column 0. Let the constants $f_{:,0}$ be determined by the property that $X(\phi_{:,1}) = 0$. Now replace all the other columns $\phi_{:,j}$ by $X(\phi_{:,j}),\ j = 2, 3, \ldots, n$. In other words: apply exactly the same extrapolation formula to each column of the matrix { that you have applied to the sequence s_k. Then continue using the new column of extrapolated values instead of s_k and the modified matrix { instead of the original, etc.

In the table S, each column $s_{k,j},\ k = 1, 2, 3, \ldots$, should converge faster than the previous one if the model is appropriate. The rows of the array $s_{k,j},\ j = 0, 1, 2, \ldots$, should in that case converge even faster. This behavior, of rows converging faster than columns, is conspicuous; its absence indicates that the model is bad.

Euler's Transformation

Euler's method is one of the rare instances of a genuine series-to-series transformation. The derivation of this grandfather of all extrapolation methods is a typical piece of Eulery. Let I be the identity ($Ia_k = a_k$) and Δ the forward difference ($\Delta a_k = a_{k+1} - a_k$) operator on a sequence. Note that $(I+\Delta)a_k = a_{k+1}$. Let $b_k = a_k/r^{k-1}$, with $r \neq 1$. Then

$$\begin{aligned}
\sum_{k=1}^{\infty} a_k &= \sum_{k=0}^{\infty} r^k b_{k+1} = \sum_{k=0}^{\infty} (r(I+\Delta))^k b_1 \\
&= (I - r(I+\Delta))^{-1} b_1 = \frac{1}{1-r}\left(I - \frac{r\Delta}{1-r}\right)^{-1} b_1 \\
&= \frac{1}{1-r}\sum_{j=0}^{\infty}\left(\frac{r\Delta}{1-r}\right)^j b_1 = \frac{1}{1-r}\sum_{j=0}^{\infty}\left(\frac{r}{1-r}\right)^j \Delta^j b_1.
\end{aligned}$$

We get our two-dimensional table by not starting at the first term, but using the transformation only to estimate the tail. This gives

$$s_{k,n} = \sum_{i=1}^{k-1} a_i + \frac{1}{1-r} \sum_{j=0}^{n} \left(\frac{r}{1-r}\right)^j \Delta^j b_k. \tag{A.10}$$

A little algebra shows that we can generate these values recursively columnwise by

$$s_{k,j+1} = s_{k+1,j} + \frac{r}{1-r}(s_{k+1,j} - s_{k,j}). \tag{A.11}$$

So Euler's transformation is seen to be the simplest possible case of (A.8), since $f_{k,l}$ is constant.

Although we have not used an explicit model in the derivation, the form of the final equation (A.10) reveals the underlying model: $s_{k,n}$ is constant in n for $n \geqslant m$ if b_k is a polynomial of degree no more than $m-1$ in k. In other words, the model is

$$a_k = r^k (c_0 + c_1 k + c_2 k^2 + \cdots + c_{n-1} k^{n-1}).$$

It is obvious that a_k has this form when

$$s_k = s + r^k (c_0 + c_1 k + c_2 k^2 + \cdots + c_n k^n).$$

While no doubt such sequences do occur, they are not very typical. More usually, in cases where we do happen to know the correct r, the first column $s_{k,1}$ converges substantially faster than s_k, but the first row $s_{0,k}$, which should be the really fast-converging one, at best has only approximately geometric convergence with factor $r/(1-r)$.

The most common application of Euler's transformation is to an alternating series converging slower than geometrically, when $r = -1$. It is fairly efficient in that case, not because the model is appropriate (it seldom is), but because $r/(1-r) = -1/2$, which is not too bad. For example, the series

$$\sum_{k=1}^{\infty} (-1)^{k-1} k^{-1} = \log 2 \doteq 0.693147180559945$$

is accelerated reasonably well by Euler's transformation, despite the fact that the high-order differences $\Delta^k c_1$ fail to approach zero any faster than a_k does. (It is a nice little exercise to prove that for this series, $\Delta^k c_1 = a_{k+1}$.) In Table A.4 and later in the chapter, we have made a column out of row 0 to facilitate comparison.

Table A.4 shows that, as expected, $s_{0,k}$ converges approximately geometrically with ratio $\frac{1}{2}$: after 10 steps we have three digits, and $2^{-10} \approx 0.001$. It is tempting to exploit that behavior through another application of Euler's transformation with $r = 1/2$, but only the first column is useful. (Guess what the first row will look like if you go all the way. Try it out. Shouldn't you have expected it?) The entries in the first column of this repeated transformation are given in Table A.4 by t_k: one extra digit—not worth the effort.

We have paid a good deal of attention to Euler's transformation, only to find that it is not spectacularly effective. But we now have the shoulders of a giant to stand on.

Table A.4. *Euler's transformation applied to* $\sum_{k=1}^{\infty}(-1)^{k-1}k^{-1}$.

k	s_k	$s_{0,k}$	t_k
1	1.000000000000000	**0.**500000000000000	1.000000000000000
2	**0.**500000000000000	**0.6**25000000000000	**0.**750000000000000
3	**0.**833333333333333	**0.6**66666666666667	**0.**708333333333333
4	**0.**583333333333333	**0.6**82291666666667	**0.69**7916666666667
5	**0.**783333333333333	**0.6**88541666666667	**0.69**4791666666667
6	**0.6**16666666666667	**0.69**1145833333333	**0.693**750000000000
7	**0.**759523809523809	**0.69**2261904761905	**0.693**377976190476
8	**0.6**34523809523809	**0.69**2750186011905	**0.693**238467261905
9	**0.**745634920634921	**0.69**2967199900793	**0.6931**84213789682
10	**0.6**45634920634921	**0.693**064856150793	**0.6931**62512400794

Modified Euler Extrapolation

An obvious modification of (A.11) is to allow r to have a different value for each column, giving

$$s_{k,j+1} = s_{k+1,j} + \frac{r_j}{1-r_j}(s_{k+1,j} - s_{k,j}). \tag{A.12}$$

It is not hard to show that this formula gives $s_{k,n} = s$ when

$$s_k = s + \sum_{j=1}^{n} c_j r_j^k. \tag{A.13}$$

It can be used, therefore, when the sequence is well approximated by a sum of geometric sequences with known decay rates. Repeated values of r_j have the same effect as in the case of the ordinary Euler transformation.

Richardson Extrapolation

Richardson extrapolation is appropriate when the sequence behaves like a polynomial in some sequence h_k.

In the original application, the model is (A.4) and (A.5) with

$$\phi_j(k) = h_k^{p_j}, \tag{A.14}$$

where the numbers h_k are step sizes in a finite-difference method and the exponents p_j are known in advance from an analysis involving Taylor series. In the most general case, we need the E-algorithm, but in the two most commonly encountered cases, the parameters $r_{k,j}$ in (A.9) can be found more simply.

1. If the step sizes are in geometric progression, that is, $h_{k+1}/h_k = r$ (the most common case being $r = 1/2$), then we put $r_{k,j} = r^{p_j}$.

2. If the exponents are a constant multiple of $1, 2, 3, \ldots$, that is, they are given by $p_j = cj$ for some constant c, then we put $r_{k,j} = (h_{k+1}/h_k)^c$.

Case 1 can be recognized as equivalent to modified Euler extrapolation, whereas case 2 is an application of Neville–Aitken interpolation.

We have already, in §A.3, seen a spectacular application of Richardson extrapolation, so we need no further example of a case where it is successful. Instead, we show what can happen when it is inappropriately used.

Richardson extrapolation is the engine inside Romberg integration. In that application, s_k is formed from trapezoidal or midpoint rules with the step size being continually halved, for example, using midpoint sums,

$$\int_0^1 f(x)\,dx \approx s_k = h \sum_{j=1}^{2^k} f((j - \tfrac{1}{2})h), \quad \text{where } h = 2^{-k}.$$

For smooth functions, the same model with even powers of h is applicable as in the case of the calculation of π, but Table A.5 shows what happens for

$$\int_0^1 -\log(x)\,dx = 1.$$

The first row converges no better than the first column, which is a sure sign that the model is bad.

A careful analysis of the error expansion will reveal which powers of h are introduced by the logarithmic singularity, but it is also quite easy to diagnose it numerically; see §A.5.2. Since the step sizes are in geometric progression, we can choose p to knock out precisely those powers. If we take the p_j from the sequence $\{1, 2, 4, 6, \ldots\}$, we obtain the results of Table A.6. Note that $s_{k,1}$ converges much faster than before, but that is not the real point. The first row converges much more quickly still, and that is what we are looking for to reassure us that the proper model has been used.

Table A.5. *Romberg integration applied to* $\int_0^1 -\log(x)\,dx = 1$*, assuming an expansion in even powers.*

h_k^{-1}	$s_{k,1}$	$s_{0,k}$
2	**0.9**42272533258662	**0.9**42272533258662
4	**0.9**71121185130247	**0.9**73044428588352
8	**0.9**85559581182433	**0.9**86736072861005
16	**0.99**2779726126725	**0.99**3394043936872
32	**0.99**6389859013850	**0.99**6700250685739
64	**0.99**8194929253506	**0.99**8350528242663

Table A.6. *Romberg integration applied to* $\int_0^1 -\log(x)\,dx = 1$, *assuming an expansion in the powers* 1, 2, 4, 6, ...

h_k^{-1}	$s_{k,1}$	$s_{0,k}$
2	**0.99**4914691495074	**0.99**4914691495074
4	**0.99**8706050625142	**0.9999**69837001832
8	**0.999**674995582249	**0.999999**853250138
16	**0.9999**18652198826	**0.999999999**613708
32	**0.9999**79656975437	**0.999999999999**550
64	**0.99999**4913863731	**0.999999999999999**

Salzer's Extrapolation

Salzer's extrapolation [Sal55] is appropriate when s_k is well approximated by a rational function in k.

The model is (A.4), (A.5) with $\phi_j(k) = (k+k_0)^{-j}$, where k_0 is a fixed number chosen in advance (usually $k_0 = 0$). This is a special case of Richardson extrapolation, obtained by putting $h_k = (k + k_0)^{-1}$ and $p_j = j$ in (A.14). So the extrapolated values can be obtained by case 2 of the Richardson extrapolation.

There are two other ways of solving the equations. Multiplying (A.7) by $(i + k_0)^n$, we obtain

$$(i + k_0)^n s_i = (i + k_0)^n s_{k,n} + \sum_{j=1}^{n} c_j (i + k_0)^{n-j}, \quad i = k, k+1, \ldots, k+n.$$

Taking the nth difference at $i = k$, everything under the sum vanishes and we are left with

$$s_{k,n} = \frac{\Delta^n((k + k_0)^n s_k)}{\Delta^n (k + k_0)^n}. \tag{A.15}$$

This formula is not recursive (which is close to Salzer's own [Sal55] formulation), and the preliminary multiplication by a different power of $(k + k_0)$ for each column $s_{:,l}$ is undesirable. Still, it gives an interesting view of the extrapolation: it is a discrete analogue of n applications of L'Hospital's rule. Suppose we wanted to find $\lim_{x\to\infty} f(x)$. One way of doing it is to use an auxiliary function g such that $g(x) \to_{x\to\infty} 0$ and to calculate

$$\lim_{x\to\infty} \frac{\frac{d^n}{dx^n}(f(x)/g(x))}{\frac{d^n}{dx^n}(1/g(x))}.$$

The third way is discussed in the next section.

For the series $s = \sum_{k=1}^{\infty} k^{-2} = \pi^2/6 \doteq 1.6449340668482$ we get the result of Table A.7 in IEEE double precision. The final result has about 11 correct digits, which is the best we can hope for because of round-off (see §A.5.3).

Table A.7. *Salzer's extrapolation applied to* $\sum_{k=1}^{\infty} k^{-2}$.

k	$s_{k,1}$	$s_{0,k}$
1	**1**.50000000000000	**1**.50000000000000
2	**1**.58333333333333	**1.6**2500000000000
3	**1.6**1111111111111	**1.64**351851851852
4	**1.6**2361111111111	**1.6449**6527777778
5	**1.6**3027777777778	**1.6449**5138888888
6	**1.6**3424603174603	**1.64493**518518521
7	**1.6**3679705215420	**1.64493**394341858
8	**1.6**3853316326531	**1.644934**04116995
9	**1.6**3976773116654	**1.644934066**24713
10	**1.64**067682207563	**1.644934067**14932
11	**1.64**136552730979	**1.6449340668**8417
12	**1.64**189971534398	**1.64493406684**468
13	**1.64**232236961279	**1.6449340668**3127

Modified Salzer's Extrapolation

Like all linear methods, Salzer extrapolation fails dismally on sequences that fail to conform to the underlying model. A simple modification goes a long way to remedy this.

The modified model is (A.4), (A.5) with $\phi_j(k) = \psi(k)(k + k_0)^{1-j}$ for some nonzero auxiliary function $\psi(k)$. Obviously, best results are obtained when $\psi(k)$ approximates $(s - s_k)$. This model no longer falls in the Richardson framework, and we have to find another computational procedure. The E-algorithm can of course be used, but there is a more economical alternative.

Dividing (A.7) by $\psi(i)$ we obtain

$$\frac{s_i}{\psi(i)} = \frac{s_{k,n}}{\psi(i)} + \sum_{j=0}^{n-1} c_{j+1}(i + k_0)^{-j}, \quad i = k, k+1, \ldots, k+n.$$

Now apply not ordinary differences as before, but divided differences, thinking of s_k as a function of t evaluated at $t_k = (k + k_0)^{-1}$. The divided differences of a sequence f_k given at those abscissas are defined recursively by

$$\delta^0 f_k = f_k;$$

$$\delta^n f_k = \frac{\delta^{n-1} f_{k+1} - \delta^{n-1} f_k}{t_{k+n} - t_k}. \tag{A.16}$$

Table A.8. *Salzer's and modified Salzer's extrapolation applied to* $\sum_{k=1}^{\infty} k^{-3/2}$.

k	$s_{0,k}$ with $\psi = 1$	$s_{0,k}$ with $\psi(k) = k^{-1/2}$
1	1.70710678118655	**2**.20710678118655
2	**2**.04280209908108	**2**.55223464247637
3	**2**.20053703479084	**2.6**0809008399373
4	**2**.28756605115163	**2.612**55796998662
5	**2**.34336074156378	**2.612**44505799459
6	**2**.38255729176400	**2.61237**916796721
7	**2**.41169662004271	**2.61237**468295396
8	**2**.43423616494222	**2.612375**22998939
9	**2**.45220141289729	**2.61237534**769965
10	**2**.46686223192588	**2.6123753**5041131
11	**2**.47905652271137	**2.61237534**876941
12	**2**.48936011369519	**2.61237534**899252
13	**2**.49818208203943	**2.61237534**782506

We obtain

$$s_{k,n} = \frac{\delta^n \left(\frac{s_k}{\psi(k)} \right)}{\delta^n \left(\frac{1}{\psi(k)} \right)}. \tag{A.17}$$

This can be implemented economically by forming two tables of divided differences, one for the denominator and one for the numerator.[88]

One can often guess a good ψ by integration; for example, if ϕ can be extended to a function defined on $(0, \infty)$, then it is plausible to choose

$$\psi(x) = \int_x^\infty \phi(x)\, dx. \tag{A.18}$$

As an example we take $s = \sum_{k=1}^{\infty} k^{-3/2} = \zeta(3/2) \doteq 2.61237534868549$. We show in Table A.8 the first row of the Salzer algorithm and of the modified Salzer algorithm with $\psi(k) = k^{-1/2}$, obtained by the method of integration.

Operator Polynomial Extrapolation

A rich family of linear extrapolation methods arises from yet another generalization of Euler's transformation. Express (A.11) in terms of the shift operator E (defined by

[88]This procedure is in fact equivalent to the E-algorithm if one takes advantage of the specially simple form of the dependence of the functions ϕ_j on j.

$Es_k = s_{k+1}$) as

$$s_{k,j+1} = \frac{(E - rI)s_{k,j}}{1 - r} = \frac{p_1(E)s_{k,j}}{p_1(1)},$$

where $p_1(t) = (t - r)$. From this we deduce that

$$s_{k,n} = \frac{p_n(E)s_k}{p_n(1)}, \tag{A.19}$$

where $p_n(t) = (t - r)^n$. We get the modified Euler formula when $p_n(t) = \prod_{j=1}^n (t - r_j)$. In general, when p_n is a polynomial of degree n, we call (A.19) an *operator polynomial extrapolation* formula. Note that the modified Salzer extrapolation (A.17) can also be viewed as a special case of (A.19).

To understand how to stand on Euler's shoulders, we need the following concept:

> A sequence a_k is *totally monotonic* if every difference $\Delta^j a_k$, $j = 0, 1, 2, \ldots$ (including a_k itself, $j = 0$), has the constant sign $(-1)^j$. A sequence a_k is *totally alternating* if $(-1)^k a_k$ is totally monotonic.

Totally monotonic sequences have many pleasant properties, the most important of which is Hausdorff's theorem:

> The null sequence a_k is totally monotonic if and only if there exists a weight function w, nonnegative over $(0, 1)$, such that
>
> $$a_k = \int_0^1 t^{k-1} w(t)\, dt, \quad k = 1, 2, \ldots .$$

Two corollaries are:

1. The sequence a_k is totally alternating if and only if there exists a weight function w, nonnegative over $(-1, 0)$, such that
$$a_k = \int_0^{-1} t^{k-1} w(t)\, dt, \quad k = 1, 2, \ldots .$$
2. If the support of w is the interval $[0, r]$ for some $0 < r < 1$, or in the alternating case $[r, 0]$ for some $-1 < r < 0$, then a_k converges geometrically with ratio r.

In view of these corollaries, we will henceforth use the notation $[0, r]$ for the closed interval with endpoints 0 and r even when $r < 0$.

Inspired by Hausdorff's theorem, let us suppose that $a_k = \int_0^r t^{k-1} w(t)\, dt$ with w nonnegative over the integration interval. Then

$$s_k = \sum_{j=1}^{k} \int_0^r t^{j-1} w(t)\, dt = \int_0^r \frac{(1 - t^k) w(t)\, dt}{1 - t}.$$

This allows us to represent (A.19) as

$$s_{k,n} = \frac{1}{p_n(1)} \int_0^r \frac{(p_n(1) - t^k p_n(t)) w(t)\, dt}{1 - t}.$$

Since $s = \lim s_k = \int_0^r \frac{w(t)\,dt}{1-t}$, we get the error formula

$$s - s_{k,n} = \frac{1}{p_n(1)} \int_0^r \frac{t^k p_n(t) w(t)\,dt}{1-t}. \tag{A.20}$$

The error formula suggests several strategies.

1. If we know nothing about the sequence a_k except that it is totally monotonic or totally alternating, there is the (fairly crude) error bound

$$\left|1 - \frac{s_{0,n}}{s}\right| \leqslant \frac{\max_{t\in[0,r]} |p_n(t)|}{p_n(1)}.$$

In the case of Euler's method for an alternating series with $r = -1$, we have $p_n(t) = (t+1)^n$, so that the crude bound is $O(2^{-n})$. A very good polynomial to use from the point of view of this bound is the Chebyshev polynomial of degree n, shifted to the interval $[-1, 0]$, which gives the bound $O(\lambda^{-n})$ with $\lambda = 3+2\sqrt{2} \doteq 5.828$ [CRZ00]. This formula is Algorithm 1 in [CRZ00].

2. If we know something about the weight function w, it is possible to optimize the polynomials to take advantage of the fact. For example, Cohen, Rodriguez Villegas, and Zagier [CRZ00] (hereafter referred to as CRVZ) derived polynomials that are good to use when w is analytic in certain regions of the complex plane that include the interval $[0, r]$. The resulting method can in the most favorable case be guaranteed to achieve $O(\lambda^{-n})$ with $\lambda \doteq 17.9$. The simplest family of polynomials in this class is defined by the identity

$$A_n(\sin^2\theta) = \frac{d^n(\sin^n\theta\cos^n\theta)}{d\theta^n}. \tag{A.21}$$

The polynomials A_n do not satisfy a three-term recursion and it seems to require $O(n^3)$ operations to generate all the coefficients of $A_0, A_1, \ldots, A_n$. Of course, one could precompute and store those.

3. If we assume that $w(t)/(1-t)$ is $(2n)$ times continuously differentiable, then a promising choice is to take the Legendre polynomials.

4. If $w(t)$ has an $O(t^\beta)$ singularity at $t = 0$, then the Jacobi polynomials suggest themselves. For example, for series of the form

$$\eta(\beta, r) = \frac{1}{\beta} - \frac{r}{1+\beta} + \frac{r^2}{2+\beta} - \frac{r^3}{3+\beta} + \cdots, \tag{A.22}$$

use the Jacobi polynomials $J^{(0,\beta-1)}$, shifting from the interval $x \in [-1, 1]$ to $t \in [0, r]$ by the transformation $t = r(x+1)/2$.

5. For monotonic series with $r = 1$, this approach has not so far been spectacularly successful. One reason is that, since $1 \in [0, r]$, the crucial factor $\max_{t\in[0,r]} |p_n(t)|/|p_n(1)|$ cannot be made smaller than 1.

Since so many of the above possibilities involve orthogonal polynomials, it is worthwhile to show in detail how these would be applied. Let the required orthogonal polynomials, shifted to the interval $[0, r]$, satisfy the recursion

$$\begin{aligned} p_0(t) &= 1; \\ p_1(t) &= (t - a_0); \\ p_{n+1}(t) &= (t - a_n)p_n(t) - b_n p_{n-1}(t), \qquad n = 1, 2, \ldots . \end{aligned}$$

Then the extrapolation algorithm is given by

$$\begin{aligned} \hat{s}_{k,0} &= s_{k,0}; \\ \hat{s}_{k,1} &= s_{k+1,0} - a_0 s_{k,0}; \\ \hat{s}_{k,n+1} &= \hat{s}_{k+1,n} - a_n \hat{s}_{k,n} - b_n \hat{s}_{k,n-1}, \; n = 1, 2, \ldots; \\ s_{k,n} &= \frac{\hat{s}_{k,n}}{p_n(1)}. \end{aligned}$$

One example must suffice here. The series $\eta(\beta, r)$ defined in (A.22) is to be evaluated for $r = 0.94$, $\beta = 0.125$. We do it four times, the last two times using more and more information. In Table A.9, each column is the transposed first row of the extrapolation table. First, the polynomials A_n of CRVZ (A.21), taken over $(0, -1)$; second, the Legendre polynomials shifted to $[0, -1]$; third, the CRVZ polynomials shifted to $[0, -0.94]$; last,

Table A.9. *The CRVZ method applied to $\eta(\beta, r)$ with $r = 0.94$, $\beta = 0.125$.*

k	CRVZ(−1)	Legendre(−1)	CRVZ(−0.94)	Jacobi(−0.94)
1	8.00000000000000	8.00000000000000	8.00000000000000	8.00000000000000
2	7.44296296296296	7.44296296296296	7.43159486016629	7.24346076458753
3	7.40927089580931	7.42063107088989	7.41224315975913	7.41890823284965
4	7.42291549578755	7.42333978080714	7.42286010429028	7.42300312779417
5	7.42312861386679	7.42305165138502	7.42312700068607	7.42310843651716
6	7.42311090693716	7.42311564776528	7.42311289945816	7.42311121107761
7	7.42311207370828	7.42311055990758	7.42311135754321	7.42311128485659
8	7.42311117324569	7.42311127821820	7.42311128787958	7.42311128682738
9	7.42311130057045	7.42311129133954	7.42311128684365	7.42311128688016
10	7.42311128604961	7.42311128388467	7.42311128688337	7.42311128688157
11	7.42311128677714	7.42311128731700	7.42311128688239	7.42311128688161
12	7.42311128692742	7.42311128679323	7.42311128688171	7.42311128688161
13	7.42311128687268	7.42311128688948	7.42311128688162	7.42311128688161
14	7.42311128688278	7.42311128688073	7.42311128688161	7.42311128688161
15	7.42311128688154	7.42311128688153	7.42311128688161	7.42311128688161
16	7.42311128688161	7.42311128688164	7.42311128688161	7.42311128688161

the Jacobi polynomials $J^{(0,-0.875)}$ shifted to $[0, -0.94]$. There is no meaningful difference between the first two methods (apart from the greater convenience of the three-term recursion in the Legendre case), but the third method is distinctly better. Not shown here is Legendre (-0.94), which also shows improvement, but slightly less so than CRVZ. The last column shows very impressive convergence, which even the various nonlinear algorithms to be discussed later cannot match. This example demonstrates the great value of having analytical information about the sequence.

A.4.2 Semilinear Algorithms

This family of algorithms is due to Levin [Lev73]. They are modified Salzer's algorithms with the following auxiliary functions (the labels come from Levin's own notation):

T $\psi(k) = a_k$.

U $\psi(k) = (k + k_0)a_k$.

W $\psi(k) = a_k^2/(a_{k+1} - a_k)$.

Note that we do not know continuous functions $\psi(x)$ for all x. The implementation goes exactly as described for the modified Salzer's algorithm.

The U-algorithm, in particular, is amazingly effective over a large class of sequences. When the T-algorithm works, so does the U-algorithm, although it may need one term more to achieve the same accuracy. Since round-off error cannot be ignored (see §A.5.3), it is better to use the T-algorithm in the cases where both work. The W-algorithm has the disadvantage of always needing one term more than the U-algorithm to form a particular $s_{k,l}$, and it is hard to find cases where the W-algorithm works but the U-algorithm does not.

Unfortunately it is also quite difficult to characterize the sequences for which any of these algorithms is exact. One reason for that is that semilinear algorithms do not possess the linear invariance property (A.6). In particular, we cannot expect that it will be exact for a sum of two functions if it happens to be exact for either separately.

To understand the startling effectiveness of the algorithms, it is useful to think of the analogue with integration: by (A.18), ψ is a plausible choice of auxiliary function when $\psi' = \phi$. Any nonzero multiple of ψ will do just as well, so the three transformations correspond, respectively, to

T $c\psi(x) = \psi'(x)$.

U $c\psi(x) = (x + x_0)\psi'(x)$.

W $c\psi(k) = (\psi'(x))^2/\psi''(x)$.

These differential equations have solutions

T $\psi(x) = e^{cx}$.

U $\psi(x) = (x + x_0)^c$.

W The general solution includes the other two as special cases.

Therefore, we expect the T-algorithm to be effective when $s_k \sim r^k$, the U-algorithm when $s_k \sim k^{-j}$, and the W-algorithm in either case. Actually the U-algorithm is nearly as effective as the T-algorithm when $s_k \sim r^k$, since its model has as its $(j+1)$st auxiliary function what the T-algorithm has as its jth. In practice, it is usually enough to use the U-algorithm all the time and ignore the others, since it is quite hard to find a function that suits the W-algorithm but not the U-algorithm, and in any case the W-algorithm is more susceptible to round-off error.

A.4.3 Strongly Nonlinear Methods

A typical feature of strongly nonlinear methods is that an approximation $s_{k,l}$ that uses precisely the values $s_{k:k+l}$ and no others is not available for all possible combinations of k and l.

Aitken's Method

Although more recent than Euler's transformation, "Aitken's Δ^2-method" is the best-known of all convergence acceleration procedures. It takes three members of the sequence and returns one number. Since it uses the elements s_k, s_{k+1}, s_{k+2} we call that number $s_{k,2}$. The basic formula resembles (A.9):

$$s_{k,2} = s_{k+2} + \frac{r_k}{1 - r_k}(s_{k+2} - s_{k+1}), \tag{A.23}$$

$$\text{where } r_k = \frac{s_{k+2} - s_{k+1}}{s_{k+1} - s_k}. \tag{A.24}$$

The formula for r_k is motivated by the model

$$a_k = cr^k,$$

the same model as for the first column of Euler's transformation but with r regarded as unknown. The nickname derives from another way of writing (A.23):

$$s_{k,2} = s_{k+2} - \frac{(\Delta s_{k+1})^2}{\Delta^2 s_k}. \tag{A.25}$$

Aitken's method is contained in two other methods:

1. Aitken's formula applied to $s_0, s_1, s_2, \ldots, s_n$ gives the same values for $s_{k,2}$ that Levin's T-transform gives for $s_{k,1}$ when applied to $s_1, s_2, \ldots, s_n$.

2. The second colum $s_{k,2}$ of the epsilon algorithm (see the next section) is identical to that of Aitken's method.

It is often effective to apply Aitken's method again on the transformed sequence, etc., but there is then no longer a simple model to tell us when the transformation is exact.

The Epsilon Algorithm

The model for Wynn's epsilon algorithm [Wyn56a] is the same (A.13) as for the modified Euler method, except that the ratios r_j are unknown rather than known. In other words, the epsilon algorithm can deliver the exact limit, using $2n+1$ members of a sequence, whenever the sequence can be written as a sum of n geometric sequences.

It is without doubt the most elegant of all convergence acceleration methods, with a marvellously simple recursion formula,

$$s_{k,j} = s_{k+1,j-2} + \frac{1}{s_{k+1,j-1} - s_{k,j-1}},$$

which contains no extraneous multipliers at all. To start off the recursion, we need $s_{k,-1} = 0$ as well as the usual $s_{k,0} = s_k$. The subscripts hide the pretty geometric picture. If we denote the four entries in the table by geographic letters of the alphabet and use a table in which each column is offset by half a step, we get

$$\begin{pmatrix} & s_{k,j-1} & \\ s_{k+1,j-2} & & s_{k,j} \\ & s_{k+1,j-1} & \end{pmatrix} = \begin{pmatrix} & N & \\ W & & E \\ & S & \end{pmatrix}, \qquad (N-S)(W-E) = 1.$$

The extrapolated values are found in the columns $s_{:,j}$ when j is even. In the case when the even columns converge (the usual case), the odd columns diverge to infinity.

Since only the even-numbered columns matter, a useful alternative formulation [Wyn66] eliminates the odd-numbered columns. Once again we get a pretty picture:

$$\begin{pmatrix} & s_{k,j-2} & \\ s_{k+2,j-4} & s_{k+1,j-2} & s_{k,j} \\ & s_{k+2,j-2} & \end{pmatrix} = \begin{pmatrix} & N & \\ W & C & E \\ & S & \end{pmatrix},$$

$$\frac{1}{C-N} + \frac{1}{C-S} = \frac{1}{C-W} + \frac{1}{C-E}.$$

The column $s_{:,2m}$ of the epsilon algorithm is exact when the sequence $s - s_k$ satisfies a linear difference equation of order m, that is, when constants $c_0, c_1, \ldots, c_m$ exist such that

$$c_0 s_k + c_1 s_{k+1} + \cdots + c_m s_{k+m} \text{ is constant for all } k.$$

It is therefore suitable for the same sequences as the modified Euler transformation, but with two important differences: on the plus side, it is not necessary to know the factors r_j in (A.12), and on the minus side, twice as many terms are required.

The Rho Algorithm

The rho algorithm [Wyn56b] is nearly as simple as the epsilon algorithm, having the model

$$s_{k,j} = s_{k+1,j-2} + \frac{j}{s_{k+1,j-1} - s_{k,j-1}}. \tag{A.26}$$

The extrapolated values appear in the same columns as in the case of the epsilon algorithm.

The column $s_{:,2m}$ of the rho algorithm is exact when

$$s_k = p(k)/q(k), \tag{A.27}$$

where p and q are polynomials of degree not more than m. It is therefore suitable for the same sequences as Salzer's transformation.

The simplicity of (A.26) arises from the hypothesis (A.27). It may sometimes be the case that a better model is

$$s_k = p(x_k)/q(x_k),$$

where the sequence x_k is known. The rho algorithm then becomes

$$s_{k,j} = s_{k+1,j-2} + \frac{x_k - x_{k-j}}{s_{k+1,j-1} - s_{k,j-1}}. \tag{A.28}$$

The Modified Rho Algorithm

Like Salzer's algorithm, the rho algorithm is not very effective when the sequence does not conform to the model of a rational function, but can be modified to certain other functions. The idea is to view the epsilon algorithm as the case $\theta = 0$, and the rho algorithm as the case $\theta = 1$, of the following algorithm:

$$s_{k,j} = s_{k+1,j-2} + \frac{1 + \theta(j-1)}{s_{k+1,j-1} - s_{k,j-1}}. \tag{A.29}$$

If s_k is well modelled by $s - s_k \approx k^{-1/\theta}\psi(k)$, where $\psi(k)$ is a rational function, the modified rho algorithm can be very effective.

The Theta Algorithm

The theta algorithm [Bre71] takes the guesswork out of choosing θ in the modified rho algorithm. To derive it, we write the first two stages of (A.29) as

$$\begin{aligned} s_{k,1} &= \frac{1}{s_{k+1,0} - s_{k,0}}, \\ s_{k,2} &= s_{k+1,0} + \frac{t}{s_{k+1,1} - s_{k,1}}, \end{aligned} \tag{A.30}$$

where $t = 1 + \theta$, and then ask ourselves, Why can't we allow t to depend on k? Since the ultimate in convergence acceleration is to reach a constant sequence, we throw caution overboard and choose t_k such that $s_{k+1,2} = s_{k,2}$ in (A.30), that is,

$$s_{k+1,0} + \frac{t_k}{s_{k+1,1} - s_{k,1}} = s_{k+2,0} + \frac{t_k}{s_{k+2,1} - s_{k+1,1}}.$$

This can be written as

$$t_k = -\frac{\Delta s_{k+1,0}}{\Delta^2 s_{k,1}}.$$

The theta algorithm is then continued to further columns by analogy to the epsilon and rho algorithms as

$$s_{k,2j+1} = s_{k+1,2j-1} + \frac{1}{\Delta s_{k,2j}},$$

$$s_{k,2j+2} = s_{k+1,2j} + \frac{\Delta s_{k+1,2j}\Delta s_{k,2j+1}}{\Delta^2 s_{k,2j}}.$$

The theta algorithm is extremely versatile in the sense that it can accelerate the convergence of a large class of sequences, which is, however, difficult to characterize. In this sense it is reminiscent of the Levin W-algorithm. On the negative side, it uses up $3n + 1$ terms of the original sequence to produce a row of n accelerated values against, respectively, $2n + 1$ terms for the other strongly linear algorithms discussed here, and only $n + 1$ terms for the linear and semilinear methods. It is also more prone to the effects of round-off error.

A.5 Practical Issues

A.5.1 Using a Subsequence

A trivial way of accelerating a sequence is to form a subsequence $t_k = s_{n_k}$, with $1 \leqslant n_1 < n_2 < n_3 < \cdots$. This technique is sometimes a useful preconditioning step before applying a convergence acceleration algorithm.

For example, suppose that in Problem 3 $s_k = \|A_k\|$ is obtained by using MATLAB to find the norm of the $k \times k$ submatrix A_k of the infinite matrix A (see §3.1). The best we can do with Levin's U-algorithm before numerical instability (see §A.5.3) dominates is about seven correct digits, using 16 terms. Using more terms makes matters worse, not better.

Now suppose we still use 16 terms, but start later: $t_k = s_{k+16}$. One would think that since these terms are closer to the limit, a better extrapolated value could be obtained. In fact, we still get about seven digits, but we already get them using 10 terms. Numerical instability sets in earlier. If we still use no term further than s_{32}, but take a larger stride, we do a little better. For example, stride 2 involves working with the subsequence s_{2k}; we obtain nine digits. However, taking strides 3, 4, etc., does not significantly improve the accuracy any further.

The real advantage of a subsequence is achieved, though, when the index sequence n_k grows faster than linearly. For example, if $s - s_k \equiv k^{-1}$ and $n_k = 2^k$, then $s - t_k \equiv 2^{-k}$: sublinear convergence has been turned into linear convergence. Similarly, linear convergence is turned into quadratic convergence. In Problem 3, one can get about 12 digits in floating-point arithmetic using $n_k = 2^{k-1}$, $k = 1, 2, \ldots, 10$. Since the subsequence already converges linearly with $\rho \approx \frac{1}{2}$, there is very little build-up of round-off error, and it is in principle possible to get fairly close to machine accuracy using this technique.

The extrapolated values for this example are given in Table A.10. Since we have been gaining a steady one to two digits per extrapolation step, and round-off error is insignificant, it requires no great leap of faith to accept the value $s = \|A\| \doteq 1.27422415282$, which is correct to 12 digits.

The epsilon algorithm also delivers contest accuracy (10 digits, almost 11) when applied to this sequence t_k (see Table 3.1 in Chapter 3).

Table A.10. *Levin's U-algorithm applied to the sequence $\|A_{n_k}\|$ of Problem 3.*

n_k	$t_k = s_{n_k}$	$t_{1,k}$
1	**1**.00000000000000	**1**.00000000000000
2	**1**.18335017655166	**1**.28951567784715
4	**1**.25253739751680	**1**.36301342016060
8	**1**.27004630585408	**1**.26782158984849
16	**1.27**352521545013	**1.274**45564643953
32	**1.274**11814436915	**1.27422**101365494
64	**1.2742**0913129766	**1.274224**05834917
128	**1.27422**212003778	**1.2742241**6013221
256	**1.27422**388594855	**1.274224152**91307
512	**1.2742241**1845808	**1.27422415282**063

A.5.2 Diagnosing the Nature of a Sequence

To estimate ρ, one can form the auxiliary sequence $\rho_k = \lim(s_{k+2} - s_{k+1})/(s_{k+1} - s_k)$, whose limit, if it exists, must equal ρ. It is in general very risky to base opinions on the nature of a slowly convergent sequence by examining a finite number of terms. The one exception arises when ρ is known to have one of a finite set of values, and the only uncertainty is which to take.

In the example of §A.4.1, we get the following values of ρ_k:

0.500487721799137
0.500065583337346
0.500008367557636
0.500001051510722

If we know that the asymptotic expansion of the error contains negative powers of 2, it is easy to decide that $\rho = 1/2$ and not $1/4$ or 1.

It must be stressed that this kind of test is of limited use.

A feature of convergence acceleration methods of which users must be aware is that in the case of a divergent alternating sequence, the algorithm usually delivers an approximation to a so-called *antilimit*. There are situations where the antilimit can be rigorously defined by analytical continuation; for example,

$$1 + r + r^2 + r^3 + \cdots = \frac{1}{1-r}$$

has a right-hand side that is defined for all $r \neq 1$ and that is the antilimit of the series on the left when $r \leqslant -1$ or $r > 1$.

This behavior can be very useful—an example follows below—but it does mean that divergence cannot be diagnosed. For example, the following sequence was shown to me by a researcher who wished to know whether it converges to 0:

$$\begin{array}{r} -1.000000000000000 \\ 0.732050807568878 \\ -0.630414938191809 \\ 0.576771533743575 \\ -0.543599590009618 \\ 0.521053669642427 \\ -0.504731494506494 \\ 0.492368058352063 \\ -0.482678712072860 \\ 0.474880464055156 \\ -0.468468857699484 \\ 0.463104220153723 \\ -0.458549452588645 \\ 0.454634026719303 \end{array}$$

On these data, Euler's transformation gives -0.0001244, Levin's T-algorithm -3.144×10^{-17}, the epsilon algorithm 1.116×10^{-10}, and the theta algorithm -4.737×10^{-12}. But 0 is not the the limit of the sequence; it is an antilimit. The sequence is divergent. However, $|s_k|$ is a convergent sequence. Salzer's transformation gives 0.402759395, Levin's U-algorithm 0.4027594, the rho algorithm 0.4027593957, and the theta algorithm 0.4027594. The evidence is overwhelming that the limit is not 0.

As an example of the usefulness of an antilimit, take, for example, $s_k = \sum_{n=0}^{k}(-1)^n n!$. Using 16 terms, Levin's T-algorithm gives 0.59634736, the epsilon algorithm 0.596, and the theta algorithm 0.596347. The series is, in fact, the asymptotic series for $f(z) = \int_z^\infty e^{z-x}x^{-1}\,dx$ in negative powers of x, evaluated at $x = 1$. By other methods we find $f(1) \doteq 0.596347362323194$. This shows that some violently divergent series can be summed, if not to great accuracy, by nonlinear acceleration methods. In fact, inputting a sequence of random deviates uniformly distributed over (0, 1) to the epsilon algorithm is quite likely to produce "accelerated" values clustering around 0.5.

Strongly nonlinear acceleration methods are so powerful that they require great care in their handling.

Here is a cautionary example. We define the two sequences

$$s_k = \sum_{n=1}^{k}(0.95)^n/n,$$

$$t_k = s_k + 19(0.95)^k/k.$$

Since $t_k - s_k$ is a null sequence, both sequences converge to the same limit, namely, $\log 20 \doteq 2.99573227355399$. And indeed, the Levin U-transformation in IEEE double precision approximates that limit to a terminal accuracy of five and seven digits, respectively. We now form $u_k = \sqrt{s_k t_k}$. This sequence obviously has the same limit as the other two. Yet Levin's algorithm, using 30 terms, insists that the limit is 2.9589224422146, with entries 25 to 30 of the first row all within half a digit in the last place of each other. The theta algorithm insists that the limit is 2.958919941, also in agreement with all these decimals between entries 21, 24, and 27 of the first row.

The sequence u_k is not sufficiently well behaved for extrapolation to work. It is neither monotonic nor alternating. The sequence decreases from over 4 to 2.95887231295868, then increases again. Both algorithms seem to treat u_k as an asymptotic sequence and to produce a spurious pseudolimit close to the place where u_k changes least. What saves us is that although both algorithms seem to produce a limit accurate to 10 digits, the two limits only agree to 6 digits. Linear algorithms like Salzer's do not produce any spurious limits at all.

Whenever possible, use more than one extrapolation method.

In the case of linear extrapolation methods, such as Richardson extrapolation, it is usually possible to prove that the accelerated sequence converges to the same limit as the original sequence. In the case of nonlinear methods, such a proof is seldom available, and when available, requires hypotheses that are unverifiable in practice. And, as we have seen, internal consistency among the entries of one nonlinear acceleration method does not guarantee anything.

In finite-precision arithmetic, "convergence" does not mean that one can in principle get arbitrarily close to the desired limit. It means that sooner or later (in the case of a good extrapolation method, sooner rather than later) a stage is reached where taking into account more terms does not further improve the approximation. Like an asymptotic series, a numerically accelerated sequence in finite precision has a certain terminal accuracy. The factors that determine terminal accuracy are discussed in the following section.

A.5.3 Condition and Stability

We have left this question to the last, but it is in fact an extremely important one. In any numerical computation, there are two main sources of error: truncation error, which arises because only a finite number of terms of an infinite sequence are taken into account; and round-off error, which arises because the computer cannot perform calculations exactly. Usually round-off error is only an important issue when the original sequence is monotonic.

In the case of acceleration methods, we typically start from a sequence in which truncation error is much larger than round-off error. The effect of the extrapolation is to reduce the truncation error, but (particularly in the case of monotonic sequences) to increase the round-off error.

A useful visual tool is to plot the absolute value of the differences in the extrapolated sequence $s_{1,k}$ on a logarithmic scale. For the example of §A.4.1, this graph is shown in Figure A.1. Also on the graph is shown an estimate of the accumulated round-off error (we will explain in a moment how to obtain that). It is conspicuous that the differences decrease steadily, and then start to rise, thereafter closely following the estimated round-off error.

The rule of thumb here is that when a monotonic sequence is accelerated numerically, the sequence thus obtained should be thought of in much the same way as an asymptotic series: the best place to stop is just before the first difference starts to increase. That is the point at which the truncation and round-off errors are both approximately equal to terminal accuracy.

There are also two main sources of round-off error itself. The first arises from the original data, which are obtained by an inexact procedure, such as the rounding of an exact

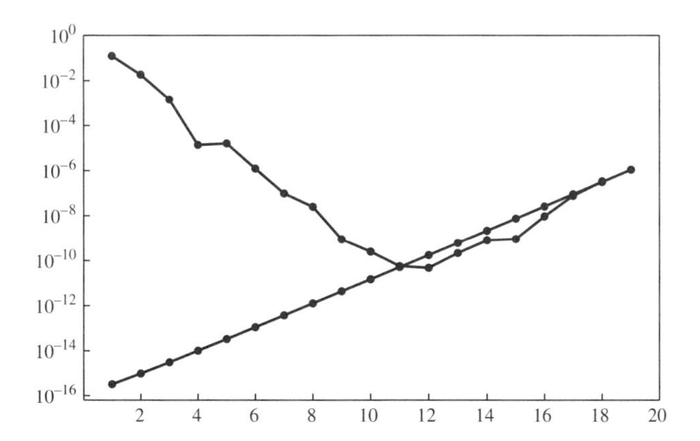

Figure A.1. *The V-shaped line shows $|s_{1,j+1} - s_{1,j}|$ for Salzer's extrapolation applied to the sequence $s_k = \sum_{n=1}^{k} n^{-2}$. The rising line shows a bound for the accumulated round-off error in $|s_{1,j}|$. This graph shows the typical behavior of a linear or semilinear method applied to a monotonic sequence. The two lines meet near the cusp of the V, allowing one to assess confidently the terminal accuracy of the extrapolation.*

value to floating point. The second arises from the further computation in finite-precision arithmetic. In a good implementation, convergence acceleration algorithms are backward stable, which means that the numerics do not introduce any error significantly larger than what would have been caused by perturbing the data at round-off error level. It is therefore sufficient in practice to consider only the propagation of initial round-off error.

All the acceleration algorithms we have considered can be thought of as having the form

$$s_{k,j} = f(s_k, s_{k+1}, \ldots, s_j),$$

although the function is never written out explicitly, but built up by recursion. We define the condition number of the formula by

$$\kappa[f] = \sum_{i=k}^{j} \left| \frac{\partial}{\partial s_i} f(s_k, s_{k+1}, \ldots, s_j) \right|.$$

Roughly speaking, if each s_i is perturbed by round-off error of size μ, then we can expect $s_{k,j}$ to be contaminated by round-off error of size $\mu\kappa[f]$.

In the case of linear transformations, it is easy to estimate the condition number. Let $\kappa_{k,j}$ be the value of $\kappa[f]$ for $s_{k,j}$. Clearly $\kappa_{k,0} = 1$ for all k. From (A.8), we find that

$$\kappa_{k,j} \leqslant \kappa_{k+1,j-1} + |f_{k,j}|(\kappa_{k+1,j-1} + \kappa_{k,j-1}).$$

This estimate is only an inequality, but it is a sharp inequality, and in practice the round-off errors do not tend to cancel out. The estimated round-off error in Figure A.1 was obtained as $2^{-53}\eta_{k,j}$, where

$$\eta_{k,0} = 1,$$

$$\eta_{k,j} = \eta_{k+1,j-1} + |f_{k,j}|(\eta_{k+1,j-1} + \eta_{k,j-1}).$$

We earlier made the remark that from the point of view of accuracy, there is not much to be gained by working with a series rather than a sequence. Even if we had the terms of the series available exactly but the partial sums were still in floating point arithmetic, we would get $\eta_{k,1} = 1$ instead of $\eta_{k,1} = 1 + |f_{k,1}|$. The recursion thereafter proceeds as usual. It is easy to show that the best that can happen is that $\eta_{k,j}$ is reduced by a factor of $(1 + c)$, where $c = \max\{|f_{k,1}|, |f_{k+1,1}|, \dots, |f_{j-1,1}|\}$.

In the case of the modified Salzer transformation, it is a nuisance to obtain the form (A.8), so we derive a similar procedure for the divided difference formulation. To go with formula (A.16), define

$$\alpha^0 f_k = f_k;$$

$$\alpha^n f_k = \frac{\alpha^{n-1} f_{k+1} + \alpha^{n-1} f_k}{t_{k+n} - t_k};$$

$$\eta_{k,n} = \frac{\alpha^n \left(\frac{1}{\psi(k)}\right)}{\delta^n \left(\frac{1}{\psi(k)}\right)}.$$

For semilinear transformations, one can to a first-order approximation ignore the dependence of the auxiliary sequence on the data. The same procedure as for the modified Salzer transformation remains valid.

The strongly nonlinear transformations are not so easy to analyze. We can no longer, as in the case of the semilinear methods, ignore the dependence of the multipliers $f_{k,j}$ on the data when the methods are written in the form (A.8). This is because the multipliers themselves depend on differences and even second differences of computed quantities, and it is not reasonable to assume that they are accurate enough to be dropped from the analysis.

But in the case of strongly nonlinear methods, this substantial analytical effort is useless. Unlike linear and semilinear methods, bounds on the round-off error tend to be highly pessimistic. The reason for this is that as round-off error starts playing a role, extrapolated values that should in exact arithmetic be very close to each other (since both are close to the limit) turn out to be not that close. The large round-off–contaminated multipliers that might arise because of division by the tiny difference between supposedly close quantities do not occur. The observed behavior for a strongly nonlinear method is that when round-off error and truncation error become comparable, the extrapolated values do not get steadily worse, but tend to vary in a random-looking way around the limit, with amplitude at the accumulated round-off level.

An example will make this clear. Table A.11 shows the rho algorithm in action on the $s_k = \|A_k\|$ values from Problem 3 (see §3.2). The differences of these numbers are graphed in Figure A.2. Note that in both cases at about $j = 10$ the algorithm stops delivering further correct digits, but does not immediately start deteriorating as do the linear and semilinear algorithms; it just mills around without getting much better or much worse. Be warned: the slight tendency for the differences to get smaller does not mean that another digit of accuracy is gained once the initial stage of fast convergence is past. It merely shows that a nonlinear method can produce antilimits even out of random sequences.

Finally, bear in mind that mathematically equivalent formulae are not numerically equivalent. A couple of examples should make this point clear. The formula (A.8) can also

Table A.11. *The rho algorithm applied to the sequence $\|A_{k_j}\|$ of Problem 3.*

j	$k_j = j$	$k_j = 5j$
1	**1**.00000000000000	**1.2**6121761618336
2	**1**.32196073923600	**1.27**547538244135
3	**1.27**123453514403	**1.274**14260105494
4	**1.27**393314066972	**1.27422**328312872
5	**1.27422**017029519	**1.27422410**947073
6	**1.2742**1718653107	**1.27422414**923215
7	**1.27422**358265440	**1.27422414**886405
8	**1.27422**398509675	**1.27422415**124899
9	**1.27422414**504466	**1.2742241528**1792
10	**1.27422416**342250	**1.2742241528**1284
11	**1.27422**409923149	**1.274224152**67127
12	**1.27422410**946623	**1.274224152**66073
13	**1.27422412**252998	**1.274224152**58201
14	**1.2742241**1949585	**1.274224152**65168
15	**1.27422415**828802	**1.274224152**67954
16	**1.27422413**800383	**1.274224152**66834
17	**1.27422414**441947	**1.274224152**67514

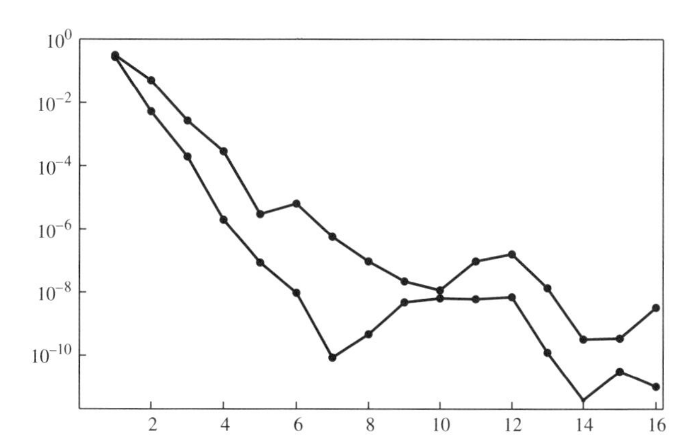

Figure A.2. *The upper line shows $|s_{1,2j+2} - s_{1,2j}|$ for the rho algorithm applied to the sequence $s_k = \|A_k\|$ arising from Problem 3. The lower line shows the same, but applied to s_{5k} instead of s_k. This graph exhibits the typical behavior of a strongly nonlinear method applied to a monotonic sequence. There is no V-shape as in the case of a linear or semilinear method, but at the stage where round-off error becomes significant, the magnitude of the differences settles down at the terminal accuracy level.*

be written as

$$s_{k,j} = (1 + f_{k,j})s_{k+1,j-1} - f_{k,j}s_{k,j-1}. \tag{A.31}$$

Suppose that $s_{k,j} \approx 1$, $s_{k+1,j-1} - s_{k,j-1} \approx 0.01$, $f_k \approx 0.1$. Then the correction term in (A.8) is approximately 0.001, so we can afford the loss of three digits in the calculation of f_k. But the correction term in (A.31) is approximately 0.1, so we can afford only the loss of one digit in the calculation of f_k. A more extreme case comes from Aitken's formula: instead of (A.23), (A.24), or (A.25), one might be tempted to use the "more elegant" formula

$$s_{k,2} = \frac{s_k s_{k+2} - s_{k+1}^2}{s_k - 2s_{k+1} + s_{k+2}}.$$

But this would produce a value of $s_{k,2}$ with the same number of correct digits as the second difference of s_k, which clearly is a quantity subject to a lot of cancellation.

A.5.4 Stopping Rules

Let me warn you, right at the outset, that any pure mathematician who feels squeamish about any of the preceding material will recoil in horror at what is about to follow. In fact, I myself feel uncomfortable with some of it.

Consider the following situation: the sequence s_k is so expensive to compute that, at each stage, before computing another term, a decision must be taken as to the necessity or indeed the utility of doing so. For example, one may wish to know s to 10 digits only, and one would like to exit (reporting success) if that accuracy has already been achieved, or to abort (reporting failure) if no further increase of accuracy can be expected from the current algorithm at the current precision.

A stopping rule is an algorithmic procedure by which that decision is automated. We give some examples of such rules, with a strong injunction that it is preferable by far to use the competent human eye.

When accelerating an alternating sequence by any of the methods discussed above, a plausible stopping rule is:

- Check whether $|s_{1,n-1} - s_{1,n-2}| \geqslant |s_{1,n-2} - s_{1,n-3}|$. If so, exit with best extrapolated value $s_{1,n-2}$ and error estimate $c|s_{1,n-1} - s_{1,n-2}|/(1-c)$, where $c = |s_{1,n-1} - s_{1,n-2}|/|s_{1,n-2} - s_{1,n-3}|$. (This is in fact often a fairly good error estimate even when the terminating condition is not yet satisfied.)

For the linear and semilinear methods, a plausible stopping rule for monotonic sequences is (having just calculated $s_{1,n-1}$ and $\eta_{1,n-1}$):

- Check whether $|\eta_{1,n-1}|\mu \geqslant |s_{1,n-1} - |s_{1,n-2}|$, where μ is the typical error in the s_k values that were used. If so, exit with best extrapolated value $s_{1,n-1}$ and error estimate $|\eta_{1,n-1}|\mu$. If not, apply the same test as for alternating sequences.

For the strongly nonlinear methods, there is no very satisfactory stopping rule for monotonic sequences. My suggestion is to use the same method as for alternating sequences and to multiply the error estimate by 10.

For the human eye, there is a very valuable visual aid. It is this: print a table of the number of digits to which $s_{k,j}$ agrees with $s_{k-1,j}$, that is, the rounded value of $-\log_{10}|1 - s_{k-1}/s_k|$. Such small integers fit nicely in one line and the behavior of the acceleration table is conspicuous.

For example, here is the digit agreement table for the modified Salzer extrapolation example that we have met before:

1	2	3	4	5	6	7	7	9	9	10	9	9	9	8	8	7
1	2	4	5	5	7	7	8	9	11	9	10	9	8	8	7	6
2	3	4	5	6	8	8	9	10	10	10	9	8	8	7	6	
2	3	5	5	7	8	8	10	10	10	9	8	8	7	6		
2	3	5	6	7	8	9	10	11	9	8	8	7	6			
2	4	5	6	7	8	9	10	9	9	9	7	7				
3	4	6	6	8	9	10	9	9	9	8	7					
3	4	6	7	8	9	10	9	10	8	7						
3	4	6	7	8	9	10	10	8	7							
3	4	6	7	9	10	10	9	8								
3	4	7	7	9	10	9	8									
3	4	7	7	9	9	9										
3	5	7	7	9	9											
3	5	7	8	9												
3	5	7	8													
3	5	7														
3	5															
3																

It is obvious that 10 digits, maybe 11, is terminal accuracy here. The ridge of terminal accuracy, flanked on both sides by less accurate values, is typical of linear and semilinear methods. For the modified rho algorithm with $\theta = 2$, on the same s_k values, we get:

3	5	7	9	11	11	11	11	12
3	5	8	10	10	10	11	12	
4	6	8	10	10	11	10	10	
4	6	9	10	11	11	11		
4	7	9	10	10	10	11		
4	7	9	10	10	10			
4	7	10	10	10	11			
5	7	10	10	10				
5	8	10	10	10				
5	8	10	10					
5	8	10	11					
5	8	10						
5	8	10						
5	8							
5	8							
5								
6								

As is typical of nonlinear methods, we observe a plateau at terminal accuracy. Terminal accuracy is also 10, maybe 11 digits; it would be unwise to trust the values that agree to 12 digits.

A.5.5 Conclusion

We have presented many convergence acceleration algorithms, so here is a short guide to which one to use.

1. If you know very precisely what the behavior of the sequence is, use an appropriate linear transformation; for example, if the sequence arises from a finite difference method for differential equations, use Richardson extrapolation or, when that is not applicable, the E-algorithm. One of the ways of solving Problem 10 does just this (see §10.3.2). Problem 8 can also be solved with the aid of a finite difference method, accelerated by Richardson extrapolation (see §8.2).

2. If the sequence is alternating, try the Levin T-transformation; if it is monotonic, try the Levin U-transformation. The T-transformation works very well for the alternating series arising from Problems 1 and 9.

3. If the sequence is alternating, confirm the Levin T result by also trying the epsilon algorithm; if it is monotonic and you know a little about its behavior, confirm the Levin U result by also trying the modified rho algorithm with a suitable value of θ.

4. If all else fails, try the Levin W-transformation and the theta algorithm.

5. If none of the previous is particularly satisfactory, try the effect of applying the transformations to a suitably chosen subsequence, with $k_j = 2^{j-1}$ being the obvious first choice.

6. Irregular sequences and series are not suitable to convergence acceleration. For example, the dependence on the sequence of primes makes the approach of extrapolation from small matrices that works so well in Problem 3 totally unsuitable for Problem 7 (see §7.2.2). Another example with even more erratic behavior is the series

$$\frac{1}{\zeta(s)} = 1 - 2^{-s} - 3^{-s} - 5^{-s} + 6^{-s} - 7^{-s} + \cdots$$

   (the general factor of n^{-s} is $\mu(n)$, the *Möbius function*. This is 0 if n has a multiple prime factor; otherwise it is $(-1)^k$, where k is the number of prime factors of n.)

7. Some sequences and series can be accelerated only with the aid of detailed information which would not be available for a series obtained numerically. For example, series in which a_k can be represented as a product of two slowly convergent null sequences, each by itself well modelled by a simple function of n, but with different asymptotic behavior, can be surprisingly intractable. One such series is $g(x, k_0) = \sum_{k=0}^{\infty} x^k (k + k_0)^{-1}$, which for $k_0 = 0.125$ and $x \approx 0.94$ behaves similarly to a series that arises in the solution to Problem 6 (see §6.3). None of the general methods we have discussed do better than six digits. It is fortunate that this series does not really require convergence acceleration, since for $(k + k_0)(1 - x) > 1$, we have the bound $|s - s_k| < x^k$; thus 600 unaccelerated terms suffice for 16 digits. In this context, $0.94 \ll 1$.

A.6 Notes and References

Although convergence acceleration is mainly an experimental science, there are some theorems (for a selection of those, see [BZ91]) saying that a given class of sequences will be accelerated by a given algorithm (for example, linear sequences are accelerated by the Aitken process).

The use of extrapolation to accelerate the Archimedes sequence is mentioned by Bauer, Rutishauser, and Stiefel [BRS63], who point out that Huygens found a_3' in 1654 and Kommerell found a_2'' in 1936.

The term "quasi-linear" is due to Brezinski [Bre88].

The E-algorithm was independently discovered by no fewer than five authors in the 1970s and has also been called by grander names, such as generalized Mühlbach–Neville–Aitken extrapolation, or the Brezinski–Håvie protocol. For its history, see [Bre00].

The literature on convergence acceleration abounds with names like E-algorithm, g-algorithm, ϵ-algorithm, etc. They usually reflect notation used in an early paper describing the algorithm.

Equation (A.12) can be generalized: if any factor r_j is repeated m times, terms of the form $s_k = r_j^k p(k)$, where p is a polynomial of degree not more than $m-1$, are annihilated.

Although the retrieval of the exact limit when the model (A.14) is satisfied is in general an $O(n^3)$ process, there exist $O(n^2)$ versions of the Richardson process that will transform the sequence to one that, though not constant, does converge faster than any of the modelling functions ϕ_j. Over half of a recent book [Sid03] is devoted to the detailed analysis of many variations of the Richardson process.

In books on difference calculus, the notation δ is usually reserved for centered differences, not for divided differences.

Our description of operator polynomial methods is based on the article by Cohen, Rodriguez, Villegas, and Zagier [CRZ00]. However, the suggestion to use Legendre and Jacobi polynomials is not mentioned there.

The epsilon algorithm has some other marvellous properties: for example, it can accelerate all totally monotonic and totally accelerating sequences, and when applied to a power series, it produces a certain subset of the Padé approximants of the function defined by that power series.

The term "stride" was coined by Carl DeVore in his write-up of Problem 3 [DeV02]. The method he uses for convergence acceleration is not covered here: it is a special-purpose nonlinear extrapolation based on the model

$$s_{k+1} - s_k \approx c_k(k + b_k)^{-d_k}.$$

The parameters c_k, b_k, and d_k are defined by assuming exactness at four consecutive k values. It is easy to eliminate c_k, leading to two nonlinear equations for b_k and d_k. Having obtained the parameters, one exploits the fact that

$$\zeta(d, b) = \sum_{n=1}^{\infty} (n + b)^{-d}$$

is a fairly well-known function (in the Maple implementation called the Hurwitz zeta function). This gives the acceleration formula

$$s_{k,3} = s_k + c_k\left(\zeta(d_k, k + b_k) - \zeta(d_k, b_k)\right).$$

In cases where the two nonlinear equations prove troublesome to solve, DeVore suggested taking $b_k = 0$, in which case three terms suffice to determine c_k and d_k. The acceleration formula becomes

$$s_{k,2} = s_k + c_k \left(\zeta(d_k, k) - \zeta(d_k, 0) \right).$$

DeVore obtained 12 digits of $\|A\|$ in 20-digit arithmetic, using the latter formula on $t_k = s_{3k}$, $k = 1, 2, \ldots, 14$.

For the numerical sequence in §A.5.2, s_k is given by the value of the $(k+1)$-st Laguerre polynomial at its first extremum in $[0, \infty)$. It was shown to me in 1973 by Syvert Nørsett. We both worked on it over a weekend, in my case to find a numerical value for the limit on a mechanical desk calculator, in his to show that the analytical limit of $|s_k|$ equals $|J_0(x_0)|$, where x_0 is the first extremum in $[0, \infty)$ of the Bessel function J_0. Thus the correct value in this case is $s \doteq 0.402759395702553$.

Appendix B

Extreme Digit-Hunting

If you add together our heroic numbers the result is $\tau =$ 1.497258836 . . . . *I wonder if anyone will ever compute the ten thousandth digit of this fundamental constant?*
—Lloyd N. Trefethen [Tre02, p. 3]

I am ashamed to tell you to how many figures I carried these computations, having no other business at the time.
—Isaac Newton, after computing 15 digits of π in 1666 [BB87, p. 339]

Since we understood all the relevant algorithms, we tried our best to find 10,000 digits for each of the 10 Challenge problems. We succeeded with 9 of them and summarize our efforts here. In all the even-numbered cases, the task was quite simple. But each odd-numbered problem provided new challenges. As one might expect, a technique that works beautifully to get 10 or 20 digits might well fail badly when trying for hundreds or thousands of digits.

All timings are on a Macintosh G4 laptop with 1 GHz CPU speed, and were done in *Mathematica* unless specified otherwise. The high-precision arithmetic in version 5 of *Mathematica* is especially fast, as it is in PARI/GP, which was used in some cases. Code for each of these is on the web page for this book. The problems divide naturally into two groups, according to whether their solution requires a lot of arithmetic at 10,000-digit precision, or not. Problems 1, 3, and 7 do require many high-precision operations. The others do not. Thus, even though Problem 5 requires much more time than Problem 9, most of the time is spent on initializing the gamma-function computations; once that is done, the rest can be done quickly.

While such an exercise might seem frivolous, the fact is that we learned a lot from the continual refinement of our algorithms to work efficiently at ultrahigh precision. The reward is a deeper understanding of the theory, and often a better algorithm for low-precision cases. A noteworthy example is Problem 9, which first required over 20 hours using a complicated series for Meijer's G function that involved the gamma function. But we were able to eliminate the very expensive gamma function from the series, and then differentiate the result symbolically, with the result that our current best time is now 34 minutes.

As always, the question of correctness arises. We have interval validations of correctness of 10,000 digits for Problems 2, 4, 6, 7, 8, and 10. For the others, we rely on the traditional techniques: comparing different algorithms, machines, software, programmers, and programming languages. And we work up, always comparing digits with past computations. Thus we have ample evidence to believe in the correctness of the 10,002 digits given here for each of nine of the challenge problems.

Problem 1. Using the method discussed in §1.4, this can be done using the trapezoidal rule with 91,356 terms to get the real part of $\int_C z^{i/z-1}\,dz$, where C is the contour given by

$$z = \frac{\pi e^t}{\pi e^t + 2 - 2t} + \frac{2i}{\cosh t},$$

and with t running from -8.031 to 10.22. The reason this simple method and small interval of integration work to such high precision is the double exponential decay of the integrand in combination with an exponential convergence rate of the trapezoidal rule. Still, it takes 22.6 hours and involves over 2 million operations at the full 10,000-digit precision. This computation was done independently by two of us, using different languages (*Mathematica*, Maple, and C++), different operating systems and computers, and different algorithms (the Ooura–Mori method).

0.32336 74316 77778 76139 93700 <<9950 digits>> 42382 81998 70848 26513 96587 27

Problem 2. Simply using high-enough precision on the initial conditions is adequate. In a fixed-precision environment one would do various experiments to estimate k so that $d + k$ digits of precision in the starting values will give d digits of the answer. But it is elegant, fast, and more rigorous to use the interval algorithm of §2.3, which will provide validated digits. It takes only two seconds to get the answer, starting with an interval of diameter 10^{-10038} around the initial conditions. This time excludes the experiments necessary to learn that 38 extra digits will be adequate.

0.99526 29194 43354 16089 03118 <<9950 digits>> 39629 06470 50526 05910 39115 30

Problem 3. For this problem we have only 273 digits, obtained in a month-long run by the method of §3.6.2. We chose the contour (3.29) with parameter $\sigma = 3/4$, the parametrization

$$t = \Phi(\tau) = \sinh\left(2\sinh\left(\frac{1}{2}\operatorname{arcsinh}(\tau)\right)\right) = \sinh\left(\frac{\tau}{\sqrt{(1+\sqrt{1+\tau^2})/2}}\right),$$

the truncation point

$$T = \frac{1}{2}\left(\log\log\left(\frac{8}{\epsilon}\right)\right)^2, \quad \epsilon = 10^{-275},$$

and the reciprocal step size $1/h = 516.2747$. In analogy to (3.32) an empirical model of the accuracy was fitted to runs in lower precisions. Here, with d denoting the number of correct digits, such a model takes the form

$$h^{-1} \approx 0.2365 \left(\frac{d}{\log d} - 2.014 \right)^2 \quad \text{for } d > 120$$

and predicts the correctness of 273 digits.

```
1.2742 24152 82122 81882 12340 <<220 digits>> 75880 55894 38735 33138 75269 029
```

Problem 4. Once the location of the minimum is obtained to a few digits, one can use Newton's method on the gradient to zero in on the corresponding critical point of f. It is only necessary to get the critical point to 5000 digits. Then use 10,000 digits of precision to evaluate the function at the critical point. This takes 8 seconds, where Newton's method is used in a way that increases the precision at each step: start with machine precision and then repeatedly double the working precision, using each result as the seed for the next. One can validate the answer in 44 seconds using the ϵ-inflation method of §4.6. Just inflate the high-precision critical point to a square of side 10^{-3} and verify Krawczyk's condition.

```
-3.3068 68647 47523 72800 76113 <<9950 digits>> 46888 47570 73423 31049 31628 61
```

Problem 5. Using the six nonlinear equations (5.2) to get the values of θ_1 and θ_2 to 5000 digits, and then maximizing $\epsilon(\theta_1, \theta_2)$, as discussed in §5.6, to 10,000 digits works nicely. It takes six hours for an initial gamma function computation to 10,000 digits. Once that is done, subsequent gamma values are very fast. Then it takes just an hour to solve the six-equation system, and only four minutes for the ϵ evaluation. Using only the six-dimensional system takes much longer, but yields the same result. And a completely independent computation using only the approach of finding a maximum to ϵ also yielded the same 10,000 digits.

```
0.21433 52345 90459 63946 15264 <<9950 digits>> 68023 90106 82332 94081 32745 91
```

Problem 6. Because of the simple form (6.1) of the answer, namely,

$$\epsilon = \frac{1}{4}\sqrt{1-\eta^2}, \quad \text{where } M\left(\sqrt{4-(\eta-1)^2}, \sqrt{4-(\eta+1)^2}\right) = 1,$$

with M being the arithmetic-geometric mean, 10,000 digits can be obtained very quickly, using fast algorithms for the arithmetic-geometric mean and the secant method to find η. The total time needed is about half a second. The result can be quickly validated using interval methods as explained in §6.5.1.

```
0.061913 95447 39909 42848 17521 <<9950 digits>> 92584 84417 28628 87590 08473 83
```

Problem 7. Using the rational form of the answer obtained by Dumas, Turner, and Wan yields 10,000 digits in no time, but sidesteps the many days needed to obtain that exact rational (see §7.5). Thus we carried out a purely numerical approach, using the conjugate gradient method with preconditioning to compute the entire first column of the inverse. This required maintaining a list of 20,000 10,010-digit reals, and took 129 hours and about $4 \cdot 10^9$

arithmetic operations at 10,010-digit precision. The result can be validated by combining knowledge of the relationship of the error to the size of the residual, which can be computed by interval methods as discussed in §7.4.2.

```
0.72507 83462 68401 16746 86877 <<9950 digits>> 52341 88088 44659 32425 66583 88
```

Problem 8. The method of equation (8.8) yields the answer by simple root-finding. The answer, t, satisfies $\theta(e^{-\pi^2 t}) = \pi/2\sqrt{5}$. The series defining θ can safely be truncated after 73 terms. Hence, the derivative of this is easy and Newton's method can be used yielding 10,000 digits in eight seconds. *Mathematica*'s `FindRoot` is efficient in the sense that, when started with a machine precision seed, it will start its work at machine precision and work up, doubling the precision at each iteration. The result can be quickly validated by the ϵ-inflation method of interval arithmetic as explained in §8.3.2.

```
0.42401 13870 33688 36379 74336 <<9950 digits>> 34539 79377 25453 79848 39522 53
```

Problem 9. The integral $I(\alpha)$ can be represented in terms of the gamma function and Meijer's G function (see §9.6), and this leads to a representation using an infinite series. Moreover, the series can be differentiated term by term, allowing us to approximate $I'(\alpha)$ as a partial sum of an infinite series. Thus we need only find the correct solution to $I'(\alpha) = 0$, which can be done by the secant method. Formula (9.13) shows how the gamma function can be eliminated, which saves a lot of time (see §5.7). That formula also yields to term-by-term differentiation, with the result that 10,000 digits can be obtained in 34 minutes. The differentiation can also be done numerically, but then one must work to 12,500-digit precision, which slows things down. The two methods, run on different machines and with different software (*Mathematica* and PARI/GP), lead to the same result. Using interval methods (bisection on an interval version of $I'(\alpha)$) validated the first 1000 digits.

```
0.78593 36743 50371 45456 52439 <<9950 digits>> 63138 27146 32604 77167 80805 93
```

Problem 10. Given the symbolic work in §10.7 that presents the answer as

$$\frac{2}{\pi} \arcsin\left((3 - 2\sqrt{2})^2 (2 + \sqrt{5})^2 (\sqrt{10} - 3)^2 (5^{1/4} - \sqrt{2}\,)^4\right),$$

this is the easiest of all: we simply compute π, three square roots, and an arcsine to 10,010 digits. This takes 0.4 seconds. An interval computation then validates correctness.

```
0.00000038375 87979 25122 61034 07133 <<9950 digits>> 65284 03815 91694 68620 19870 94
```

Appendix C
Code

We may say most aptly, that the Analytical Engine weaves algebraical patterns just as the Jacquard-loom weaves flowers and leaves.
— Augusta Ada Byron King, Countess of Lovelace
[Men43, Note A, p. 696]

Here we collect, for the convenience of the reader, all the small code that was used in the chapters but not displayed there in order not to distract from the flow of the arguments. This code, and more elaborate versions with all the bells and whistles, can be downloaded from the web page for this book:

`www.siam.org/books/100digitchallenge`

C.1 C

Monte Carlo Simulation of Brownian Motion

This small C program was used to obtain Table 10.1 in §10.1.

```
#include <math.h>
#include <stdio.h>
#include <stdlib.h>

void main()                              /* walk on spheres */
{
    const float pi2 = 8.*atan(1.)/RAND_MAX;
    const float a = 10., b = 1., h = 1.e-2; /* geometry */
    const int n = 1e8;                 /* number of samples */
    float x,y,r,phi,p;
    int k,count,hit;

    count = 0;
    for (k=0; k<n; k++) {                /* statistics loop */
        x = 0.; y = 0.;
```

```
        do {                              /* a single run */
            r = min(min(a-x,a+x),min(b-y,b+y));
            phi = pi2*rand();
            x += r*cos(phi); y += r*sin(phi);
        } while ((fabs(x)<a-h) & (hit=(fabs(y)<b-h)));
        count += hit;
    }
    p = ((float)count)/n;
    printf("%e \n",p);
}
```

C.2 PARI/GP

PARI/GP is an interactive programming environment that was originally developed by Karim Belabas and Henri Cohen for algorithmic number theory. It has many powerful features of symbolic algebra and of high-precision arithmetic, including basic linear algebra. PARI/GP (we used version 2.1.3) is free for private or academic purposes and can be downloaded from

http://pari.math.u-bordeaux.fr/download.html

C.2.1 Problem-Dependent Functions and Routines

Operator Norm with Power Method (§3.3)

Used in session on p. 55.

```
{OperatorNorm(x,tol) =
    n=length(x); b=vector(2*n,l, 1+(l^2-l)/2);
    lambda=0; lambda0=1;
    while(abs(lambda-lambda0)>tol,
        xhat=x/sqrt(sum(k=-n,-1, x[-k]^2));
        y=vector(n,k, sum(l=-n,-1, xhat[-l]/(b[k-l]+l)));
        x=vector(n,j, sum(k=-n,-1,  y[-k]/(b[j-k]-j)));
        lambda0=lambda; lambda=sum(k=-n,-1,xhat[-k]*x[-k]);
    );
    [sqrt(lambda),x]
}
```

C.3 MATLAB

All the MATLAB code was tested to run under version 6.5 (R13).

C.3.1 Problem-Dependent Functions and Routines

Operator Norm with Power Method (§3.3)

MATLAB version of the PARI/GP procedure used in session on p. 55.

```
function [s,v,u] = OperatorNorm(x,tol)
```

```
% Input:    x       initial right singular vector
%           tol     desired absolute tolerance
% Output:   s       maximal singular value (operator norm)
%           v       right singular vector
%           u       left  singular vector

n = length(x); y = zeros(n,1);
k = 1:n;
lambda = 0; lambda0 = 1;
while abs(lambda-lambda0) > tol
    xhat = x/norm(x);
    for j=1:n, y(j)=(1./((j+k-1).*(j+k)/2-(k-1)))*xhat; end
    for j=1:n, x(j)=(1./((k+j-1).*(k+j)/2-(j-1)))*y;    end
    lambda0 = lambda; lambda   = x'*xhat;
end
s = sqrt(lambda);
v = x/norm(x);
u = y/norm(y);

return
```

Return Probability (§6.2)

Used in session on p. 124.

```
function p = ReturnProbability(epsilon,n)

pE = 1/4 + epsilon; pW = 1/4 - epsilon; pN = 1/4; pS = 1/4;

m = 2*n+1; ctr = sub2ind([m,m],n+1,n+1);
A_EW = spdiags(ones(m,1)*[pW pE],[-1 1],m,m);
A_NS = spdiags(ones(m,1)*[pS pN],[-1 1],m,m);
A = kron(A_EW,speye(m)) + kron(speye(m),A_NS);
r = A(:,ctr); A(:,ctr) = 0; q = (speye(m^2)-A)\r;
p = q(ctr);

return
```

Occupation Probabilities (§6.3.1)

Used in session on p. 128 and the function `ExpectedVisitsExtrapolated`.

```
function p = OccupationProbability(epsilon,K)

global pE pW pN pS;
pE = 1/4 + epsilon; pW = 1/4 - epsilon; pN = 1/4; pS = 1/4;

p = zeros(K,1); p(1) = 1; Pi = 1;
for k=1:K-1
    Pi = step(step(Pi)); p(k+1) = Pi(k+1,k+1);
end

return
```

```
function PiNew = step(Pi)

global pE pW pN pS;
[k,k] = size(Pi); PiNew = zeros(k+1,k+1); i=1:k;
PiNew(i+1,i+1) =                     pE*Pi;
PiNew(i  ,i  ) = PiNew(i  ,i  ) + pW*Pi;
PiNew(i  ,i+1) = PiNew(i  ,i+1) + pN*Pi;
PiNew(i+1,i  ) = PiNew(i+1,i  ) + pS*Pi;

return
```

Extrapolation of the Expected Number of Visits (§6.3.2)

Used in session on p. 129.

```
function val = ExpectedVisistsExtrapolated(epsilon,K,extraTerms)

p = OccupationProbability(epsilon,K+extraTerms);
steps = (extraTerms-1)/2;
val = sum(p(1:K))+WynnEpsilon(p(end-2*steps:end),steps,'series','off');

return
```

C.3.2 General Functions and Routines

Strebel's Summation Formula

Used in session on p. 64.

```
function [w,c] = SummationFormula(n,alpha)

% Calculates weights w and nodes c for a summation formula that
% approximates sum(f(k),k=1..infinity) by sum(w(k)*f(c(k)),k=1..n).
% Works well if f(k) is asymptotic to k^(-alpha) for large k.
% Put alpha = 'exp' to use an exponential formula.

switch alpha
    case 'exp'
        k=n:-1:2;
        u=(k-1)/n;
        c1 = exp(2./(1-u).^2-1./u.^2/2); c = k + c1;
        w = 1 + c1.*(4./(1-u).^3+1./u.^3)/n;
        kinf = find(c==Inf);
        c(kinf)=[]; w(kinf)=[];
        c = [c 1]; w = [w 1];
    otherwise
        n = ceil(n/2); a1 = (alpha-1)/6; a6 = 1+1/a1;
        k2 = n-1:-1:0;
        w2 = n^a6./(n-k2).^a6-a6*k2/n;
        c2 = n+a1*(-n+n^a6./(n-k2).^(1/a1))-a6*k2.^2/2/n;
        k1 = n-1:-1:1;
        w = [w2 ones(size(k1))];
        c = [c2            k1  ];
end

return
```

Trapezoidal Sums with sinh-Transformations

Used in session on p. 71.

```
function [s,steps] = TrapezoidalSum(f,h,tol,level,even,varargin)

% [s,steps] = TrapezoidalSum(f,h,tol,level,even,varargin)
%
% applies the truncated trapezoidal rule with sinh-transformation
%
%  f          integrand, evaluates as f(t,varargin)
%  h          step size
%  tol        truncation tolerance
%  level      number of recursive applications of sinh-transformation
%  even       put 'yes' if integrand is even
%  varargin   additional arguments for f
%
%  s          value of the integral
%  steps      number of terms in the truncation trapezoidal rule

[sr,kr] = TrapezoidalSum_(f,h,tol,level,varargin{:});
if isequal(even,'yes')
    sl = sr; kl = kr;
else
    [sl, kl] = TrapezoidalSum_(f,-h,tol,level,varargin{:});
end
s = sl+sr; steps = kl+kr+1;

return

function [s,k] = TrapezoidalSum_(f,h,tol,level,varargin)

t = 0; F0 = TransformedF(f,t,level,varargin{:})/2;
val = [F0]; F = 1; k = 0;
while abs(F) >= tol
    t = t+h; F = TransformedF(f,t,level,varargin{:});
    k = k+1; val = [F val];
end
s = abs(h)*sum(val);

return

function val = TransformedF(f,t,level,varargin)

dt = 1;
for j=1:level
    dt = cosh(t)*dt; t = sinh(t);
end
val = feval(f,t,varargin{:})*dt;

return
```

Wynn's Epsilon Algorithm

Used in function `ExpectedVisistsExtrapolated`.

```
function val = WynnEpsilon(a,steps,object,display)

% function S = WynnEpsilon(a,steps,object,display)
%
% extrapolates a sequence or series by using Wynn's epsilon algorithm
%
%   a           vector of terms
%   steps       number of extrapolation steps
%   object      'sequence': extrapolation to the limit of a
%               'series'  : extrapolation to the limit of cumsum(a)
%   display     'on':   plot the absolute value of differences in the
%               first row of the extrapolation table for diagnosis
%
%   val         extrapolated value

S = zeros(length(a),2*steps+1);
switch object
    case 'sequence'
        S(:,1) = a;
    case 'series'
        S(:,1) = cumsum(a);
    otherwise
        error('MATLAB:badopt','%s: no such object known',object);
end
for j=2:2*steps+1
    if j==2
        switch object
            case 'sequence'
                S(1:end-j+1,j) = 1./diff(a);
            case 'series'
                S(1:end-j+1,j) = 1./a(2:end);
        end
    else
        S(1:end-j+1,j) = S(2:end-j+2,j-2)+1./diff(S(1:end-j+2,j-1));
    end
end
S = S(:,1:2:end);
if isequal(display,'on')
    h=semilogy(0:length(S(1,:))-2,abs(diff(S(1,:))));
    set(h,'LineWidth',3);
end
val = S(1,end);

return
```

Richardson Extrapolation with Error Control

Used in sessions on pp. 173, 174, and 208.

```
function [val,err,ampl] = richardson(tol,p,nmin,f,varargin)
```

```
% [val,err,ampl] = richardson(tol,p,nmin,f,varargin)
%
% richardson extrapolation with error control
%
%   tol     requested relative error
%   p       f(...,h) has an asymptotic expansion in h^p
%   nmin    start double harmonic sequence at h=1/2/nmin
%   f       function, evaluates as f(varargin,h)
%
%   val         extrapolated value (h -> 0)
%   err         estimated relative error
%   ampl        amplification of relative error in the
%                   evaluation of f

% initialize tableaux
j_max = 9; T = zeros(j_max,j_max);
n = 2*(nmin:(j_max+nmin));     % double harmonic sequence
j=1; h=1/n(1);
T(1,1) = feval(f,varargin{:},h);
val=T(1,1); err=abs(val);

% do Richardson until convergence
while err(end)/abs(val) > tol & j<j_max
    j=j+1; h=1/n(j);
    T(j,1) = feval(f,varargin{:},h);
    A(j,1) = abs(T(j,1));
    for k=2:j
       T(j,k)=T(j,k-1)+(T(j,k-1)-T(j-1,k-1))/((n(j)/n(j-k+1))^p-1);
       A(j,k)=A(j,k-1)+(A(j,k-1)+A(j-1,k-1))/((n(j)/n(j-k+1))^p-1);
    end
    val = T(j,j);
    err = abs(diff(T(j,1:j)));   % subdiagonal error estimates
    if j > 2              % extrapolate error estimate once more
        err(end+1) = err(end)^2/err(end-1);
    end
end
ampl = A(end,end)/abs(val);
err = max(err(end)/abs(val),ampl*eps);

return
```

Approximation of the Heat Equation on Rectangles

Used in session on p. 173.

```
function u_val = heat(pos,rectangle,T,u0,f,bdry,h);

% u_val = heat(pos,rectangle,T,u0,f,bdry,h)
%
% solves the heat equation u_t - \Delta u(x) = f with dirichlet
% boundary conditions on rectangle [-a,a] x [-b,b] using a five-point
% stencil and explicit Euler time-stepping
%
```

```
%   pos          [x,y] a point of the rectangle
%   rectangle    [a,b]
%                2a   length in x-direction
%                2b   length in y-direction
%   T            final time
%   u0           intial value                        (constant)
%   f            right-hand side                     (constant)
%   bdry         Dirichlet data [xl,xr,yl,yu]
%                xl  boundary value on {-a} x [-b,b]    (constant)
%                xr  boundary value on { a} x [-b,b]    (constant)
%                yl  boundary value on [-a,a] x {-b}    (constant)
%                yu  boundary value on [-a,a] x { b}    (constant)
%   h   h = 1/n  suggested approximate grid-size
%   u_val        solution at pos = [x,y]

% the grid
a = rectangle(1); b = rectangle(2);
n = 2*ceil(1/2/h); % make n even
nx = ceil(2*a)*n+1; ny = ceil(2*b)*n+1;
dx = 2*a/(nx-1); dy = 2*b/(ny-1);
nt = ceil(4*T)*ceil(1/h^2); dt = T/nt;
x = 1:nx; y=1:ny;

% initial and boundary values
u = u0*ones(nx,ny);
u(x(1),y)  = bdry(1); u(x(end),y)  = bdry(2);
u(x,y(1))  = bdry(3); u(x,y(end))  = bdry(4);

% the five-point stencil
stencil = [-1/dy^2 -1/dx^2 2*(1/dx^2+1/dy^2) -1/dx^2 -1/dy^2];

% the time-stepping
x = x(2:end-1); y = y(2:end-1);
for k=1:nt
    u(x,y) = u(x,y) + dt*(f - stencil(1)*u(x,y-1)...
                            - stencil(2)*u(x-1,y)...
                            - stencil(3)*u(x,y)...
                            - stencil(4)*u(x+1,y)...
                            - stencil(5)*u(x,y+1));
end

% the solution
u_val = u(1+round((pos(1)+a)/dx),1+round((pos(2)+b)/dy));

return
```

Poisson Solver on Rectangle with Constant Data

Used in sessions on pp. 205 and 206.

```
function u_val = poisson(pos,rectangle,f,bdry,solver,h);

% u_val = poisson(pos,rectangle,f,bdry,solver,h)
%
```

```
% solves poisson equation with dirichlet boundary conditions
% on rectangle [-a,a] x [-b,b] using a five-point stencil
%
%   pos         [x,y] a point of the rectangle
%   rectangle   [a,b]
%               2a   length in x-direction
%               2b   length in y-direction
%   f           right-hand side of -\Delta u(x) = f   (constant)
%   bdry        Dirichlet data [xl,xr,yl,yu]
%               xl  boundary value on {-a} x [-b,b]   (constant)
%               xr  boundary value on { a} x [-b,b]   (constant)
%               yl  boundary value on [-a,a] x {-b}   (constant)
%               yu  boundary value on [-a,a] x { b}   (constant)
%   solver      'Cholesky'  sparse Cholesky solver
%               'FFT'       FFT based fast solver
%   h           h discretization parameter
%
%   u_val       solution at pos = [x,y]

% the grid
n = ceil(1/h);
a = rectangle(1); b = rectangle(2);
nx = 2*ceil(a)*n-1; ny = 2*ceil(b)*n-1; x=1:nx; y=1:ny;
dx = (2*a)/(nx+1); dy = (2*b)/(ny+1);

% the right hand side
r = f*ones(nx,ny);
r(1,y)  = r(1,y)+bdry(1)/dx^2; r(nx,y)  = r(nx,y)+bdry(2)/dx^2;
r(x,1)  = r(x,1)+bdry(3)/dy^2; r(x,ny)  = r(x,ny)+bdry(4)/dy^2;

% solve it
switch solver
    case 'Cholesky' % [Dem87,Sect. 6.3.3]
        Ax = spdiags(ones(nx,1)*[-1 2 -1]/dx^2,-1:1,nx,nx);
        Ay = spdiags(ones(ny,1)*[-1 2 -1]/dy^2,-1:1,ny,ny);
        A  = kron(Ay,speye(nx)) + kron(speye(ny),Ax);
        u = A\r(:); u = reshape(u,nx,ny);
    case 'FFT'      % [Dem87,Sect. 6.7]
        u = dst2(r);
        d = 4*(sin(x'/2*pi/(nx+1)).^2*ones(1,ny)/dx^2+ ...
                        ones(nx,1)*sin(y/2*pi/(ny+1)).^2/dy^2);
        u = u./d;
        u = dst2(u);
    otherwise
        error('MATLAB:badopt','%s: no such solver known',solver);
end

% the solution
u_val = u(round((pos(1)+a)/dx),round((pos(2)+b)/dy));

return

% subroutines for 1D and 2D fast sine transform [Dem97,p.324]

function y = dst(x)
```

```
n = size(x,1); m = size(x,2);
y = [zeros(1,m);x]; y = imag(fft(y,2*n+2));
y = sqrt(2/(n+1))*y(2:n+1,:);
return

function y = dst2(x)
y = dst(dst(x)')';
return
```

C.4 INTLAB

INTLAB is a MATLAB toolbox for self-validating algorithms written by Siegfried Rump [Rum99a, Rum99b]. For portability the toolbox is written entirely in MATLAB, making heavy use of BLAS routines. INTLAB (we used version 4.1.2) is free for private or academic purposes and can be downloaded from

`www.ti3.tu-harburg.de/~rump/intlab/`.

A tutorial can be found in the master's thesis of Hargreaves [Har02].

C.4.1 Utilities

Gradient of `hull` Command

Needed for applying `IntervalNewton` *to the function* `theta`.

Automatic differentiation of the command `'hull'` needs some short code to be put in a file named `'hull.m'` to a subdirectory `'@gradient'` of the working directory:

```
function a = hull(a,b)

a.x  = hull(a.x,b.x);
a.dx = hull(a.dx,b.dx);

return
```

Subdivision of Intervals and Rectangles

Used in the commands `IntervalBisection` *and* `IntervalNewton`.

```
function x = subdivide1D(x)

% subdivides the intervals of a list (row vector) by bisection

x1 = infsup(inf(x),mid(x));
x2 = infsup(mid(x),sup(x));
x = [x1 x2];

return

function x = subdivide2D(x)
```

```
% subdivides the rectangles of a list of
% rectangles (2 x k matrix of intervals)

x1_ = x(1,:); x2_ = x(2,:);
l1 = infsup(inf(x1_),mid(x1_));
r1 = infsup(mid(x1_),sup(x1_));
l2 = infsup(inf(x2_),mid(x2_));
r2 = infsup(mid(x2_),sup(x2_));
x = [l1 l1 r1 r1; l2 r2 l2 r2];

return
```

C.4.2 Problem-Dependent Functions and Routines

Objective Function (Chapter 4)

```
function f = fun(x)

f.x      = exp(sin(50*x(1,:)))+sin(60*exp(x(2,:)))+ ...
           sin(70*sin(x(1,:)))+sin(sin(80*x(2,:)))- ...
               sin(10*(x(1,:)+x(2,:)))+(x(1,:).^2+x(2,:).^2)/4;

f.dx(1,:) = 50*cos(50*x(1,:)).*exp(sin(50*x(1,:)))+ ...
               70*cos(70*sin(x(1,:))).*cos(x(1,:))- ...
               10*cos(10*x(1,:)+10*x(2,:))+1/2*x(1,:);

f.dx(2,:) = 60*cos(60*exp(x(2,:))).*exp(x(2,:))+ ...
               80*cos(sin(80*x(2,:))).*cos(80*x(2,:))- ...
               10*cos(10*x(1,:)+10*x(2,:))+1/2*x(2,:);

return
```

Proof of Rigorous Bound for $\lambda_{\min}(A_{1142})$ (§7.4.2)

The results of the following session are used in the proof of Lemma 7.1.

```
>> n = 1142; p_n = 9209;
>> A = spdiags(primes(p_n)',0,n,n); e = ones(n,2);
>> for k=2.^(0:floor(log2(n))), A = A + spdiags(e,[-k k],n,n); end
>> [V,D] = eig(full(A)); V = intval(V);
>> R = V*D-A*V;
>> for i=1:n, lambda(i) = midrad(D(i,i),norm(R(:,i))/norm(V(:,i))); end
>> [lambda0,j] = min(inf(lambda));
>> lambda_min = infsup(lambda(j))

lambda_min = [  1.120651470854673e+000,  1.120651470922646e+000]

>> lambda0 = intval(1.120651470854673);
>> alpha2 = 11*100; lambda1 = 9221-100;
>> lambdaF = infsup(2*(lambda0*lambda1-alpha2)/...
                    (lambda0+lambda1+sqrt(4*alpha2+(lambda1-lambda0)^2)))

lambdaF = [  1.000037435271862e+000,  1.000037435271866e+000]
```

Interval Theta-Function (§8.3.2)

Used in session on p. 179.

```
function val = theta(q,k)

val = hull(theta_(q,k-1),theta_(q,k));

return

function val = theta_(q,k)

j=0:k; a=(-1).^j./(2*j+1).*q.^(j.*(j+1));
val=2*q^(1/4).*sum(a);

return
```

C.4.3 General Functions and Routines

Two-Dimensional Minimization

As an alternative to using Mathematica in the session on p. 85 one can call:

```
>> [minval,x]=LowestCriticalPoint(@fun,infsup([-1;-1],[1;1]),...
                         infsup(-inf,-3.24),5e-11)

intval minval = -3.306868647_____
x =   -0.02440307969482
       0.21061242715405
```

Implementation of Algorithm 4.2

```
function [minval,x] = LowestCriticalPoint(fun,x,minval,tol)

% [minval,x] = LowestCriticalPoint(fun,x,minval,tol)
%
% solves (interior) global minimization problem on a list
% of rectangles using interval arithmetic.
%
%   fun          objective function. f = fun(x) should give
%                for a (2 x n)-vector of input intervals x
%                a cell structure f.x, f.dx containing the
%                (1 x n)-vector of f-intervals f.x and the
%                (2 x n)-vector of df-intervals f.dx
%
%   x            input: list of rectangles, i.e. (2 x n)-
%                vector of intervals, specifying search region
%                output: midpoint of final enclosing rectangle
%
%   minval       interval enclosing global minimum
%
%   tol          relative tolerance (for minima below 1e-20,
%                absolute tolerance)
```

```
while relerr(minval) > tol
    x = subdivide2D(x);
    f = feval(fun,x);
    upper = min(minval.sup,min(f.x.sup));
    rem = (f.x > upper) | any(not(in(zeros(size(f.dx)),f.dx)));
    f.x(rem) = []; x(:,rem) = [];
    minval = infsup(min(f.x.inf),upper);
end
x = mid(x);

return
```

Interval Bisection

Used in session on p. 138.

```
function x = IntervalBisection(fun,x,tol,varargin)

% x = IntervalBisection(f,x,tol,varargin)
%
% applies the interval bisection method for root-finding
%
%   f           interval function
%   x           at input:  search interval
%               at output: interval enclosing the roots
%   tol         relative error
%   varargin    additional arguments for f

while max(relerr(x))>tol
    x = subdivide1D(x);
    f = x;
    for k=1:length(x)
        f(k) = feval(fun,x(k),varargin{:});
    end
    rem = not(in(zeros(size(f)),f));
    f(rem) = []; x(rem) = [];
end
x=infsup(min(inf(x)),max(sup(x)));

return
```

Interval Newton Iteration

Used in session on p. 179.

```
function X = IntervalNewton(f,X1,varargin)

% X = IntervalNewton(f,X1,varargin)
%
% applies interval Newton method until convergence
%
%   f           interval function, must be enabled for automatic
%                   differentiation, call f(x,varargin)
%   X1          initial interval
```

```
%   varargin     additional arguments for f
%
%   X            converged interval

X = intval(inf(X1));
while X ~= X1
    X = X1;
    x = intval(mid(X));
    F = feval(f,gradientinit(X),varargin{:});
    fx = feval(f,x,varargin{:});
    X1 = intersect(x-fx/F.dx,X);
end

return
```

Interval Arithmetic-Geometric Mean

Used in session on p. 138.

```
function m = AGM(a,b)

rnd = getround;
if isa(a,'double'), a = intval(a); end
if isa(b,'double'), b = intval(b); end
minf = inf(EnclosingAGM(a.inf,b.inf));
msup = sup(EnclosingAGM(a.sup,b.sup));
m = infsup(minf,msup);
setround(rnd);

return

function m = EnclosingAGM(a,b)

a1 = -inf; b1 = inf;
while (a > a1) | (b < b1)
    a1 = a; b1 = b;
    setround(-1); a = sqrt(a1*b1);
    setround( 1); b = (a1+b1)/2;
end
m = infsup(a,b);

return
```

C.5 *Mathematica*

All the *Mathematica* code is for version 5.0.

C.5.1 Utilities

Kronecker Tensor Product of Matrices

Used in the function `ReturnProbability` *on p. 278.*

```
SetAttributes[KroneckerTensor, OneIdentity];
KroneckerTensor[u_?MatrixQ, v_?MatrixQ] :=
 Module[{w = Outer[Times, u, v]}, Partition[
   Flatten[Transpose[w, {1, 3, 2, 4}]], Dimensions[w][[2]] Dimensions[w][[4]]]];
KroneckerTensor[u_, v_, w__] := Fold[KroneckerTensor, u, {v, w}];
CircleTimes = KroneckerTensor;
```

Supporting Interval Functions

Used in various interval functions such as `ReliableTrajectory`, `LowestCritical-Point`, `IntervalBisection`, *and* `IntervalNewton`.

```
mid[X_] := (Min[X] + Max[X]) / 2;
diam[X_] := Max[X] - Min[X];
diam[{X__Interval}] := Max[diam /@ {X}];
extremes[X_] := {Min[X], Max[X]};
hull[X_] := Interval[extremes[X]];
MidRad[x_, r_] := x + Interval[{-1, 1}] r;
IntervalMin[{Interval[{a_, b_}], Interval[{c_, d_}]}] :=
  Interval[{Min[a, c], Min[b, d]}];
IntervalMin[{}] := ∞;
```

Subdivision of Intervals and Rectangles

Used in various interval functions such as `LowestCriticalPoint`, `IntervalBi-section`, *and* `IntervalNewton`.

```
subdivide1D[X_] := Interval /@ {{Min[X], mid[X]}, {mid[X], Max[X]}};
subdivide2D[{X_, Y_}] := Distribute[{subdivide1D[X], subdivide1D[Y]}, List];
```

Count of Digit Agreement and Form for Pretty Interval Output

Used in sessions on pp. 42, 139, 161, 179, and 279.

```
DigitsAgreeCount[a_, b_] := (prec = Ceiling@Min[Precision /@ {a, b}];
   {{ad, ae}, {bd, be}} = RealDigits[#, 10, prec] & /@ {a, b};
   If[ae ≠ be ∨ a b ≤ 0, Return[0]]; If[ad == bd, Return@Length[ad]];
   {{com}} = Position[MapThread[Equal, {ad, bd}], False, 1, 1] - 1; com);
DigitsAgreeCount[Interval[{a_, b_}]] := DigitsAgreeCount[a, b];
IntervalForm[Interval[{a_, b_}]] :=
 (If[(com = DigitsAgreeCount[a, b]) == 0, Return@Interval[{a, b}]];
  start = Sign[a] N[FromDigits@{ad[[Range@com]], 1}, com];
  {low, up} = SequenceForm @@ Take[#, {com + 1, prec}] & /@ {ad, bd};
  If[ae == 0, start /= 10; ae++];
  SequenceForm[DisplayForm@SubsuperscriptBox[NumberForm@start, low, up],
   If[ae ≠ 1, Sequence @@ {" × ", DisplayForm@SuperscriptBox[10, ae - 1]}, ""]])
```

C.5.2 Problem-Dependent Functions and Routines

Reliable Photon Trajectory (§2.3)

Used in session on p. 42.

```
Options[ReliableTrajectory] :=
  {StartIntervalPrecision → Automatic, AccuracyGoal → 12};
ReliableTrajectory[p_, v_, tMax_, opts___Rule] :=
  Module[{error = ∞, ϵ, s0, s, P, V, t, Trem, path, S, T},
   {ϵ, s, s0} = {10.^(-AccuracyGoal), AccuracyGoal, StartIntervalPrecision} /.
      {opts} /. Options[ReliableTrajectory];
   If[s0 === Automatic, s0 = s];
   s = s0; wp = Max[17, s + 2];
   While[error > ϵ,
    P = N[(MidRad[#, 10^-s] &) /@ p, wp];
    V = N[(MidRad[#, 10^-s] &) /@ v, wp];
    path = {P};
    Trem = Interval[tMax];
    While[Trem > 0, M = Round[P + 2 V / 3 ];
     S = t /. (Solve[(P + t V - M).(P + t V - M) == 1 / 9, t]);
     If[FreeQ[S, Power[Interval[{_?Negative, _?Positive}], _]],
          T = IntervalMin[Cases[S, _?Positive]], Break[]];
     Which[
      T ≤ Trem, P += T V; V = H[P - M].V; Trem -= T,
      T > Trem && Trem ≥ 2 / 3, P += 2 V / 3; Trem -= 2 / 3,
      T > Trem && Trem < 2 / 3, P += Trem V; Trem = 0,
      True, Break[]];
     AppendTo[path, P];
     If[Precision[{Trem, P, V, T}] < ag, Break[]]];
    wp = Max[17, (++s) + 2];
    error = diam[P + Table[MidRad[-Max[Abs[Trem]], Max[Abs[Trem]]], {2}]]];
   Print[StringForm["Initial condition interval radius is 10^-``.", s]];
   path];
```

Return Probability (§6.2)

Mathematica version of the MATLAB function on p. 265.

```
Matrices[n_] := Matrices[n] = (m = 2 n + 1; pE = 1/4 + ϵ0;
   pW = 1/4 - ϵ0; pN = 1/4; pS = 1/4; Id = SparseArray[{i_, i_} → 1, {m, m}];
   PEW = SparseArray[{{i_, j_} /; j == i + 1 → pE, {i_, j_} /; j == i - 1 → pW}, {m, m}];
   PNS = SparseArray[{{i_, j_} /; j == i + 1 → pN, {i_, j_} /; j == i - 1 → pS}, {m, m}];
   {PEW ⊗ Id + Id ⊗ PNS, Id ⊗ Id});
ReturnProbability[ϵ_Real, n_Integer] := ({A, Id} = Matrices[n];
   Block[{ϵ0 = ϵ}, m = 2 n + 1; ctr = n m + n + 1; r = A〚All, ctr〛;
    A〚All, ctr〛 = 0; q = LinearSolve[Id - A, r]; q〚ctr〛]);
```

Multiplication by the Matrix A_n (§7.3)

Black-box definition of the sparse matrix A_n on p. 149.

```
n = 20000;
BitLength := Developer`BitLength;
indices[i_] := indices[i] =
  i + Join[2^(Range@BitLength[n - i] - 1), -2^(Range@BitLength[i - 1] - 1)]
diagonal = Prime[Range@n];
A[x_] := diagonal * x + MapIndexed[Plus @@ x[[indices[#2[[1]]]]] &, x];
```

Proof of the Rigorous Bound for $\lambda_{\min}(A_{1142})$ (§7.4.2)

The proof of Lemma 7.1 can be based on the following session.

```
n = 1142;
A = SparseArray[{{i_, i_} → Prime[i]}, n] + (# + Transpose[#]) &@
   SparseArray[Flatten@Table[{i, i + 2^j} → 1, {i, n - 1}, {j, 0, Log[2., n - i]}], n];
{λ, V} = Eigensystem[Normal@N[A]];
r = (λMin = λ[[n]]) (x = Interval /@ V[[n, All]]) - x.A;
(λMin += Interval[{-1, 1}] Norm[r] / Norm[x]) // IntervalForm
```

$1.1206514708^{96148}_{84156}$

```
λ1 = Prime[n + 1] - 100; α2 = 11 × 100;
```

$$\left(\text{λFMin} = \frac{2\,((\text{λ0} = \text{Interval[Min[λMin]]})\ \text{λ1} - \text{α2})}{\text{λ0} + \text{λ1} + \sqrt{4 \times \text{α2} + (\text{λ1} - \text{λ0})^2}}\right) \text{ // IntervalForm}$$

$1.000037435301345^{8}_{5}$

Interval Theta Function (§8.3.2)

Used in session on p. 179.

```
SetAttributes[{θ0, θ1}, Listable];
```

$$\text{θ0[q\_, K\_]} := 2\, q^{1/4} \sum_{k=0}^{K} \frac{(-1)^k}{2k+1} q^{k(k+1)};$$

$$\text{θ1[q\_, K\_]} := \frac{q^{-3/4}}{2} \sum_{k=0}^{K} (-1)^k (2k+1)\, q^{k(k+1)};$$

```
θ[q_Interval, K_Integer] := hull[θ0[q, {K - 1, K}]];
θ^(1,0)[q_Interval, K_Integer] := hull[θ1[q, {K - 1, K}]];
```

C.5.3 General Functions and Routines

Two-Dimensional Minimization

Used in session on p. 85.

```
LowestCriticalPoint[f_, {x_, a_, b_}, {y_, c_, d_}, upperbound_, tol_] :=
  (rects = N[{ {Interval[{a, b}], Interval[{c, d}]}}];
   fcn[{xx_, yy_}] := f /. {x → xx, y → yy};
   gradf[{xx_, yy_}] := Evaluate[{D[f, x], D[f, y]} /. {x → xx, y → yy}];
   {low, upp} = {-∞, upperbound};
   While[(upp - low) > tol,
    rects = Join @@ subdivide2D /@ rects;
    fvals = fcn /@ rects;
    upp = Min[upp, Min[Max /@ fvals]];
    pos = Flatten[Position[Min /@ fvals, _?(# ≤ upp &)]];
    rects = rects[[pos]]; fvals = fvals[[pos]];
    pos = Flatten[
      Position[Apply[And, IntervalMemberQ[gradf /@ rects, 0], {1}], True]];
    rects = rects[[pos]]; low = Min[fvals[[pos]]]];
   {{low, upp}, Map[mid, rects, {2}]});
```

Interval Arithmetic-Geometric Mean

Used in session on p. 139.

```
AGMStep[{a_, b_}] := {√(a b), 1/2 (a + b)};
EnclosingAGM[{a_Real, b_Real}] :=
  Interval@FixedPoint[extremes@AGMStep[Interval /@ #] &, {a, b}];
Unprotect[ArithmeticGeometricMean];
ArithmeticGeometricMean[A_Interval, B_Interval] :=
  Block[{Experimental`$EqualTolerance = 0, Experimental`$SameQTolerance = 0},
   Interval@extremes[EnclosingAGM /@ {Min /@ #, Max /@ #} &@{A, B}]];
Protect[ArithmeticGeometricMean];
```

Interval Bisection

Used in session on p. 139.

```
IntervalBisection[f_, {a_, b_}, tol_] :=
 (X = Interval[{a, b}]; pos = {1}; While[Max[diam /@ X] > tol,
   pos = Flatten@Position[f /@ (X = Join @@ subdivide1D /@ X),
      _?(IntervalMemberQ[#, 0] &)]; X = X[[pos]]]; IntervalUnion @@ X)
```

Interval Newton Iteration

Used in session on p. 179.

```
IntervalNewton[f_, {a_, b_}] := Block[{Experimental`$EqualTolerance = 0},
  X1 = Interval[{a, b}]; X = Interval[a]; While[X =!= X1, X = X1;
   x = Interval[mid[X]]; X1 = IntervalIntersection[x - f[x]/f'[X], X]]; X]
```

Appendix D
More Problems

Whatever the details of the matter, it finds me too absorbed by numerous occupations for me to be able to devote my attention to it immediately.

—John Wallis, upon hearing about a problem posed by Fermat in 1657 [Hav03, p. 92]

While realizing that the solution of problems is one of the lowest forms of Mathematical research, and that, in general, it has no scientific value, yet its educational value can not be over estimated. It is the ladder by which the mind ascends into the higher fields of original research and investigation. Many dormant minds have been aroused into activity through the mastery of a single problem.

—Benjamin Finkel and John Colaw on the first page of the first issue of the *American Mathematical Monthly*, 1894

To help readers experience first-hand the excitement, frustration, and joy of working on a challenging numerical problem, we include here a selection in the same style as Trefethen's 10. Of these 22, the two at the end can be considered research problems in the sense that the proposer does not know even a single digit of the answer.

If you solve one of these and wish to share your solution, we will be happy to receive it. We will post, on the web page of this book, solutions that are submitted to us.

1. What is $\sum_n 1/n$, where n is restricted to those positive integers whose decimal representation does not contain the substring 42? (*Folkmar Bornemann*)

2. What is the sum of the series $\sum_{n=1}^{\infty} 1/f(n)$, where $f(1) = 1$, $f(2) = 2$, and if $n > 2$, $f(n) = nf(d)$, with d the number of base-2 digits of n? (*David Smith*)

Remark. Problem A6 of the 2002 Putnam Competition asked for the integers $b \geqslant 2$ such that the series, when generalized to base-b digits, converges.

3. Let $m(k) = k - k/d(k)$ where $d(k)$ is the smallest prime factor of k. What is

$$\lim_{x \uparrow 1} \frac{1}{1-x} \prod_{k=2}^{\infty} \left(1 - \frac{x^{m(k)}}{k+1}\right) ?$$

(Arnold Knopfmacher)

Remark. This problem arose from the work of Knopfmacher and Warlimont [KW95]. It is a variation of functions that arise in the study of probabilities related to the irreducible factors in polynomials over Galois fields.

4. If N point charges are distributed on the unit sphere, the potential energy is

$$E = \sum_{j=1}^{N-1} \sum_{k=j+1}^{N} |x_j - x_k|^{-1},$$

where $|x_j - x_k|$ is the Euclidean distance between x_j and x_k. Let E_N denote the minimal value of E over all possible configurations of N charges. What is E_{100}? *(Lloyd N. Trefethen)*

5. Riemann's prime counting function is defined as

$$R(x) = \sum_{k=1}^{\infty} \frac{\mu(k)}{k} \operatorname{li}(x^{1/k}),$$

where $\mu(k)$ is the Möbius function, which is $(-1)^{\rho}$ when k is a product of ρ different primes and zero otherwise, and $\operatorname{li}(x) = \int_0^x dt/\log t$ is the logarithmic integral, taken as a principal value in Cauchy's sense. What is the largest positive zero of R? *(Jörg Waldvogel)*

Remark. The answer to this problem is truly shocking.

6. Let A be the 48×48 Toeplitz matrix with -1 on the first subdiagonal, $+1$ on the main diagonal and the first three superdiagonals, and 0 elsewhere, and let $\|\cdot\|$ be the matrix 2-norm. What is $\min_p \|p(A)\|$, where p ranges over all monic polynomials of degree 8?

(Lloyd N. Trefethen)

7. What is the value of

$$\int_{-1}^{1} \exp\left(x + \sin e^{e^{e^{x+1/3}}}\right) dx \ ?$$

(Lloyd N. Trefethen)

8. What is the value of

$$\int_0^{\infty} x\, J_0(x\sqrt{2}) J_0(x\sqrt{3}) J_0(x\sqrt{5}) J_0(x\sqrt{7}) J_0(x\sqrt{11})\, dx,$$

where J_0 denotes the Bessel function of the first kind of order zero?

(Folkmar Bornemann)

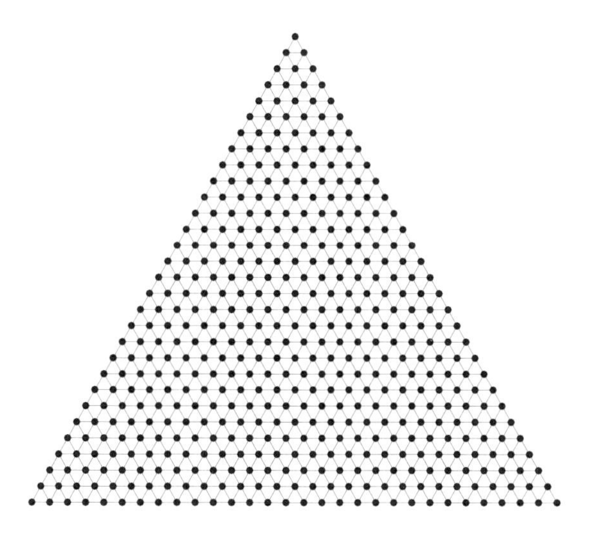

Figure D.1. *A triangular lattice.*

9. If $f(x, y) = e^{-(y+x^3)^2}$ and $g(x, y) = \frac{1}{32}y^2 + e^{\sin y}$, what is the area of the region of the x-y plane in which $f > g$? *(Lloyd N. Trefethen)*

10. The square $c_m(\mathbb{R}^N)^2$ of the least constant in the Sobolev inequality for the domain $\mathbb{R}^N$ is given by the multidimensional integral

$$c_m(\mathbb{R}^N)^2 := (2\pi)^{-N} \int_{\mathbb{R}^N} \left(\sum_{|k| \leqslant m} x^{2k} \right)^{-1} dx,$$

where $k = (k_1, k_2, \ldots, k_N)$ is a multi-index with nonnegative integer elements, and

$$|k| := \sum_{j=1}^{N} k_j, \quad x^k := \prod_{j=1}^{N} x_j^{k_j}.$$

For example, we have $c_3(\mathbb{R}^1) = 0.5$. What is $c_{10}(\mathbb{R}^{10})$? *(Jörg Waldvogel)*

11. A particle starts at the top vertex of the array shown in Figure D.1 with 30 points on each side, and then takes 60 random steps. What is the probability that it ends up in the bottom row? *(Lloyd N. Trefethen)*

12. The random sequence x_n satisfies $x_0 = x_1 = 1$ and the recursion $x_{n+1} = 2x_n \pm x_{n-1}$, where each $\pm$ sign is chosen independently with equal probability. To what value does $|x_n|^{1/n}$ converge for $n \to \infty$ almost surely? *(Folkmar Bornemann)*

13. Six masses of mass 1 are fixed at positions $(2,-1)$, $(2,0)$, $(2,1)$, $(3,-1)$, $(3,0)$, and $(3,1)$. Another particle of mass 1 starts at $(0,0)$ with velocity 1 in a direction θ (counter-clockwise from the x-axis). It then travels according to Newton's laws, feeling an inverse-square force of magnitude r^{-2} from each of the six fixed masses. What is the shortest time in which the moving particle can be made to reach position $(4,1)$? *(Lloyd N. Trefethen)*

14. Suppose a particle's movement in the x–y plane is governed by the kinetic energy $T=\frac{1}{2}(\dot{x}^2+\dot{y}^2)$ and the potential energy

$$U = y + \frac{\epsilon^{-2}}{2}(1+\alpha x^2)(x^2+y^2-1)^2.$$

The particle starts at the position $(0,1)$ with the velocity $(1,1)$. For which parameter α does the particle hit $y=0$ first at time 10 in the limit $\epsilon \to 0$? *(Folkmar Bornemann)*

15. Let $u=(x,y,z)$ start at $(0,0,z_0)$ at $t=0$ with $z_0 \geqslant 0$ and evolve according to the equations

$$\begin{aligned}
\dot{x} &= -x + 10y + \|u\|(-0.7y - 0.03z),\\
\dot{y} &= -y + 10z + \|u\|(0.7y - 0.1z),\\
\dot{z} &= -z + \|u\|(0.03x + 0.1y),
\end{aligned}$$

where $\|u\|^2 = x^2+y^2+z^2$. If $\|u(50)\| = 1$, what is z_0? *(Lloyd N. Trefethen)*

16. Consider the Poisson equation $-\Delta u(x) = \exp(\alpha\|x\|^2)$ on a regular pentagon inscribed to the unit circle. On four sides of the pentagon there is the Dirichlet condition $u=0$, while on one side u satisfies a Neumann condition; that is, the normal derivative of u is zero. For which α does the integral of u along the Neumann side equal e^{α}? *(Folkmar Bornemann)*

17. At what time t_∞ does the solution of the equation $u_t = \Delta u + e^u$ on a 3×3 square with zero boundary and initial data blow up to ∞? *(Lloyd N. Trefethen)*

18. Figure D.2 shows the Daubechies scaling function $\phi_2(x)$ (see [Dau92]) drawn as a curve in the x–y plane. Suppose the heat equation $u_t = u_{xx}$ on the interval $[0,3]$ is solved with initial data $u(x,0)=\phi_2(x)$ and boundary conditions $u(0)=u(3)=0$. At what time does the length of this curve in the x–y plane become 5.4? *(Lloyd N. Trefethen)*

19. Let u be an eigenfunction belonging to the third eigenvalue of the Laplacian with Dirichlet boundary conditions on an L-shaped domain that is made up from three unit squares. What is the length of the zero-level set of u? *(Folkmar Bornemann)*

20. The Koch snowflake is a fractal domain defined as follows. Start with an equilateral triangle with side length 1, and replace the middle third of each side by two sides of an outward-pointing equilateral triangle of side length $1/3$. Now replace the middle third of each of the 12 new sides by two sides of an outward-pointing equilateral triangular of side

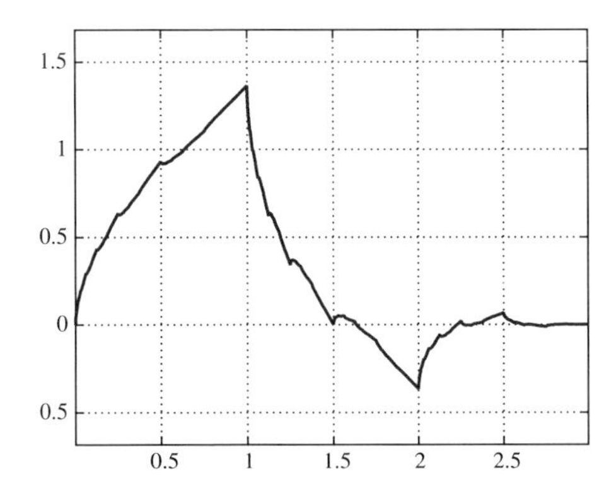

Figure D.2. *Daubechies scaling function $\phi_2(x)$.*

length $1/9$; and so on ad infinitum. What is the smallest eigenvalue of the negative of the Laplacian on the Koch snowflake with zero boundary conditions? *(Lloyd N. Trefethen)*

21. Consider the Poisson equation $-\text{div}(c(x)\text{grad}\,u(x)) = 1$ on the unit square with homogeneous Dirichlet boundary conditions. What is the supremum of the integral of u over the square if $a(x)$ is allowed to vary over all measurable functions that are 1 on half of the area of the square, and 100 on the rest? *(Folkmar Bornemann)*

22. Let $h(z)$ be that solution to the functional equation $\exp(z) = h(h(z))$ which is analytic in a neighborhood of the origin and increasing for real z. What is $h(0)$? What is the radius of convergence of the Maclaurin series of h? *(Dirk Laurie)*

Remark. There are additional properties needed to make h unique. One such simple property has to be found before solving this problem; none is known in the literature right now.

References

[Apo74] Tom M. Apostol, *Mathematical analysis*, second ed., Addison-Wesley, Reading, MA, 1974. (Cited on p. 185.)

[AS84] Milton Abramowitz and Irene A. Stegun (eds.), *Handbook of mathematical functions with formulas, graphs, and mathematical tables*, Wiley, New York, 1984. (Cited on pp. 23, 56, 102, 103, 116, 117, 135.)

[AW97] Victor Adamchik and Stan Wagon, *A simple formula for π*, Amer. Math. Monthly **104** (1997), no. 9, 852–855. (Cited on p. viii.)

[Bai00] David H. Bailey, *A compendium of BPP-type formulas for mathematical constants*, manuscript, November 2000, `http://crd.lbl.gov/~dhbailey/dhbpapers/bbp-formulas.pdf`. (Cited on p. viii.)

[Bak90] Alan Baker, *Transcendental number theory*, second ed., Cambridge University Press, Cambridge, UK, 1990. (Cited on p. 219.)

[Bar63] Vic D. Barnett, *Some explicit results for an asymmetric two-dimensional random walk*, Proc. Cambridge Philos. Soc. **59** (1963), 451–462. (Cited on pp. 13, 130, 137.)

[BB87] Jonathan M. Borwein and Peter B. Borwein, *Pi and the AGM. A study in analytic number theory and computational complexity*, Wiley, New York, 1987. (Cited on pp. 137, 138, 176, 216, 217, 259.)

[BB04] Jonathan M. Borwein and David H. Bailey, *Mathematics by experiment: Plausible reasoning in the 21st century*, A. K. Peters, Wellesley, 2004. (Cited on p. viii.)

[BBG04] Jonathan M. Borwein, David H. Bailey, and Roland Girgensohn, *Experimentation in mathematics: Computational paths to discovery*, A. K. Peters, Wellesley, 2004. (Cited on p. viii.)

[BBP97] David Bailey, Peter Borwein, and Simon Plouffe, *On the rapid computation of various polylogarithmic constants*, Math. Comp. **66** (1997), no. 218, 903–913. (Cited on p. viii.)

[Ber89] Bruce C. Berndt, *Ramanujan's notebooks. Part II*, Springer-Verlag, New York, 1989. (Cited on p. 211.)

[Ber98] ———, *Ramanujan's notebooks. Part V*, Springer-Verlag, New York, 1998. (Cited on pp. 221.)

[Ber01] Michael Berry, *Why are special functions special?*, Physics Today **54** (2001), no. 4, 11–12. (Cited on p. 121.)

[BF71] Paul F. Byrd and Morris D. Friedman, *Handbook of elliptic integrals for engineers and scientists*, 2nd rev. ed., Springer-Verlag, New York, 1971. (Cited on p. 142.)

[BR95] Bruce C. Berndt and Robert A. Rankin, *Ramanujan, Letters and commentary*, American Mathematical Society, Providence, 1995. (Cited on p. 220.)

[Bre71] Claude Brezinski, *Accélération de suites à convergence logarithmique*, C. R. Acad. Sci. Paris Sér. A-B **273** (1971), A727–A730. (Cited on p. 246.)

[Bre76] Richard P. Brent, *Fast multiple-precision evaluation of elementary functions*, J. Assoc. Comput. Mach. **23** (1976), no. 2, 242–251. (Cited on p. 30.)

[Bre88] Claude Brezinski, *Quasi-linear extrapolation processes*, Numerical Mathematics, Singapore 1988, Birkhäuser, Basel, 1988, pp. 61–78. (Cited on p. 257.)

[Bre00] ———, *Convergence acceleration during the 20th century*, J. Comput. Appl. Math. **122** (2000), no. 1–2, 1–21. (Cited on pp. 230, 231, 257.)

[BRS63] Friedrich L. Bauer, Heinz Rutishauser, and Eduard Stiefel, *New aspects in numerical quadrature*, Proc. Sympos. Appl. Math., Vol. XV, Amer. Math. Soc., Providence, R.I., 1963, pp. 199–218. (Cited on pp. 68, 257.)

[BT99] B. Le Bailly and J.-P. Thiran, *Computing complex polynomial Chebyshev approximants on the unit circle by the real Remez algorithm*, SIAM J. Numer. Anal. **36** (1999), no. 6, 1858–1877. (Cited on p. 119.)

[BY93] Folkmar Bornemann and Harry Yserentant, *A basic norm equivalence for the theory of multilevel methods*, Numer. Math. **64** (1993), no. 4, 455–476. (Cited on p. 157.)

[BZ91] Claude Brezinski and Michela Redivo Zaglia, *Extrapolation methods*, North-Holland, Amsterdam, 1991. (Cited on pp. 225, 230, 257.)

[BZ92] Jonathan M. Borwein and I. John Zucker, *Fast evaluation of the gamma function for small rational fractions using complete elliptic integrals of the first kind*, IMA J. Numer. Anal. **12** (1992), no. 4, 519–526. (Cited on p. 143.)

[Cau27] Augustin-Louis Cauchy, *Sur quelques propositions fondamentales du calcul des résidus*, Exerc. Math. **2** (1827), 245–276. (Cited on pp. 211, 212.)

[CDG99] David W. Corne, Marco Dorigo, and Fred Glover (eds.), *New ideas in optimization*, McGraw-Hill, Berkshire, 1999. (Cited on p. 106.)

[CGH+96] Robert M. Corless, Gaston H. Gonnet, David E. G. Hare, David J. Jeffrey, and Donald E. Knuth, *On the Lambert W function*, Adv. Comput. Math. **5** (1996), no. 4, 329–359. (Cited on pp. 23, 24.)

[CM01] Nikolai Chernov and Roberto Markarian, *Introduction to the ergodic theory of chaotic billiards*, Instituto de Matemática y Ciencías Afines (IMCA), Lima, 2001. (Cited on p. 44.)

[Col95] Courtney S. Coleman, CODEE Newsletter (spring 1995), cover. (Cited on p. 91.)

[Cox84] David A. Cox, *The arithmetic-geometric mean of Gauss*, Enseign. Math. (2) **30** (1984), no. 3–4, 275–330. (Cited on p. 216.)

[Cox89] ———, *Primes of the form $x^2 + ny^2$. Fermat, class field theory and complex multiplication*, Wiley, New York, 1989. (Cited on p. 220.)

[CP01] Richard Crandall and Carl Pomerance, *Prime numbers, A computational perspective*, Springer-Verlag, New York, 2001. (Cited on pp. 156, 202.)

[CRZ00] Henri Cohen, Fernando Rodriguez Villegas, and Don Zagier, *Convergence acceleration of alternating series*, Experiment. Math. **9** (2000), no. 1, 3–12. (Cited on pp. 241, 257.)

[Dau92] Ingrid Daubechies, *Ten lectures on wavelets*, Society for Industrial and Applied Mathematics (SIAM), Philadelphia, 1992. (Cited on p. 284.)

[DB02] Peter Deuflhard and Folkmar Bornemann, *Scientific computing with ordinary differential equations*, Springer-Verlag, New York, 2002, Translated by Werner C. Rheinboldt. (Cited on pp. 170, 172, 207.)

[Dem97] James W. Demmel, *Applied numerical linear algebra*, Society for Industrial and Applied Mathematics (SIAM), Philadelphia, 1997. (Cited on pp. 124, 205.)

[DeV02] Carl DeVore, 2002, A Maple worksheet on Trefethen's Problem 3, `http://groups.yahoo.com/group/100digits/files/Tref3.mws`. (Cited on p. 257.)

[DH03] Peter Deuflhard and Andreas Hohmann, *Numerical analysis in modern scientific computing. An introduction*, second ed., Springer-Verlag, New York, 2003. (Cited on pp. 148, 154.)

[Dix82] John D. Dixon, *Exact solution of linear equations using p-adic expansions*, Numer. Math. **40** (1982), no. 1, 137–141. (Cited on p. 167.)

[DJ01] Richard T. Delves and Geoff S. Joyce, *On the Green function for the anisotropic simple cubic lattice*, Ann. Phys. **291** (2001), 71–133. (Cited on p. 144.)

[DKK91] Eusebius Doedel, Herbert B. Keller, and Jean-Pierre Kernévez, *Numerical analysis and control of bifurcation problems. I. Bifurcation in finite dimensions*, Internat. J. Bifur. Chaos Appl. Sci. Engrg. **1** (1991), no. 3, 493–520. (Cited on p. 88.)

[DR84] Philip J. Davis and Philip Rabinowitz, *Methods of numerical integration*, second ed., Academic Press, Orlando, FL, 1984. (Cited on pp. 68, 70.)

[DR90] John M. DeLaurentis and Louis A. Romero, *A Monte Carlo method for Poisson's equation*, J. Comput. Phys. **90** (1990), no. 1, 123–140. (Cited on pp. 200, 201.)

[DS00] Jack Dongarra and Francis Sullivan, *The top 10 algorithms*, IEEE Computing in Science and Engineering **2** (2000), no. 1, 22–23. (Cited on p. viii.)

[DT02] Tobin A. Driscoll and Lloyd N. Trefethen, *Schwarz–Christoffel mapping*, Cambridge University Press, Cambridge, UK, 2002. (Cited on p. 223.)

[DTW02] Jean-Guillaume Dumas, William Turner, and Zhendong Wan, *Exact solution to large sparse integer linear systems*, Abstract for ECCAD'2002, May 2002. (Cited on p. 164.)

[Dys96] Freeman J. Dyson, *Review of "Nature's Numbers" by Ian Stewart*, Amer. Math. Monthly **103** (1996), 610–612. (Cited on p. 147.)

[EMOT53] Arthur Erdélyi, Wilhelm Magnus, Fritz Oberhettinger, and Francesco G. Tricomi, *Higher transcendental functions. Vols. I, II*, McGraw-Hill, New York, 1953. (Cited on p. 135.)

[Erd56] Arthur Erdélyi, *Asymptotic expansions*, Dover, New York, 1956. (Cited on p. 70.)

[Eva93] Gwynne Evans, *Practical numerical integration*, Wiley, Chichester, UK, 1993. (Cited on p. 18.)

[Fel50] William Feller, *An introduction to probability theory and its applications. Vol. I*, Wiley, New York, 1950. (Cited on pp. 125, 126.)

[FH98] Samuel P. Ferguson and Thomas C. Hales, *A formulation of the Kepler conjecture*, Tech. report, ArXiv Math MG, 1998, `http://arxiv.org/abs/math.MG/9811072`. (Cited on p. 95.)

[FLS63] Richard P. Feynman, Robert B. Leighton, and Matthew Sands, *The Feynman lectures on physics. Vol. 1: Mainly mechanics, radiation, and heat*, Addison-Wesley, Reading, MA, 1963. (Cited on p. 230.)

[Fou78] Joseph Fourier, *The analytical theory of heat*, Cambridge University Press, Cambridge, UK, 1878, Translated by Alexander Freeman. Reprinted by Dover Publications, New York, 1955. French original: "Théorie analytique de la chaleur," Didot, Paris, 1822. (Cited on pp. 169, 174, 175, 176.)

[GH83] John Guckenheimer and Philip Holmes, *Nonlinear oscillations, dynamical systems, and bifurcations of vector fields*, Springer-Verlag, New York, 1983. (Cited on p. 40.)

[GL81] Alan George and Joseph W. H. Liu, *Computer solution of large sparse positive definite systems*, Prentice-Hall, Englewood Cliffs, NJ, 1981. (Cited on pp. 149, 150, 151.)

[GL96] Gene H. Golub and Charles F. Van Loan, *Matrix computations*, third ed., Johns Hopkins Studies in the Mathematical Sciences, Johns Hopkins University Press, Baltimore, 1996. (Cited on pp. 49, 51, 53.)

[Goo83] Nicolas D. Goodman, *Reflections on Bishop's philosophy of mathematics*, Math. Intelligencer **5** (1983), no. 3, 61–68. (Cited on p. 147.)

[Grif90] Peter Griffin, *Accelerating beyond the third dimension: Returning to the origin in simple random walk*, Math. Sci., **15** (1990), 24–35. (Cited on p. 146.)

[GW01] Walter Gautschi and Jörg Waldvogel, *Computing the Hilbert transform of the generalized Laguerre and Hermite weight functions*, BIT **41** (2001), no. 3, 490–503. (Cited on p. 70.)

[GZ77] M. Lawrence Glasser and I. John Zucker, *Extended Watson integrals for the cubic lattices*, Proc. Nat. Acad. Sci. U.S.A. **74** (1977), no. 5, 1800–1801. (Cited on p. 143.)

[Hac92] Wolfgang Hackbusch, *Elliptic differential equations. Theory and numerical treatment*, Springer-Verlag, Berlin, 1992. (Cited on p. 205.)

[Had45] Jacques Hadamard, *The psychology of invention in the mathematical field*, Princeton University Press, Princeton, NJ, 1945. (Cited on p. 17.)

[Han92] Eldon Hansen, *Global optimization using interval analysis*, Monographs and Textbooks in Pure and Applied Mathematics, Vol. 165, Marcel Dekker, New York, 1992. (Cited on pp. 84, 94, 95, 98.)

[Har40] Godfrey H. Hardy, *Ramanujan. Twelve lectures on subjects suggested by his life and work*, Cambridge University Press, Cambridge, UK, 1940. (Cited on p. 221.)

[Har02] Gareth I. Hargreaves, *Interval analysis in MATLAB*, Master's thesis, University of Manchester, December 2002, Numerical Analysis Report No. 416. (Cited on p. 272.)

[Hav03] Julian Havil, *Gamma*, Princeton University Press, Princeton, NJ, 2003. (Cited on p. 281.)

[Hen61] Ernst Henze, *Zur Theorie der diskreten unsymmetrischen Irrfahrt*, ZAMM **41** (1961), 1–9. (Cited on pp. 13, 137.)

[Hen64] Peter Henrici, *Elements of numerical analysis*, Wiley, New York, 1964. (Cited on p. 19.)

[Hen74] ———, *Applied and computational complex analysis. Vol. 1: Power series—integration—conformal mapping—location of zeros*, Wiley, New York, 1974. (Cited on pp. 67, 211, 212, 213.)

[Hen77] ———, *Applied and computational complex analysis. Vol. 2: Special functions—integral transforms—asymptotics—continued fractions*, Wiley, New York, 1977. (Cited on pp. 62, 70.)

[Hen86] ———, *Applied and computational complex analysis. Vol. 3: Discrete Fourier analysis—Cauchy integrals—construction of conformal maps—univalent functions*, Wiley, New York, 1986. (Cited on pp. 203, 204, 213, 214, 223.)

[Her83] Joseph Hersch, *On harmonic measures, conformal moduli and some elementary symmetry methods*, J. Analyse Math. **42** (1982/83), 211–228. (Cited on p. 213.)

[Hig96] Nicholas J. Higham, *Accuracy and stability of numerical algorithms,* 2nd ed., Society for Industrial and Applied Mathematics (SIAM), Philadelphia, 2002. `http://www.ma.man.ac.uk/~higham/asna` (Cited on pp. 5, 40, 48, 49, 54, 151, 152, 159, 163, 164, 173, 192, 205, 208.)

[HJ85] Roger A. Horn and Charles R. Johnson, *Matrix analysis*, Cambridge University Press, Cambridge, UK, 1985. (Cited on pp. 53, 155, 162, 163, 165.)

[Hof67] Peter Hofmann, *Asymptotic expansions of the discretization error of boundary value problems of the Laplace equation in rectangular domains*, Numer. Math. **9** (1966/1967), 302–322. (Cited on p. 206.)

[Hug95] Barry D. Hughes, *Random walks and random environments. Vol. 1: Random walks*, Oxford University Press, New York, 1995. (Cited on pp. 121, 122, 130, 139, 140, 143.)

[IMT70] Masao Iri, Sigeiti Moriguti, and Yoshimitsu Takasawa, *On a certain quadrature formula (Japanese)*, Kokyuroku Ser. Res. Inst. for Math. Sci. Kyoto Univ. **91** (1970), 82–118; English translation: J. Comput. Appl. Math. 17, 3–20 (1987). (Cited on p. 68.)

[Jac29] Carl Gustav Jacob Jacobi, *Fundamenta nova theoriae functionum ellipticarum*, Bornträger, Regiomontum (Königsberg), 1829. (Cited on p. 218.)

[JDZ03] Geoff S. Joyce, Richard T. Delves, and I. John Zucker, *Exact evaluation of the Green functions for the anisotropic face-centred and simple cubic lattices*, J. Phys. A: Math. Gen. **36** (2003), 8661–8672. (Cited on p. 144.)

[Joh82] Fritz John, *Partial differential equations*, fourth ed., Springer-Verlag, New York, 1982. (Cited on p. 176.)

[Kea96] R. Baker Kearfott, *Rigorous global search: Continuous problems*, Nonconvex Optimization and Its Applications, Vol. 13, Kluwer Academic Publishers, Dordrecht, 1996. (Cited on pp. 84, 93, 94, 95, 97.)

[Kno56] Konrad Knopp, *Infinite sequences and series*, Dover, New York, 1956. (Cited on p. 135.)

[Knu81] Donald E. Knuth, *The art of computer programming. Vol. 2: Seminumerical algorithms*, second ed., Addison-Wesley, Reading, 1981. (Cited on p. 29.)

[Koe98] Wolfram Koepf, *Hypergeometric summation. An algorithmic approach to summation and special function identities*, Vieweg, Braunschweig, 1998. (Cited on pp. 131, 132.)

[Kol48] Andrey N. Kolmogorov, *A remark on the polynomials of P. L. Čebyšev deviating the least from a given function*, Uspehi Matem. Nauk (N.S.) **3** (1948), no. 1(23), 216–221. (Cited on p. 118.)

[KS91] Erich Kaltofen and B. David Saunders, *On Wiedemann's method of solving sparse linear systems*, Applied algebra, algebraic algorithms and error-correcting codes (New Orleans, LA, 1991), Lecture Notes in Comput. Sci., Vol. 539, Springer, Berlin, 1991, pp. 29–38. (Cited on p. 166.)

[Küh82] Wilhelm O. Kühne, *Huppel en sy maats*, Tafelberg, Kaapstad, 1982. (Cited on p. 101.)

[KW95] Arnold Knopfmacher and Richard Warlimont, *Distinct degree factorizations for polynomials over a finite field*, Trans. Amer. Math. Soc. **347** (1995), no. 6, 2235–2243. (Cited on p. 282.)

[Lan82] Oscar E. Lanford, III, *A computer-assisted proof of the Feigenbaum conjectures*, Bull. Amer. Math. Soc. (N.S.) **6** (1982), no. 3, 427–434. (Cited on p. 95.)

[LB92] John Lund and Kenneth L. Bowers, *Sinc methods for quadrature and differential equations*, Society for Industrial and Applied Mathematics (SIAM), Philadelphia, 1992. (Cited on p. 188.)

[Lev73] David Levin, *Development of non-linear transformations of improving convergence of sequences*, Internat. J. Comput. Math. **3** (1973), 371–388. (Cited on p. 243.)

[Lon56] Ivor M. Longman, *Note on a method for computing infinite integrals of oscillatory functions*, Proc. Cambridge Philos. Soc. **52** (1956), 764–768. (Cited on p. 18.)

[Luk75] Yudell L. Luke, *Mathematical functions and their approximations*, Academic Press, New York, 1975. (Cited on pp. 117, 195.)

[LV94] Dirk P. Laurie and Lucas M. Venter, *A two-phase algorithm for the Chebyshev solution of complex linear equations*, SIAM J. Sci. Comput. **15** (1994), no. 6, 1440–1451. (Cited on pp. 111, 115.)

[Lyn85] James N. Lyness, *Integrating some infinite oscillating tails*, Proceedings of the International Conference on Computational and Applied Mathematics (Leuven, 1984), Vol. 12/13, 1985, pp. 109–117. (Cited on p. 19.)

[Men43] Luigi Frederico Menabrea, *Sketch of the analytical engine invented by Charles Babbage, Esq. With notes by the translator (A.A.L.)*, Taylor's Scientific Memoirs **3** (1843), no. 29, 666–731. (Cited on p. 263.)

[Mil63] John Milnor, *Morse theory*, Based on lecture notes by M. Spivak and R. Wells. Annals of Mathematics Studies, No. 51, Princeton University Press, Princeton, NJ, 1963. (Cited on p. 89.)

[Mil94] Gradimir V. Milovanović, *Summation of series and Gaussian quadratures*, Approximation and Computation (West Lafayette, IN, 1993), Birkhäuser, Boston, 1994, pp. 459–475. (Cited on p. 66.)

[Mon56] Elliot W. Montroll, *Random walks in multidimensional spaces, especially on periodic lattices*, J. Soc. Indust. Appl. Math. **4** (1956), 241–260. (Cited on p. 145.)

[Moo66] Ramon E. Moore, *Interval analysis*, Prentice-Hall, Englewood Cliffs, NJ, 1966. (Cited on p. 92.)

[Mor78] Masatake Mori, *An IMT-type double exponential formula for numerical integration*, Publ. Res. Inst. Math. Sci. Kyoto Univ. **14** (1978), no. 3, 713–729. (Cited on p. 68.)

[MS83] Guriĭ I. Marchuk and Vladimir V. Shaĭdurov, *Difference methods and their extrapolations*, Springer-Verlag, New York, 1983. (Cited on p. 206.)

[MS01] Masatake Mori and Masaaki Sugihara, *The double-exponential transformation in numerical analysis*, J. Comput. Appl. Math. **127** (2001), no. 1–2, 287–296. (Cited on p. 188.)

[MT03] Oleg Marichev and Michael Trott, *Meijer G function*, The Wolfram Functions Site, Wolfram Research, 2003, `http://functions.wolfram.com`. (Cited on p. 194.)

[Neh52] Zeev Nehari, *Conformal mapping*, McGraw-Hill, New York, 1952. (Cited on p. 203.)

[Neu90] Arnold Neumaier, *Interval methods for systems of equations*, Encyclopedia of Mathematics and Its Applications, vol. 37, Cambridge University Press, Cambridge, U.K., 1990. (Cited on pp. 93, 94.)

[Nie06] Niels Nielsen, *Handbuch der Theorie der Gammafunktion.*, Teubner, Leipzig, 1906. (Cited on p. 56.)

[NPWZ97] István Nemes, Marko Petkovšek, Herbert S. Wilf, and Doron Zeilberger, *How to do Monthly problems with your computer*, Amer. Math. Monthly **104** (1997), no. 6, 505–519. (Cited on p. 132.)

[Olv74] Frank W. J. Olver, *Asymptotics and special functions*, Academic Press, New York, 1974. (Cited on pp. 70, 227.)

[OM99] Takuya Ooura and Masatake Mori, *A robust double exponential formula for Fourier-type integrals*, J. Comput. Appl. Math. **112** (1999), no. 1–2, 229–241. (Cited on pp. 28, 190, 191.)

[PBM86] Anatoliĭ P. Prudnikov, Yury A. Brychkov, and Oleg I. Marichev, *Integrals and series. Vol. 1: Elementary functions*, Gordon & Breach, New York, 1986. (Cited on p. 142.)

[PdDKÜK83] Robert Piessens, Elise de Doncker-Kapenga, Christoph W. Überhuber, and David K. Kahaner, *QUADPACK: A subroutine package for automatic integration*, Springer-Verlag, Berlin, 1983. (Cited on pp. 18, 183.)

[Pól21] Georg Pólya, *Über eine Aufgabe der Wahrscheinlichkeitsrechnung betreffend die Irrfahrt im Straßennetz*, Math. Ann. **83** (1921), 149–160. (Cited on pp. 122, 135, 143.)

[Pow64] Michael J. D. Powell, *An efficient method for finding the minimum of a function of several variables without calculating derivatives*, Comput. J. **7** (1964), 155–162. (Cited on p. 184.)

[PTVF92] William H. Press, Saul A. Teukolsky, William T. Vetterling, and Brian P. Flannery, *Numerical recipes in C*, second ed., Cambridge University Press, Cambridge, UK, 1992. (Cited on p. 77.)

[PW34] Raymond E. A. C. Paley and Norbert Wiener, *Fourier transforms in the complex domain*, American Mathematical Society, New York, 1934. (Cited on p. 70.)

[PWZ96] Marko Petkovšek, Herbert S. Wilf, and Doron Zeilberger, $A = B$, A. K. Peters, Wellesley, 1996. (Cited on pp. 131, 132.)

[Rai60] Earl D. Rainville, *Special functions*, Macmillan, New York, 1960. (Cited on p. 135.)

[Rau91] Jeffrey Rauch, *Partial differential equations*, Springer-Verlag, New York, 1991. (Cited on pp. 199, 209.)

[Rem34a] Eugene J. Remes (Evgeny Ya. Remez), *Sur le calcul effectif des polynômes d'approximation de Tchebichef*, C. R. Acad. Sci. Paris **199** (1934), 337–340. (Cited on p. 107.)

[Rem34b] ———, *Sur un procédé convergent d'approximation successives pour déterminer les polynômes d'approximation*, C. R. Acad. Sci. Paris **198** (1934), 2063–2065. (Cited on p. 107.)

[RS75] Michael Reed and Barry Simon, *Methods of modern mathematical physics. II. Fourier analysis, self-adjointness*, Academic Press, New York, 1975. (Cited on p. 70.)

[Rud87] Walter Rudin, *Real and complex analysis*, third ed., McGraw-Hill, New York, 1987. (Cited on p. 48.)

[Rum98] Siegfried M. Rump, *A note on epsilon-inflation*, Reliab. Comput. **4** (1998), no. 4, 371–375. (Cited on p. 97.)

[Rum99a] ——, *Fast and parallel interval arithmetic*, BIT **39** (1999), no. 3, 534–554. (Cited on p. 272.)

[Rum99b] ——, *INTLAB—interval laboratory*, Developments in Reliable Computing (Tibor Csendes, ed.), Kluwer, Dordrecht, 1999, pp. 77–104. (Cited on p. 272.)

[Rut90] Heinz Rutishauser, *Lectures on numerical mathematics*, Birkhäuser, Boston, 1990. (Cited on p. 107.)

[Sal55] Herbert E. Salzer, *A simple method for summing certain slowly convergent series*, J. Math. Phys. **33** (1955), 356–359. (Cited on p. 237.)

[Sch69] Charles Schwartz, *Numerical integration of analytic functions*, J. Computational Phys. **4** (1969), 19–29. (Cited on p. 68.)

[Sch89] Hans R. Schwarz, *Numerical analysis: A comprehensive introduction*, Wiley, Chichester, UK, 1989, With a contribution by Jörg Waldvogel, Translated from the German. (Cited on p. 68.)

[Sid03] Avram Sidi, *Practical extrapolation methods: Theory and applications*, Cambridge University Press, Cambridge, UK, 2003. (Cited on pp. 230, 257.)

[Sin70a] Yakov G. Sinaĭ, *Dynamical systems with elastic reflections. Ergodic properties of dispersing billiards*, Uspehi Mat. Nauk **25** (1970), no. 2 (152), 141–192. (Cited on p. 44.)

[Sin70b] Ivan Singer, *Best approximation in normed linear spaces by elements of linear subspaces*, Springer-Verlag, Berlin, 1970. (Cited on pp. 113, 119.)

[SL68] Vladimir I. Smirnov and N. A. Lebedev, *Functions of a complex variable: Constructive theory*, The MIT Press, Cambridge, MA, 1968. (Cited on pp. 113, 119.)

[Smi97] Frank Smithies, *Cauchy and the creation of complex function theory*, Cambridge University Press, Cambridge, UK, 1997. (Cited on p. 211.)

[Sok97] Alan D. Sokal, *Monte Carlo methods in statistical mechanics: Foundations and new algorithms*, Functional integration (Cargèse, 1996), NATO Adv. Sci. Inst. Ser. B Phys., Vol. 361, Plenum, New York, 1997, pp. 131–192. (Cited on pp. 199, 201.)

[SR97] Lawrence F. Shampine and Mark W. Reichelt, *The MATLAB ODE suite*, SIAM J. Sci. Comput. **18** (1997), no. 1, 1–22. (Cited on p. 171.)

[Ste65] Hans J. Stetter, *Asymptotic expansions for the error of discretization algorithms for non-linear functional equations*, Numer. Math. **7** (1965), 18–31. (Cited on p. 172.)

[Ste73] Frank Stenger, *Integration formulae based on the trapezoidal formula*, J. Inst. Math. Appl. **12** (1973), 103–114. (Cited on p. 68.)

[Ste84] Gilbert W. Stewart, *A second order perturbation expansion for small singular values*, Linear Algebra Appl. **56** (1984), 231–235. (Cited on p. 58.)

[Ste01] ———, *Matrix algorithms. Vol. II, Eigensystems*, Society for Industrial and Applied Mathematics (SIAM), Philadelphia, 2001. (Cited on p. 162.)

[SW97] Dan Schwalbe and Stan Wagon, *VisualDSolve, Visualizing differential equations with Mathematica*, TELOS/Springer-Verlag, New York, 1997. (Cited on p. 87.)

[Syl60] James Joseph Sylvester, *Notes on the meditation of Poncelet's theorem*, Philosophical Magazine **20** (1860), 533. (Cited on p. 181.)

[Sze75] Gábor Szegő, *Orthogonal polynomials*, fourth ed., American Mathematical Society, Providence, 1975. (Cited on p. 135.)

[Tab95] Serge Tabachnikov, *Billiards*, Société Mathématique de France, Marseille, 1995. (Cited on pp. 40, 44.)

[Tan88] Ping Tak Peter Tang, *A fast algorithm for linear complex Chebyshev approximations*, Math. Comp. **51** (1988), no. 184, 721–739. (Cited on p. 119.)

[TB97] Lloyd N. Trefethen and David Bau, III, *Numerical linear algebra*, Society for Industrial and Applied Mathematics (SIAM), Philadelphia, 1997. (Cited on pp. 4, 10, 154, 156.)

[Tho95] James W. Thomas, *Numerical partial differential equations: Finite difference methods*, Springer-Verlag, New York, 1995. (Cited on p. 172.)

[TM74] Hidetosi Takahasi and Masatake Mori, *Double exponential formulas for numerical integration*, Publ. Res. Inst. Math. Sci. Kyoto Univ. **9** (1973/74), 721–741. (Cited on p. 68.)

[Tre81] Lloyd N. Trefethen, *Near-circularity of the error curve in complex Chebyshev approximation*, J. Approx. Theory **31** (1981), no. 4, 344–367. (Cited on p. 115.)

[Tre98] ———, *Maxims about numerical mathematics, computers, science, and life*, SIAM News **31** (1998), no. 1, 4. `http://www.siam.org/siamnews/01-98/maxims.htm`. (Cited on pp. vii, 4, 34.)

[Tre00] ———, *Predictions for scientific computing 50 years from now*, Mathematics Today (April 2000), 53–57. (Cited on p. 4.)

[Tre02] ———, *The $100, 100-Digit Challenge*, SIAM News **35** (2002), no. 6, 1–3. (Cited on pp. 2, 169, 259.)

[Tse96] Ching-Yih Tseng, *A multiple-exchange algorithm for complex Chebyshev approximation by polynomials on the unit circle*, SIAM J. Numer. Anal. **33** (1996), no. 5, 2017–2049. (Cited on p. 119.)

[Tuc02] Warwick Tucker, *A rigorous ODE solver and Smale's 14th problem*, Found. Comput. Math. **2** (2002), no. 1, 53–117. (Cited on p. 95.)

[vzGG99] Joachim von zur Gathen and Jürgen Gerhard, *Modern computer algebra*, Cambridge University Press, New York, 1999. (Cited on pp. 164, 166, 167.)

[Wal88] Jörg Waldvogel, *Numerical quadrature in several dimensions*, Numerical Integration, III (Oberwolfach, 1987), Birkhäuser, Basel, 1988, pp. 295–309. (Cited on p. 68.)

[Wat39] George N. Watson, *Three triple integrals*, Quart. J. Math., Oxford Ser. **10** (1939), 266–276. (Cited on p. 143.)

[Wat88] G. Alistair Watson, *A method for the Chebyshev solution of an overdetermined system of complex linear equations*, IMA J. Numer. Anal. **8** (1988), no. 4, 461–471. (Cited on p. 111.)

[Wat00] ———, *Approximation in normed linear spaces*, J. Comput. Appl. Math. **121** (2000), no. 1–2, 1–36. (Cited on p. 119.)

[Web91] Heinrich Weber, *Elliptische Functionen und algebraische Zahlen*, Vieweg, Braunschweig, 1891. (Cited on pp. 220, 221.)

[Wen89] Ernst Joachim Weniger, *Nonlinear sequence transformations for the acceleration of convergence and the summation of divergent series*, Computer Physics Reports **10** (1989), 189–371. (Cited on pp. 225, 230.)

[Wie86] Douglas H. Wiedemann, *Solving sparse linear equations over finite fields*, IEEE Trans. Inform. Theory **32** (1986), no. 1, 54–62. (Cited on p. 165.)

[Wim81] Jet Wimp, *Sequence transformations and their applications*, Mathematics in Science and Engineering, Vol. 154, Academic Press, New York, 1981. (Cited on p. 230.)

[WW96] Edmund T. Whittaker and George N. Watson, *A course of modern analysis*, Cambridge University Press, Cambridge, UK, 1996, Reprint of the fourth (1927) edition. (Cited on pp. 101, 135, 176.)

[Wyn56a] Peter Wynn, *On a device for computing the $e_m(S_n)$ tranformation*, Math. Tables Aids Comput. **10** (1956), 91–96. (Cited on p. 245.)

[Wyn56b] ———, *On a procrustean technique for the numerical transformation of slowly convergent sequences and series*, Proc. Cambridge Philos. Soc. **52** (1956), 663–671. (Cited on p. 245.)

[Wyn66] ——, *Upon systems of recursions which obtain among the quotients of the Padé table*, Numer. Math. **8** (1966), 264–269. (Cited on p. 245.)

[You81] Laurence Chisholm Young, *Mathematicians and their times*, North-Holland, Amsterdam, 1981. (Cited on p. 48.)

[Zau89] Erich Zauderer, *Partial differential equations of applied mathematics*, second ed., Wiley, New York, 1989. (Cited on pp. 202, 209.)

[Zuc79] I. J. Zucker, *The summation of series of hyperbolic functions*, SIAM J. Math. Anal. **10** (1979), no. 1, 192–206. (Cited on p. 220.)

Index